Cellular and Molecular Mechanisms in Pathogenesis of Multiple Sclerosis

Cellular and Molecular Mechanisms in Pathogenesis of Multiple Sclerosis

Editor

Edwin Wan

MDPI • Basel • Beijing • Wuhan • Barcelona • Belgrade • Manchester • Tokyo • Cluj • Tianjin

Editor
Edwin Wan
West Virginia University School
of Medicine
USA

Editorial Office
MDPI
St. Alban-Anlage 66
4052 Basel, Switzerland

This is a reprint of articles from the Special Issue published online in the open access journal *Cells* (ISSN 2073-4409) (available at: https://www.mdpi.com/journal/cells/special_issues/pathogenesis_Multiple_Sclerosis).

For citation purposes, cite each article independently as indicated on the article page online and as indicated below:

LastName, A.A.; LastName, B.B.; LastName, C.C. Article Title. *Journal Name* **Year**, *Article Number*, Page Range.

ISBN 978-3-03943-555-5 (Hbk)
ISBN 978-3-03943-556-2 (PDF)

© 2020 by the authors. Articles in this book are Open Access and distributed under the Creative Commons Attribution (CC BY) license, which allows users to download, copy and build upon published articles, as long as the author and publisher are properly credited, which ensures maximum dissemination and a wider impact of our publications.

The book as a whole is distributed by MDPI under the terms and conditions of the Creative Commons license CC BY-NC-ND.

Contents

About the Editor vii

Edwin C. K. Wan
Cellular and Molecular Mechanisms in the Pathogenesis of Multiple Sclerosis
Reprinted from: *Cells* **2020**, *9*, 2223, doi:10.3390/cells9102223 1

Alessandra Musella, Antonietta Gentile, Livia Guadalupi, Francesca Romana Rizzo, Francesca De Vito, Diego Fresegna, Antonio Bruno, Ettore Dolcetti, Valentina Vanni, Laura Vitiello, Silvia Bullitta, Krizia Sanna, Silvia Caioli, Sara Balletta, Monica Nencini, Fabio Buttari, Mario Stampanoni Bassi, Diego Centonze and Georgia Mandolesi
Central Modulation of Selective Sphingosine-1-Phosphate Receptor 1 Ameliorates Experimental Multiple Sclerosis
Reprinted from: *Cells* **2020**, *9*, 1290, doi:10.3390/cells9051290 5

Gabriele Di Sante, Susanna Amadio, Beatrice Sampaolese, Maria Elisabetta Clementi, Mariagrazia Valentini, Cinzia Volonté, Patrizia Casalbore, Francesco Ria and Fabrizio Michetti
The S100B Inhibitor Pentamidine Ameliorates Clinical Score and Neuropathology of Relapsing—Remitting Multiple Sclerosis Mouse Model
Reprinted from: *Cells* **2020**, *9*, 748, doi:10.3390/cells9030748 23

Sonsoles Barriola, Fernando Pérez-Cerdá, Carlos Matute, Ana Bribián and Laura López-Mascaraque
A Clonal NG2-Glia Cell Response in a Mouse Model of Multiple Sclerosis
Reprinted from: *Cells* **2020**, *9*, 1279, doi:10.3390/cells9051279 39

Monokesh K. Sen, Mohammed S. M. Almuslehi, Erika Gyengesi, Simon J. Myers, Peter J. Shortland, David A. Mahns and Jens R. Coorssen
Suppression of the Peripheral Immune System Limits the Central Immune Response Following Cuprizone-Feeding: Relevance to Modelling Multiple Sclerosis
Reprinted from: *Cells* **2019**, *8*, 1314, doi:10.3390/cells8111314 55

Emanuele D'Amico, Aurora Zanghì, Alessandra Romano, Mariangela Sciandra, Giuseppe Alberto Maria Palumbo and Francesco Patti
The Neutrophil-to-Lymphocyte Ratio is Related to Disease Activity in Relapsing Remitting Multiple Sclerosis
Reprinted from: *Cells* **2019**, *8*, 1114, doi:10.3390/cells8101114 89

Itay Raphael, Rachel R. Joern and Thomas G. Forsthuber
Memory CD4+ T Cells in Immunity and Autoimmune Diseases
Reprinted from: *Cells* **2020**, *9*, 531, doi:10.3390/cells9030531 97

Tamás Biernacki, Dániel Sandi, Krisztina Bencsik and László Vécsei
Kynurenines in the Pathogenesis of Multiple Sclerosis: Therapeutic Perspectives
Reprinted from: *Cells* **2020**, *9*, 1564, doi:10.3390/cells9061564 121

Kelly L. Monaghan and Edwin C.K. Wan
The Role of Granulocyte-Macrophage Colony-Stimulating Factor in Murine Models of Multiple Sclerosis
Reprinted from: *Cells* **2020**, *9*, 611, doi:10.3390/cells9030611 157

About the Editor

Edwin Wan conducted his Ph.D. at City University of Hong Kong under supervision of Dr. David Wang-Fun Fong, focusing on multidrug resistance in cancers. Following the completion of his graduate study, he joined Dr. Xiao-ping Zhong's lab at Duke University to investigate the role of diacylglycerol kinases, a group of enzymes that terminate T-cell receptor signaling, in the development and function of T cells. He then moved to National Institutes of Health to continue his postdoctoral training under Dr. Warren J. Leonard to investigate the functional role of interleukin-21 in innate immunity. In 2016, he moved to West Virginia University to establish his research group, where his lab focuses on investigating the cellular and molecular mechanisms that promote neuroinflammatory diseases, such as multiple sclerosis and ischemic stroke-induced brain damage.

Editorial

Cellular and Molecular Mechanisms in the Pathogenesis of Multiple Sclerosis

Edwin C. K. Wan [1,2,3]

1. Department of Microbiology, Immunology, and Cell Biology, West Virginia University, Morgantown, WV 26506, USA; edwin.wan@hsc.wvu.edu
2. Department of Neuroscience, West Virginia University, Morgantown, WV 26506, USA
3. Rockefeller Neuroscience Institute, West Virginia University, Morgantown, WV 26506, USA

Received: 26 September 2020; Accepted: 30 September 2020; Published: 1 October 2020

Multiple sclerosis (MS) is one of the most common neurological disorders in young adults. The etiology of MS is not known, but it is generally accepted that it is autoimmune in nature. Our knowledge of the pathogenesis of MS has increased tremendously in the past decade through clinical studies and the use of experimental autoimmune encephalomyelitis (EAE), a model that has been widely used for MS research. Major advances in the field, such as understanding the roles of pathogenic Th17 cells, myeloid cells, and B cells in MS/EAE, as well as cytokine and chemokine signaling that controls neuroinflammation, have led to the development of potential and clinically approved disease-modifying agents (DMAs).

There are many aspects related to the initiation, relapse and remission, and progression of MS that are yet to be elucidated. For instance, what are the genetic and environmental risk factors that promote the initiation of MS and how do these factors impact the immune system? What factors drive the progression of MS and what are the roles of peripheral immune cells in disease progression? How do the CNS-infiltrated immune cells interact with the CNS-resident glial cells when the disease progresses? What is the role of the microbiome in MS? Can we develop animal models that better represent subcategories of MS? Understanding the cellular and molecular mechanisms that govern the pathogenesis of MS will help to develop novel and more specific therapeutic strategies that will ultimately improve clinical outcomes of the treatments. This Special Issue of *Cells* has published original research articles, retrospective clinical reports, and review articles that investigate the cellular and molecular basis of MS.

Most DMAs can effectively reduce the rate of relapse in patients with relapsing–remitting MS (RRMS) but these DMAs are ineffective in stopping the disease from progressing. Ozanimod is a sphingosine-1-phosphate (S1P1 and S1P5) receptor agonist that is recently approved for the treatment of clinically isolated syndrome, RRMS, and active secondary progressive MS (SPMS). Since neurodegeneration is a prominent feature for SPMS, Centonze and colleagues [1] sought to investigate the potential neuroprotective effect of ozanimod. Using corticostriatal slices prepared from EAE-induced mice, they showed a direct anti-excitotoxic effect of ozanimod and found that ozanimod reduces the expression of inflammatory marker iNOS, IL-1β, and TNF. The authors further demonstrated that ozanimod reduces the synaptotoxic effect during EAE by modulating lymphocytes. To further investigate whether the effect of ozanimod is due to the modulation of S1P1 or S1P5 activity, the authors used selective S1P1 and S1P5 agonists and showed that targeting S1P1 is more effective than targeting S1P5 in ameliorating glutamatergic dysfunction. Finally, they showed that mice treated with S1P1 agonist developed less severe EAE. Overall, this study provides evidence that targeting S1P1/S1P5 activity has both anti-inflammatory and neuroprotective effects.

Inflammation in the CNS during relapse can be caused by mediators produced from both peripheral infiltrated immune cells and CNS-resident glial cells. While most current DMAs for MS treatment target blocking the infiltration and/or function of peripheral immune cells, the potential of targeting glial cells in MS is not well explored. Michetti, Ria, and colleagues [2] showed that targeting S100B, a protein primarily produced by astrocytes, can ameliorate the relapsing–remitting EAE. EAE-induced mice treated with pentamidine isethionate (PTM), an antiprotozoal drug shown previously to block S100B activity, had reduced EAE severity score in the preclinical, onset, and remission phases during the disease course, which correlates with the reduction of *Ifng* and *Tnfa* expression in the brain, as well as the decrease of NOS activity. The numbers of CD68+ cells and demyelinating lesions were reduced in the PTM-treated EAE mice compared to EAE mice without drug treatment. Overall, this study suggests that targeting neurotoxic mediators produced by astrocytes reduces the severity of EAE. Whether the same treatment strategy applies to MS warrants further investigation.

Demyelination is a hallmark of MS pathology. NG2-glia are oligodendrocyte precursors that can differentiate into mature oligodendrocytes and thus may contribute to remyelination in patients with MS. Using a sophisticated StarTrack approach to follow the fate of NG2-glia which are derived from single progenitors from the dorsal subventricular zone, Lopez-Mascaraque and colleagues [3] investigated the phenotypic heterogenicity of NG2-glia relative to their ontogenic origin, and asked whether EAE triggers a clonal NG2-glial response. They showed that the NG2-glia from single progenitors dispersed throughout the grey and white matter in a clonal manner. In addition, the heterogeneity of NG2-glia seems established in the early embryonic development stage. Using EAE as the disease model, the authors further showed that the clonal NG2-glia are morphologically diverse relative to the distance from EAE lesions. Overall, this study improves our understanding in the nature of heterogeneity of NG2-glia under normal and pathological conditions.

Although the activation of autoreactive T cells plays a critical role in the pathogenesis of MS, the etiology of this disease remains uncertain. The "inside-out" theory proposes that MS is initiated by demyelination of the CNS, followed by the activation of adaptive immune response in the periphery. Coorssen, Mahns, and colleagues [4] investigated whether the cuprizone-mediated demyelination in mice, in the presence of pertussis toxin, stimulates T-cell infiltration to the brain. They showed that both standard-dose (0.2%) and low-dose (0.1%) treatment of cuprizone for five weeks induces oligodendrocytosis, demyelination, and gliosis in the brain. However, standard-dose treatment also induces splenic atrophy and a more significant reduction of CD4+ and CD8+ cells in the spleen compared to mice with low-dose cuprizone treatment. They further showed that demyelination in the brain does not stimulate T-cell infiltration, even in the presence of pertussis toxin, which is known to disrupt blood–brain barrier integrity. In addition, the authors determined changes in the brain proteome following cuprizone treatment using two-dimensional gel electrophoresis coupled with liquid chromatography and tandem mass spectrometry. They found changes of multiple brain proteoforms following cuprizone treatment, the majority of which are associated with mitochondrial function. Thus, their study suggests that the cuprizone-induced demyelination may be elicited by mitochondrial perturbations.

In a clinical prospective, identification of biomarkers in patients' blood may serve as a simple and rapid tool to evaluate disease activity in MS patients. D'Amico and colleagues [5] performed a retrospective study to analyze data from patients with RRMS, and asked whether the blood neutrophil-to-lymphocyte ratio (NLR) correlates with disease severity in these patients. The authors obtained NLR data from 84 newly diagnosed, naïve RRMS patients and found that NLR ratio positively correlates with high disease activity, classified as at least two relapses in the year prior to the study entry and at least one gadolinium-enhancing lesion at the time of study. This study, although with a small sample size, suggests that NLR may serve as one of the biomarkers to evaluate disease activity in MS patients. This observation should be confirmed further with multicenter, large cohort clinical studies.

In addition to the aforementioned research articles, this Special Issue of *Cells* also published three review articles that cover different research topics in MS. Forsthuber and colleagues [6] summarized the role of memory CD4+ T cells in autoimmune diseases, with specific focus on MS and EAE. Vecsei and colleagues [7] discussed the importance of kynurenines in the pathogenesis of MS. Last but not least, we [8] provided a comprehensive review on the role of granulocyte-macrophage colony-stimulating factor (GM-CSF) in mediating neuroinflammation, and discussed the potential of targeting GM-CSF in the treatment of MS.

Taken together, articles published in this Special Issue of *Cells* have covered a variety of topics in MS, which help us to better understand the underlying cellular and molecular mechanisms that drive MS pathogenesis.

Funding: E.C.K.W. is supported by NIH grant P20 GM109098 and the Innovation Award Program from Praespero.

Acknowledgments: I thank Sheran Law and David Fong for their support throughout my scientific career.

Conflicts of Interest: The author declares no conflict of interest.

References

1. Musella, A.; Gentile, A.; Guadalupi, L.; Rizzo, F.R.; De Vito, F.; Fresegna, D.; Bruno, A.; Dolcetti, E.; Vanni, V.; Vitiello, L.; et al. Central Modulation of Selective Sphingosine-1-Phosphate Receptor 1 Ameliorates Experimental Multiple Sclerosis. *Cells* **2020**, *9*, 1290. [CrossRef] [PubMed]
2. Di Sante, G.; Amadio, S.; Sampaolese, B.; Clementi, M.E.; Valentini, M.; Volonte, C.; Casalbore, P.; Ria, F.; Michetti, F. The S100B Inhibitor Pentamidine Ameliorates Clinical Score and Neuropathology of Relapsing-Remitting Multiple Sclerosis Mouse Model. *Cells* **2020**, *9*, 748. [CrossRef] [PubMed]
3. Barriola, S.; Perez-Cerda, F.; Matute, C.; Bribian, A.; Lopez-Mascaraque, L. A Clonal NG2-Glia Cell Response in a Mouse Model of Multiple Sclerosis. *Cells* **2020**, *9*, 1279. [CrossRef] [PubMed]
4. Sen, M.K.; Almuslehi, M.S.M.; Gyengesi, E.; Myers, S.J.; Shortland, P.J.; Mahns, D.A.; Coorssen, J.R. Suppression of the Peripheral Immune System Limits the Central Immune Response Following Cuprizone-Feeding: Relevance to Modelling Multiple Sclerosis. *Cells* **2019**, *8*, 1314. [CrossRef] [PubMed]
5. D'Amico, E.; Zanghi, A.; Romano, A.; Sciandra, M.; Palumbo, G.A.M.; Patti, F. The Neutrophil-to-Lymphocyte Ratio is Related to Disease Activity in Relapsing Remitting Multiple Sclerosis. *Cells* **2019**, *8*, 1114. [CrossRef] [PubMed]
6. Raphael, I.; Joern, R.R.; Forsthuber, T.G. Memory CD4(+) T Cells in Immunity and Autoimmune Diseases. *Cells* **2020**, *9*, 531. [CrossRef] [PubMed]
7. Biernacki, T.; Sandi, D.; Bencsik, K.; Vecsei, L. Kynurenines in the Pathogenesis of Multiple Sclerosis: Therapeutic Perspectives. *Cells* **2020**, *9*, 1564. [CrossRef] [PubMed]
8. Monaghan, K.L.; Wan, E.C.K. The Role of Granulocyte-Macrophage Colony-Stimulating Factor in Murine Models of Multiple Sclerosis. *Cells* **2020**, *9*, 611. [CrossRef] [PubMed]

© 2020 by the author. Licensee MDPI, Basel, Switzerland. This article is an open access article distributed under the terms and conditions of the Creative Commons Attribution (CC BY) license (http://creativecommons.org/licenses/by/4.0/).

Article

Central Modulation of Selective Sphingosine-1-Phosphate Receptor 1 Ameliorates Experimental Multiple Sclerosis

Alessandra Musella [1,2,†], Antonietta Gentile [2,3,†], Livia Guadalupi [3], Francesca Romana Rizzo [3], Francesca De Vito [4], Diego Fresegna [2,3], Antonio Bruno [3], Ettore Dolcetti [3], Valentina Vanni [2,3], Laura Vitiello [2], Silvia Bullitta [2,3], Krizia Sanna [3], Silvia Caioli [3], Sara Balletta [3], Monica Nencini [2], Fabio Buttari [4], Mario Stampanoni Bassi [4], Diego Centonze [3,4,*] and Georgia Mandolesi [1,2]

1. Department of Human Sciences and Quality of Life Promotion, University of Rome San Raffaele, 00166 Rome, Italy; alessandra.musella@sanraffaele.it (A.M.); georgia.mandolesi@sanraffaele.it (G.M.)
2. IRCCS San Raffaele Pisana, 00166 Rome, Italy; gntnnt01@uniroma2.it (A.G.); diego.fresegna@gmail.com (D.F.); valentina_vanni@hotmail.it (V.V.); laura.vitiello@sanraffaele.it (L.V.); Silvia.Bullitta@uniroma2.it (S.B.); monicanencini@hotmail.com (M.N.)
3. Department of Systems Medicine, University of Rome Tor Vergata, 00133 Rome, Italy; livia.guadalupi@gmail.com (L.G.); f.rizzo@med.uniroma2.it (F.R.R.); antonio.bruno91@yahoo.it (A.B.); ettoredolcetti@hotmail.it (E.D.); krizia.sanna@live.it (K.S.); silviacaioli@yahoo.it (S.C.); balletta.sara@gmail.com (S.B.)
4. Unit of Neurology, IRCCS Neuromed, 86077 Pozzilli (IS), Italy; f.devito.molbio@gmail.com (F.D.V.); fabio.buttari@gmail.com (F.B.); mario_sb@hotmail.it (M.S.B.)
* Correspondence: centonze@uniroma2.it; Tel.: +39-06-7259-6010; Fax: +39-06-7259-6006
† These authors contributed equally.

Received: 20 April 2020; Accepted: 18 May 2020; Published: 22 May 2020

Abstract: Future treatments of multiple sclerosis (MS), a chronic autoimmune neurodegenerative disease of the central nervous system (CNS), aim for simultaneous early targeting of peripheral immune function and neuroinflammation. Sphingosine-1-phosphate (S1P) receptor modulators are among the most promising drugs with both "immunological" and "non-immunological" actions. Selective S1P receptor modulators have been recently approved for MS and shown clinical efficacy in its mouse model, the experimental autoimmune encephalomyelitis (EAE). Here, we investigated the anti-inflammatory/neuroprotective effects of ozanimod (RPC1063), a $S1P_{1/5}$ modulator recently approved in the United States for the treatment of MS, by performing ex vivo studies in EAE brain. Electrophysiological experiments, supported by molecular and immunofluorescence analysis, revealed that ozanimod was able to dampen the EAE glutamatergic synaptic alterations, through attenuation of local inflammatory response driven by activated microglia and infiltrating T cells, the main CNS-cellular players of EAE synaptopathy. Electrophysiological studies with selective $S1P_1$ (AUY954) and $S1P_5$ (A971432) agonists suggested that $S1P_1$ modulation is the main driver of the anti-excitotoxic activity mediated by ozanimod. Accordingly, in vivo intra-cerebroventricular treatment of EAE mice with AUY954 ameliorated clinical disability. Altogether these results strengthened the relevance of $S1P_1$ agonists as immunomodulatory and neuroprotective drugs for MS therapy.

Keywords: sphingosine-1-phosphate receptors; glutamate synaptic dysfunction; microglia; T lymphocytes; experimental autoimmune encephalomyelitis (EAE); pro-inflammatory cytokines; neuroinflammation; ozanimod; AUY954; A971432; $S1P_1$; $S1P_5$

1. Introduction

Multiple sclerosis (MS) is an autoimmune disease extensively reported to be triggered by myelin-targeted T-cell infiltration into the central nervous system (CNS) and by B-cell autoantibodies. The resulting brain inflammation is regarded as the main cause of both demyelination and neurodegeneration [1–6]. Neurodegeneration combined with a lack of repair is increasingly recognized to contribute to disease progression and to occur independently of focal white matter damage [5]. Recently, several preclinical and clinical studies have demonstrated that a marked and early imbalance between glutamate excitatory and GABA inhibitory transmission affects the brain of both MS patients and mice with experimental autoimmune encephalomyelitis (EAE), a mouse model of MS [5,7–10]. Of note, pro-inflammatory cytokines released by activated microglia and lymphocytes in both EAE and MS have been shown to exert direct synaptopathic effects especially on glutamatergic transmission [11].

Together with the loss of synapses, these synaptic alterations lead to a diffuse "synaptopathy" that is driven by inflammation and, in turn, causes excitotoxicity, a well-known mechanism of neurodegeneration. Therapeutic strategies to target synaptopathy are particularly appealing, because —unlike loss of neurons—synaptic dysfunction and loss of synapses are reversible processes. In this respect, several studies in EAE have shown that disease modifying therapies (DMTs) in use for MS, beyond their peripheral immunomodulation, can target the inflammatory synaptopathy, providing neuroprotection [12–15].

Developing neuroprotective and neurorepair therapeutic strategies is required for the future treatment of MS. A simultaneous and early targeting of peripheral immune cell function and of CNS-intrinsic inflammation, potentially through combinatorial therapies designed to modulate these two immunological arms of the disease, along with the provision of neuroprotective or neuroregenerative drugs, are challenging therapeutic goals. Promising molecules with both immunological and non-immunological actions are represented by modulators of the family of sphingosine-1-phosphate (S1P) receptors [16]. S1P is a powerful bioactive sphingolipid that activates S1P receptors in an autocrine and/or paracrine way [17]. The five receptor subtypes that have been identified, $S1P_1$, $S1P_2$, $S1P_3$, $S1P_4$, and $S1P_5$, are G-protein-coupled receptors that are almost ubiquitously expressed and play important roles in cell survival, growth, and differentiation in many cell types, including cells of the immune and central nervous system [18]. Modulation of lymphocyte trafficking, mainly mediated by $S1P_1$, is the main mechanism responsible for the beneficial effects of S1P drugs, such as fingolimod (FTY720), the first nonselective S1P receptor agonist approved for relapsing remitting (RRMS) patients [19,20] and siponimod (BAF312), a $S1P_1$ and $S1P_5$-selective fingolimod-congener, one of the few MS drugs approved for secondary progressive MS (SPMS) [21]. However, the relevant observation of a reduced brain volume loss in fingolimod-treated RRMS patients [22], and the successful reduction of neurological damage accumulation in siponimod-treated SPMS patients, highlighted a "non-immunological" effect due to modulation of S1P receptors in CNS resident cells, including neurons, astrocytes, oligodendrocytes, and microglia. Studies in experimental MS, especially in the EAE model, have attempted to define the molecular mechanisms underlying these effects, highlighting the involvement of $S1P_1$ or $S1P_5$ specifically expressed on astroglia and oligodendrocytes, respectively [23–27]. In our previous paper [13], we performed a long-lasting, preventive, and central delivery of siponimod in EAE mice and we observed a late amelioration of clinical disability. Such a beneficial effect was accompanied by a reduced astrogliosis and microgliosis in the EAE striatum, a brain region particularly affected in EAE, and by neuroprotective action on GABA-ergic neurons, likely because of a $S1P_1$-mediated reduction of the central inflammation [13]. Despite a reduced microgliosis, we could not detect any amelioration of EAE glutamate-mediated excitotoxicity [5,28], suggesting a specificity of siponimod action in this experimental condition. These results also raised the possibility that different S1P modulators could elicit specific anti-synaptotoxic effects.

In the present study, by means of ex vivo and in vitro experiments we investigated the anti-excitotoxic activity of ozanimod (RPC1063), an MS oral $S1P_1/S1P_5$ modulator [29], on glutamatergic transmission alterations in EAE striatum. Given the safety and efficacy profile, ozanimod has recently been approved in the United States for clinically isolated syndrome, RRMS, and active SPMS. Ozanimod also does not require first dose heart rate monitoring as with fingolimod, nor genetic CYP2C9 screening as with siponimod [30]. Furthermore, by using ex vivo and/or central in vivo approaches, we addressed the potential neuroprotective effects of selective $S1P_1$ and $S1P_5$ agonists on EAE mice, in order to discriminate the different involvement of these S1P receptor subtypes in ameliorating synaptic alterations and clinical disability.

2. Methods

2.1. Drug Formulation

Based on EC_{50} values [29,31,32] and to achieve selectivity for S1P receptors, for ex vivo experiments we selected the following concentration in nM: ozanimod (1000), AUY954 (300); A971432 (200). Drugs were dissolved in DMSO (0.1–0.25% final concentration) for in vitro experimentation. For in vivo experiments two AUY954 dosages were tested: 2.7 and 0.55 µg/day; the drug was dissolved in a solution containing 10% Solutol/Kolliphor HS15 (BASF Pharma Solutions), final pH range between 6 and 7.

2.2. Mice

Animals employed in this study were 6–8-week-old C57BL/6 female mice, obtained from Charles-River (Milan, Italy). Mice were housed under constant conditions in an animal facility with a regular 12 h light/dark cycle. Food and water were supplied *ad libitum*. All the efforts were made to minimize the number of animals used and their suffering. In particular, when animals experienced hindlimb weakness, moistened food and water were made easily accessible to the animals on the cage floor. Mice with hindlimb paresis received glucose solution by subcutaneous injection or food by gavage during the entire procedure. In the rare presence of a tetraparalyzed animal, mice were sacrificed. Minipump-implanted mice were housed in individual cages endowed with special bedding (TEK-Fresch, Envigo, Casatenovo (LC), Italy) in order to avoid skin infections around the surgical scar. Animal experiments were performed according to the Internal Institutional Review Committee, the European Directive 2010/63/EU and the European Recommendations 526/2007, and the Italian D.Lgs 26/2014.

2.3. EAE Model

Chronic-progressive EAE was induced as previously described [8,33]. Six-eight weeks old C57BL/6 female mice were active immunized with an emulsion of mouse myelin oligodendrocyte glycoprotein peptide 35–55 (MOG35–55, 85% purity; Espikem, Prato, Italy) in Complete Freund's Adjuvant (CFA; Difco, Los Altos, CA, USA), followed by intravenous administration of pertussis toxin (500 ng; Merck, Milan, Italy) on the day of immunization and two days post immunization (dpi). Animals were scored daily for clinical symptoms of EAE according to the following scale: 0 no clinical signs; 1 flaccid tail; 2 hindlimb weakness; 3 hindlimb paresis; 4 tetraparalysis; and 5 death due to EAE; intermediate clinical signs were scored by adding 0.5.

2.4. Minipump and Surgery

One week before immunization, mice were implanted with subcutaneous osmotic minipumps allowing continuous intracerebroventricular (icv) infusion of either vehicle or AUY954 for 4 weeks as described in [8,34].

2.5. Ex Vivo Experiments

EAE mice (21–25 dpi) were killed by cervical dislocation and corticostriatal slices (200 µm) were prepared from fresh tissue blocks of the brain with the use of a vibratome (Leica VT1200 S) [28].

For molecular biology experiments, the slices were incubated for 2 h with ozanimod (1000 nM) or vehicle (DMSO 0.1% final concentration) in a chamber containing oxygenated artificial cerebrospinal fluid (ACSF). After incubation, the cortex was removed and the striatum was stored at −80 °C until use.

For the ex vivo electrophysiological experiments, fresh EAE striatal slices were incubated with AUY954 (300 nM), A971432 (200 nM), or ozanimod (1000 nM) for 2 h. A one-hour incubation of T lymphocytes isolated from EAE mice and treated with ozanimod (overnight) was performed before electrophysiological recordings when specified.

2.6. In Vitro Experiments

The BV2 immortalized murine microglial cell line was cultured as previously described [13]. BV2 cells were maintained in a humidified incubator with 5% CO_2 and cultured in DMEM supplemented with 5% FBS, 100 U/mL penicillin, and 100 mg/mL streptomycin. BV2 cells were pre-treated for 1 h with vehicle (DMSO) or ozanimod (1000 nM) and then in vitro activated for 2 h with a Mix of Th1-specific proinflammatory cytokines: 100 U/mL IL-1β (Euroclone, Milan, Italy), 200 U/mL tumor necrosis factor (TNF, Miltenyi Biotec, Bologna, Italy), and 500 U/mL interferon γ (IFNγ, Becton Dickinson, Milan, Italy) [28]

2.7. Electrophysiology

Mice were killed by cervical dislocation, and corticostriatal coronal slices (200 μm) were prepared from fresh tissue blocks of the brain with the use of a vibratome. A single slice was then transferred to a recording chamber and submerged in a continuously flowing ACSF (34 °C, 2–3 mL/min) gassed with 95% O_2–5% CO_2. The composition of the control ACSF was (in mM): 126 NaCl, 2.5 KCl, 1.2 $MgCl_2$, 1.2 NaH_2PO_4, 2.4 $CaCl_2$, 11 glucose, 25 $NaHCO_3$. To study spontaneous glutamate-mediated excitatory postsynaptic currents (sEPSCs), the recording pipettes were filled with internal solution of the following composition (mM): K^+-gluconate (125), NaCl (10), $CaCl_2$ (1.0), $MgCl_2$ (2.0), 1,2-bis (2-aminophenoxy) ethane-N,N,N,N-tetra acetic acid (BAPTA; 0.5), HEPES (19), GTP; (0.3), Mg-ATP; (1.0), adjusted to pH 7.3 with KOH. Bicuculline (10 μM) was added to the external solution to block $GABA_A$-mediated transmission. The detection threshold of sEPSCs was set at twice the baseline noise. Offline analysis was performed on spontaneous synaptic events recorded during fixed time epochs (1–2 min, three to five samplings), sampled every 5 or 10 min. Only cells that exhibited stable frequencies in control (less than 20% changes during the control samplings) were used for analysis. For kinetic analysis, events with peak amplitude between 10 and 50 pA were grouped, aligned by half-rise time, normalized by peak amplitude, and averaged to obtain rise times and decay times.

2.8. Real Time PCR (qPCR)

Total RNA was extracted from treated BV2 cells and EAE striatal slices according to the standard miRNeasy Micro kit protocol (Qiagen). The RNA quantity and purity were analyzed with the Nanodrop 1000 spectrophotometer (Thermo Scientific). The quality of RNA was assessed by visual inspection of the agarose gel electrophoresis images. Next, 700–900 ng of total RNA was reverse-transcribed using high-capacity cDNA reverse transcription kit (Applied Biosystem) according to the manufacturer's instructions and 3–24 ng of cDNA were amplified in triplicate using the Applied Biosystem 7900HT Fast Real Time PCR system. mRNA relative quantification was performed using the comparative cycle threshold ($2^{-\Delta\Delta Ct}$) method. β-actin was used as endogenous control. For the mRNA quantification of cytokines, Tgfb1, RANTES, and iNOS and FIZZ1, SensiMix II Probe Hi-Rox Kit (Bioline; Meridian Life Science) and the following TaqMan gene expression assays were used: IL-1b ID: Mm00434228_m1; Tnf ID: Mm00443258_m1; IL6 ID: Mm00446190_m1; Il10 ID: Mm01288386_m1; Tgfb1 ID: Mm01178820_m1; Ccl5 (coding for RANTES) ID:Mm01302427_m1; Nos2 (coding for iNOS) ID: Mm00440502_m1; Retnla (coding for FIZZ1) ID: Mm00445109_m1; Actb ID: Mm00607939_s1. SensiMix SYBR Hi-Rox Kit (Bioline) was utilized for the quantification of mRNA coding for ionized binding protein type-1 (IBA1) by using the following primers: Aif1 mRNA coding for IBA1 (NM_019467): forward GACAGACTGCCAGCCTAAGACAA, reverse CATTCGCTTCAAGGACATAATATCG; β-actin (NM_007393): forward CCTAGCACCATGAAGATCA AGATCA, reverse AAGCCATGCCAATGTTGTCTCT.

2.9. Immunohistochemistry and Confocal Microscopy

Striatal slices (200 µm) cut from the brain of 21–25 dpi EAE mice were incubated in oxygenated ACSF in the presence of vehicle (DMSO 0.1% final concentration) or ozanimod (1000 nM) for 2 h. Slices were fixed in 4% PFA and equilibrated with 30% sucrose before cutting 30 µm slices, to perform immunohistochemistry and confocal microscopy experiments [8]. Sections were permeabilized in PBS with Triton-X 0.25% (TPBS). All following incubations were performed in TPBS. Sections were pre-incubated with 10% normal donkey serum solution for 1 h at room temperature and incubated with the appropriate mix of following primary antibodies: rabbit anti-Iba1 (1:750, Wako, Milan, Italy), goat anti-IL-1β (1:300, R&D System, Milan, Italy), mouse anti-TNF (1:1500, Abcam, Milan, Italy) overnight at 4 °C. After three washes, 10 min each, sections were incubated with the secondary antibody Alexa-488 conjugated donkey anti-Rabbit (1:200, Invitrogen, Milan, Italy); Cy3-conjugated donkey anti-mouse or anti-goat (1:200, Jackson, Milan, Italy) for 2 h at RT, rinsed and DAPI counterstained. Sections were mounted with Vecta-shield (Vector Labs, Milan, Italy) on poly-L-lysine-coated slides, air-dried and coverslipped. Images from immunolabeled samples were acquired using a model LSM7 Zeiss confocal laser-scanner microscope with 20× objective (zoom 1.5×). The images had a pixel resolution of 1024 × 1024. The confocal pinhole was kept at 1.0, the gain and the offset were lowered to prevent saturation in the brightest signals, and sequential scanning for each channel was performed. Images were exported in Tiff format and adjusted for brightness and contrast as needed using ImageJ software. Acquisitions were made on three slices cut from three EAE mice and incubated with ozanimod or vehicle.

2.10. Murine CD3+ Cell Isolation

Mice were sacrificed at 15–25 dpi through cervical dislocation and the spleens were quickly removed and stored in sterile phosphate-buffered saline (PBS). After mechanical dissociation of the tissue, the cell suspension was passed through a 40-µm cell strainer (BD Biosciences, San Jose, CA, USA) to remove cell debris and centrifuged. The cell suspension obtained was subjected to magnetic cell sorting separation (CD3 microbeads kit; Miltenyi Biotec) to obtain a pure lymphocyte population. T cells were co-incubated overnight with ozanimod (1000 nM). About 5×10^3 pure T cells were incubated with striatal slices for 30–60 min, in a total volume of 1 mL of oxygenated ACSF before the electrophysiological recordings.

2.11. T-Cell Absolute Count

T-cell absolute count was performed on blood samples kept from the mandibular vein of the mouse. To evaluate T lymphocyte number, we used an anti-mouse CD3 PE conjugated antibody (BD Biosciences, San Jose, CA, USA, clone 145–2C11). At predetermined optimal concentrations, 100 µL of blood was stained by incubation with the antibody. Fifty microliters of Count Bright Absolute Counting Beads (Molecular Probes, Milan, Italy) were added, and erythrocytes were lysed using ACK solution (Lonza Bio Whittaker, Walkersville, MD, USA), according to standard protocols. Cell suspensions were acquired and analyzed on LSR Fortessa™ X20 SO Cell Analyzer (BD Biosciences). By comparing the ratio of bead events to cell events, absolute numbers of cells in the sample were calculated. In particular, the formula used was: (number of cells counted/number of beads counted) × (lot-specific number of beads/sample volume).

2.12. Statistical Analysis

For each type of experiment, at least three mice of each group were used. Data are presented as means ± SEM. The significance level was established at $p < 0.05$. Statistical analysis was performed using paired or unpaired Student's *t*-test. Multiple comparisons were analyzed by one-way ANOVA, followed by Tukey's HSD. Differences between groups in clinical score analysis were tested by

Mann-Whitney test. Linear regression test was used to calculate correlation between clinical score and T cell counts. Throughout the text experiments "N" refers to the number of animals, while "n" refers to the number of cells for electrophysiological experiments and to the number of slices or biological samples for molecular biology experiments.

3. Results

3.1. Ex Vivo Ozanimod Treatment Restores Normal Glutamatergic Transmission in EAE Striatum

We first explored a potential anti-excitotoxic effect of the $S1P_{1/5}$ modulator ozanimod. To this aim, we induced $MOG_{(35-55)}$ EAE in a group of mice and performed ex vivo treatments and glutamatergic transmission recording in corticostriatal slices derived from EAE mice (20–25 dpi) [28]. In details, we performed ex vivo incubation of EAE corticostriatal slices with ozanimod (1000 nM) or vehicle for two hours and performed whole-cell voltage-clamp recordings from medium spiny neurons (MSNs) to evaluate both the duration and the frequency of sEPSCs. These post- and pre-synaptic parameters, respectively, are typically enhanced in EAE condition relative to control healthy mice [28].

Ozanimod reduced the duration (decay time and half width) of glutamate-mediated spontaneous currents compared to EAE-vehicle, reaching values similar to healthy mice (dashed lines) (20–25 dpi, EAE $n = 8$, EAE + ozanimod $n = 8$: EAE vs. EAE + ozanimod decay time $p < 0.05$, half width $p < 0.01$; EAE + ozanimod vs. control $p > 0.05$ for both half width and decay time; EAE vs. control $p < 0.01$ for both half width and decay time) (Figure 1A–C). Moreover, we investigated the effect of ozanimod on other sEPSC parameters, demonstrating that ozanimod treatment also recovered the frequency without affecting the amplitude (Figure 1A,D,E).

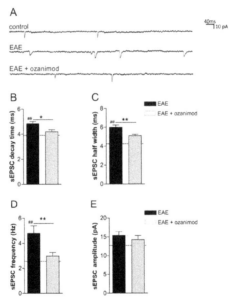

Figure 1. Ex vivo treatment of corticostriatal slices with ozanimod recovers synaptic alterations induced by experimental autoimmune encephalomyelitis (EAE). (**A**) Examples of spontaneous glutamate-mediated excitatory postsynaptic currents (sEPSC) traces recorded from medium spiny neurons (MSNs) in corticostriatal slices in the different experimental conditions (healthy control, EAE, EAE + ozanimod). Bath incubation corticostriatal slices with ozanimod (1000 nM, 2 h) recovers EAE-induced alterations of glutamatergic transmission, in terms of decay time (**B**), half width (**C**), and frequency (**D**). sEPSC amplitude is not affected by ozanimod treatment (**E**). Dotted lines refer to healthy mouse values. Data are expressed as mean ± SEM. Unpaired *t*-test, * $p < 0.05$; ** $p < 0.01$ EAE vs. EAE + ozanimod; ## $p < 0.01$ EAE vs. healthy mice.

These results highlight a direct anti-excitotoxic impact of ozanimod on EAE corticostriatal slices.

3.2. Ozanimod Treatment Exerts an Anti-Inflammatory Action on EAE Striatum and on Activated Microglial Cell Line

The observed beneficial effect of ozanimod on glutamatergic dysfunction in EAE slices suggests its potential local immunomodulatory activity. Of note, microglia, which mainly express $S1P_1$, are regarded as the main source of inflammatory mediators that contribute to synaptopathy in the EAE/MS brains [8,35]. Thus, by qPCR we analyzed mRNA levels of specific markers of microglia activation and inflammation in corticostriatal slices treated with ozanimod (as described for electrophysiological experiments).

First, we quantified the expression levels of microglia-specific transcripts coding for the binding adaptor molecule 1 (IBA-1) and M1- and M2-like markers, the inducible nitric oxide synthetase (iNOS) and FIZZ1, also known as resistin like alpha (Retnla), respectively. We observed that the M2-like marker FIZZ1 was significantly up-regulated following 2 h of ozanimod incubation (EAE 20–25 dpi, EAE n = 4 slices, EAE + ozanimod n = 5 slices; EAE vs. EAE + ozanimod $p < 0.001$), while the mRNA levels of IBA-1 and iNOS did not significantly change (EAE vs. EAE + ozanimod $p > 0.05$; Figure 2A).

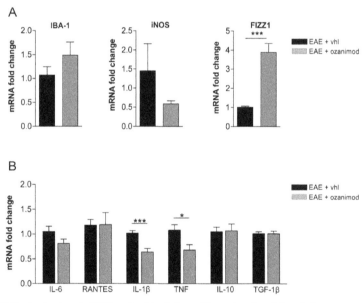

Figure 2. Ex vivo treatment of EAE corticostriatal slices with ozanimod modulates inflammatory markers related to microglial activation and lowers IL-1β and TNF mRNA levels. (**A**) qPCR quantification of microglial markers from EAE striatal slices incubated with ozanimod (1000 nM, 2 h) shows a significant upregulation of the M2 marker FIZZ1 (resistin like alpha, Retnla) without changing the expression of iNOS (inducible nitric oxide synthetase) and IBA1 (binding adaptor molecule 1). (**B**) qPCR quantification of cytokines from EAE striatal slices incubated with ozanimod (1000 nM, 2 h) shows a significant downregulation of IL-1β and TNF mRNAs. All data are expressed as mean ± SEM and as fold change of EAE vehicle samples. Unpaired t-test, * $p < 0.05$; *** $p < 0.001$.

Next, we investigated the expression levels of several pro-(TNF, IL-1β, and IL-6) and anti-(TGFβ, IL-10) inflammatory cytokines and the chemokine RANTES as a measure of the immunomodulatory activity of ozanimod. As shown in Figure 2B, treatment with ozanimod (1000 nM) downregulated TNF and IL-1β mRNAs (EAE 20–25 dpi, EAE n = 11 slices, EAE + ozanimod n = 12 slices; EAE vs. EAE + ozanimod TNF: $p < 0.001$; IL-1β: $p < 0.05$).

To support qPCR data, we performed immunofluorescence experiments on EAE striatal slices incubated with ozanimod and vehicle, focusing on TNF and IL-1β, which have been clearly shown to alter the glutamatergic transmission during EAE [8,28]. By means of confocal imaging we confirmed the remarkable effect of ozanimod on IL-1β, while the effect on TNF protein was less evident (Figure 3A,B). Indeed, we observed that ozanimod strongly attenuated IL-1β staining (Figure 3A, red) in lesioned area of EAE striatal slices, characterized by an intense labeling of the microglia/macrophage activation marker Iba1 (Figure 3A, green; counterstaining with dapi-cyan). In contrast, TNF immunolabeling was slightly reduced in ozanimod treated slices (Figure 3B, red).

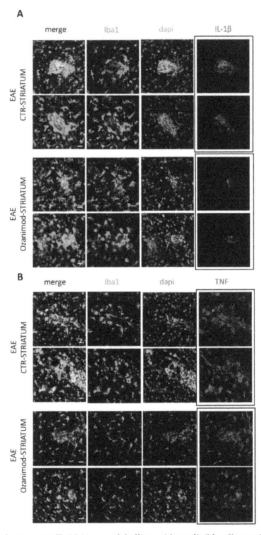

Figure 3. Ozanimod attenuates IL-1β immunolabelling with negligible effect on TNF in EAE striatal slices. Confocal images of EAE striatal slices incubated with vehicle or ozanimod (1000 nM, 2 h) stained for IBA1 (green), cell nuclei (cyan), IL-1 β (red in **A**), and TNF (red in **B**) show that ozanimod treatment leads to a significantly milder expression of IL-1 β within lesioned area, highlighted by IBA1 and dapi staining, with minor effect on TNF. Scale bars: 20 µm.

Finally, to further support a direct neuromodulatory effect of ozanimod on microglia cells, BV2 cells were in vitro treated with ozanimod and activated with a mix of Th1 cytokines known to mimic EAE condition [28]. After treatment, activated BV2 cells and the relative controls were processed for qPCR analysis to quantify mRNAs of IL-6, IL-1β, TNF, RANTES/CCL5, and IL-10. As shown in Figure 4, ozanimod treatment induced a downregulation of IL-6, RANTES/CCL5, and TNF mRNAs (n = 3 biological samples slices per condition, $p < 0.05$), with no effects on IL-1β and the anti-inflammatory cytokine IL-10 mRNAs in comparison to Th1 mix control condition (n = 3 biological samples slices for each treatment, $p > 0.05$; data not shown).

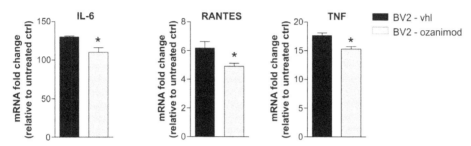

Figure 4. In vitro treatment of BV2 cell line with ozanimod modulates mRNA levels of pro-inflammatory cytokines. qPCR experiments performed on BV2 microglial cells activated by Th1 Mix for 2 h and incubated with ozanimod (1000 nM, 1 h pretreatment) or vehicle (DMSO) show a downregulation of IL-6, TNF and RANTES mRNAs. All data are expressed as mean ± SEM and as fold change of untreated controls. Unpaired t-test, * $p < 0.05$.

3.3. Ozanimod Pre-Treatment of EAE T Lymphocytes Rescues Striatal Glutamatergic Alterations In Ex Vivo EAE Model

Infiltrating T cells are another important source of cytokines in the inflamed EAE and MS brains [6,36]. Noteworthy, we previously reported that T lymphocytes isolated from the spleen of EAE mice induced a TNF-dependent enhancement of the kinetic properties of the sEPSCs when incubated for 1–2 h with healthy corticostriatal slices, mimicking EAE condition [28]. Therefore, we addressed the hypothesis that the neuroprotective effect of ozanimod could be mediated by its immunomodulatory activity on T lymphocytes, beside its well-known peripheral effect on CNS-accessing T cells. To this aim, we pre-incubated EAE lymphocytes with ozanimod overnight (1000 nM) and then we assessed their synaptotoxic effect on striatal slices of healthy mice by recording striatal glutamatergic currents. Decay time and half width parameters of sEPSCs were significantly reduced following incubation with ozanimod-treated EAE lymphocytes compared to untreated cells (ozanimod-treated EAE T cells n = 16, EAE T cells n = 5; unpaired t-test, decay time: ozanimod-treated EAE T cells vs. EAE T cells $p < 0.01$; half width ozanimod-treated EAE T cells vs. EAE T cells $p < 0.01$; Figure 5).

This experimental strategy that allows the investigation of a direct synaptotoxic role for infiltrating lymphocytes, together with the evidence that T cells predominantly express $S1P_1$ reinforce the central neuroprotective action of ozanimod mediated by $S1P_1$ modulation.

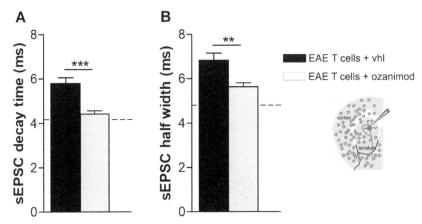

Figure 5. In vitro treatment of EAE T cells with ozanimod abolishes T cell synaptotoxicity. (**A**,**B**) The enhancement of sEPSC decay time (**A**) and sEPSC half width (**B**), typically induced by EAE lymphocytes, was significantly reduced by in vitro treatment of EAE T cells with ozanimod (1000 nM). Dotted lines refer to control condition (control T cells). Data are presented as ± SEM. Unpaired *t*-test, ** $p < 0.01$; *** $p < 0.001$.

3.4. Selective Agonists of Central $S1P_1$ and $S1P_5$ Differently Modulate EAE Striatal Glutamatergic Alterations

The above data point to a major role of $S1P_1$ in mediating the anti-synaptotoxic effects of ozanimod in EAE mice. In order to assess whether ozanimod agonism on $S1P_1$ could account for such beneficial action, we tested the effects of AUY954 and A971432, $S1P_1$, and $S1P_5$ agonist respectively, on EAE corticostriatal synaptic transmission. To this aim, we incubated EAE striatal slices for 2 h ex vivo with AUY954 (300 nM), A971432 (200 nM), or vehicle and performed whole-cell voltage-clamp recordings from MSNs to evaluate both pre- and post- synaptic parameters of sEPSCs. The statistical analysis of sEPSC kinetics, specifically decay time and half width, showed that both AUY954 and A971432 reduced, at least in part, the duration of the sEPSCs that is significantly increased in EAE striatum (Figure 6). Specifically, AUY954 was the most effective drug modulating both half width and decay time, reaching values similar to those of the control condition (EAE $n = 15$, EAE + AUY $n = 7$, control mice $n = 9$; decay time: EAE vs. control $p < 0.05$; EAE vs. EAE + AUY $p < 0.01$; half width: EAE vs. control $p < 0.001$; EAE vs. EAE + AUY $p < 0.01$; Figure 6). Treatment with A971432, instead, had a minor effect, as showed by the incomplete recovery of the decay time (EAE + A971432 $n = 12$; EAE vs. EAE + A971432: decay time $p > 0.05$; half width $p < 0.05$; Figure 6).

The frequency and the amplitude of sEPSCs were unchanged under both treatments with respect to EAE-vehicle condition (EAE $n = 15$, EAE + AUY $n = 7$, EAE + A971432 $n = 12$; control mice $n = 9$; $p > 0.05$ vs. EAE mice, data not shown).

These results show that a local and brief $S1P_1$ modulation ameliorates EAE striatal synaptic dysfunction more efficiently than $S1P_5$ agonist treatment.

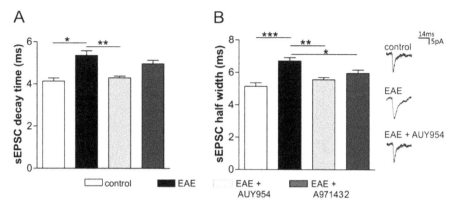

Figure 6. Ex vivo treatment with S1P$_1$ and S1P$_5$ agonists ameliorates glutamatergic dysfunction in EAE striatal slices. Whole-cell patch clamp recordings from MSNs show that the kinetics of the glutamatergic currents, decay time (**A**) and half width (**B**), were increased in EAE striatum and were completely rescued after 2 h incubation with AUY954 (300 nM, S1P$_1$ specific). A971432 (200 nM, S1P5 specific) only partially rescued the EAE sEPSC kinetics. On the right, representative peaks of electrophysiological recordings in the different experimental conditions are shown. Data are presented as mean ± SEM. ANOVA, * $p < 0.05$; ** $p < 0.01$; *** $p < 0.001$.

3.5. In Vivo Treatment with S1P$_1$ Selective Agonist Ameliorates EAE Disease

The next experiments aimed at investigating a direct neuroprotective effect of S1P$_1$ modulator in vivo in the EAE model. We preventively treated EAE mice with two different doses of AUY954 (2.7 µg/day and 0.55 µg/day) and vehicle, by means of a continuous icv infusion for four weeks. The higher dose was selected based on the maximum concentration available for this experimental condition. Since the effect on clinical score between high and low dose of AUY954 was similar (data not shown), the data were gathered together. We observed that daily central treatments with the drug ameliorated the clinical score of the disease (EAE-AUY954 icv $N = 15$, EAE-vhl $N = 6$; Mann-Whitney test $p < 0.05$; Figure 7A).

To address the possibility that the amelioration of the clinical score was mediated by a peripheral effect of the drug on lymphocytes, we counted the total number of CD3+ cells in peripheral blood samples of EAE mice that received the two different dosages of AUY954 or vehicle. As expected, EAE induced a significant increase in T lymphocytes, compared to healthy mice ($N = 4$, controls $N = 3$; $p < 0.05$; Figure 7B). Interestingly, irrespective of the dose administered, we did not observe any significant drop in T lymphocyte counts in blood samples of AUY954-treated mice compared to EAE-vehicle mice ($N = 6$, dosage 0.55 µg/day; $N = 5$, 2.7 µg/day; EAE-high and EAE-low $p < 0.05$ compared to control; Figure 7B). Of note, AUY954-treated EAE mice with zero score showed similar CD3+ counts in comparison to EAE-vhl sick mice (mean score 2.5). In accordance, there was no correlation between clinical score and T cell counts in AUY954-treated mice ($r^2 = 0.06$; Figure 7C), further supporting the idea that the beneficial effect of AUY954 treatment on clinical disability was due to a central effect of the drug.

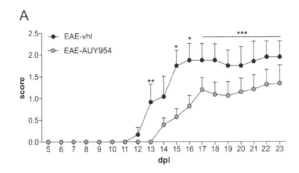
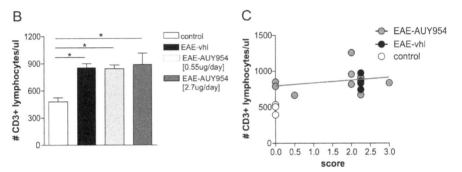

Figure 7. AUY954 icv treatment ameliorates EAE clinical disability without affecting T cell absolute count. (**A**) The graph shows representative clinical course of EAE mice treated for four weeks with two different AUY954 dosages (2.7 µg/day and 0.55 µg/day) or vehicle, preventively delivered by icv infusion. AUY954 icv treatment significantly ameliorated EAE disease progression. Mann-Whitney test on day 13, 15, 16, and from 17 to 23 days post immunization (dpi) by cumulating the data, * $p < 0.05$; ** $p < 0.01$; *** $p < 0.001$). (**B**) CD3+ lymphocytes were counted in the peripheral blood of EAE mice (21 dpi) receiving vehicle (vhl) or different AUY954 dosages, 2.7 µg/day and 0.55 µg/day. No significant reduction was observed at any dosage in comparison to EAE-vhl mice. Conversely, T cell count was significantly less in healthy mice in comparison to other EAE groups. Data are presented as mean ± SEM; ANOVA Tukey's $p < 0.05$. (**C**) Statistical analysis between clinical score and T cell count shows the lack of correlation between these two parameters in AUY954-EAE mice ($r^2 = 0.06$).

4. Discussion

In the present study, we investigated the modulatory effects of ozanimod on inflammatory glutamate-mediated excitotoxicity, a pathological feature of EAE and MS brains and we evaluated the different involvement of sphingosine receptor subtypes ($S1P_1$ and $S1P_5$) in the EAE model. Two main results emerged: first, the $S1P_1/S1P_5$ modulator ozanimod has central neuroprotective effects likely mediated by an action on microglia cells and infiltrating lymphocytes, resulting in a reduced release of proinflammatory cytokines, the main players of inflammatory synaptopathy. Second, the central delivery of a selective $S1P_1$ modulator showed neuroprotective effects, in terms of both EAE clinical score and inflammatory synaptopathy, suggesting a primary involvement of this receptor subtype in ozanimod-induced neuroprotection also in MS.

Glutamate-mediated excitotoxicity is increasingly regarded as a relevant pathogenic mechanism of neurodegeneration in MS [37,38]. Importantly, long-lasting potentiation of the glutamatergic transmission, meaning increased glutamate release from presynaptic terminals and prolonged post-synaptic action of the neurotransmitter, contributes to dendritic spine degeneration and neuronal loss in EAE [2]. In the last years, the recognition of excitotoxicity as a pathogenic mechanism in MS has brought to light a novel therapeutic target, in which the S1P receptor family may play a role. In this respect, fingolimod, the first S1P modulator developed for MS treatment, has been shown to exert its therapeutic actions via a multimodal mechanism, which includes not only the expected T cell retention in lymph nodes [39], but also an anti-inflammatory and anti-excitotoxic action in the CNS. In particular, in EAE mice, oral fingolimod was able to restore presynaptic and postsynaptic alterations of glutamatergic transmission and to promote the recovery of dendritic spines [12], probably owing to its ability to suppress T cell infiltration into the brain and to dampen microglia activation and astrogliosis [23,24]. More interestingly, fingolimod was found to reduce glutamate-mediated intracortical excitability measured by paired-pulse TMS in patients with RRMS [40]. Noteworthy, the anti-excitotoxic activity of fingolimod has been proven in models of "pure" excitotoxicity. Fingolimod was shown to protect against excitotoxic insult in cortical neuronal cultures and organotypic slices [25–27], as well as in the in vivo models of excitotoxicity induced by kainic acid (KA) or glutamate [26,27].

Fingolimod is a nonselective S1P agonist, showing affinity for $S1P_{1,3-5}$, which does not allow to discriminate which receptor subtype is involved in such synaptic activity. In the search for S1P receptor modulators with direct CNS activity, drug design studies have focused on $S1P_1$ and $S1P_5$ agonists, like siponimod that has shown remarkable neuroprotective effects in human and experimental MS [13,41]. In particular, icv delivery of siponimod in EAE mice improved clinical disability during the late phase of the disease and ameliorated the GABAergic defects of EAE striatum but not the glutamatergic counterpart [13]. Of note, an enhanced survival of GABA-ergic neurons as well as a reduced neuroinflammation accounted for such synaptic effect. The neuroprotective effect of siponimod might involve the pro-survival signaling mediated by brain-derived neurotrophic factor (BDNF), as observed with fingolimod [42]. Recently, activation of AKT and ERK kinases as well as an increase of BDNF levels have been associated with a neuroprotective effect mediated by selective $S1P_5$ agonists in other neurological disease [42–44].

To better clarify the putative mechanisms underlying the anti-synaptopathic activity of $S1P_{1/5}$ agonism at glutamatergic synapses, we focused on ozanimod, a MS drug recently approved in the United States for clinically isolated syndrome, RRMS, and active SPMS (https://packageinserts.bms.com/pi/pi_zeposia.pdf). Among the multiple beneficial effects obtained during the clinical trials, brain volume loss was significantly reduced compared to treatment with IFN-β-1a both at 12 (SUNBEAM) and 24 (RADIANCE B) months in the ozanimod arms, suggesting a protective effect on brain atrophy in both cohorts [30,45]. In accordance to direct CNS effects exerted by ozanimod, oral therapeutic administration of this drug was shown to ameliorate EAE clinical score in a dose-dependent manner [29]. This result seems to be related only in part to the restriction of autoreactive lymphocyte trafficking, as indicated by the lack of lymphopenia in mice receiving the lowest, but clinically overt effective dose used in the study. Hence, the authors argued that other direct CNS effects exerted by ozanimod may exist [29].

Here, we focused on ozanimod action on EAE synaptic alterations by studying its effects in an ex vivo experimental paradigm. First, we found that ozanimod corrected both the presynaptic and the postsynaptic alterations of the sEPSCs in EAE corticostriatal slices. Next, we investigated the $S1P_1$ agonist AUY954 and the $S1P_5$ agonist A971432 in ex vivo system and observed that the engagement of $S1P_1$ is more efficient in ameliorating the glutamatergic alterations in the EAE striatum, with respect to $S1P_5$. Altogether these ex vivo results highlight a specific and fully protective action of ozanimod relatively to AUY954 (no effect on sEPSC frequency) and A971432 (partial effect on sEPSC kinetics and no effect on frequency).

Overall, these observations further strengthen the idea that the complexity of the sphingosine system implies a fine-tuning regulation of the balance between receptor affinity of the agonist and receptor levels. It is still missing a clear picture of potential changes in the expression of S1P receptors in EAE and in MS [46], not only in the different brain area but also at cellular levels. Moreover, the trafficking of S1P receptors and the downstream signaling might be differentially modulated by each drug and by the duration of the treatment [24].

Although further investigations are necessary to define specific intracellular pathways triggered by each S1P receptor modulator, our results further support a neuroprotective effect of S1P-signaling in EAE disease, mediated by an impairment of excitotoxic damage through modulation of $S1P_1$ on microglia cells and T lymphocytes. We previously demonstrated that TNF release from activated microglia cells could mimic the glutamatergic alterations observed in EAE striatum [28]. Similarly, we observed that T lymphocytes derived either from the spleen of EAE mice [28] or from blood of RRMS patients during a relapse (chimeric ex vivo MS model), could exert a direct glutamatergic synaptotoxic effect on healthy murine striatal slices [11]. As already mentioned, $S1P_1$ appears to be the most expressed in multiple cell types in the CNS, including neurons, astrocytes, and microglia. Astrocytes have been previously demonstrated to respond to several S1P compounds both in vitro [47,48] and in vivo. In particular, Choi and colleagues [24] elegantly demonstrated a clear $S1P_1$ involvement in neuroprotection by S1P agonists (Fingolimod and AUY954) in the EAE model, by inducing EAE in conditional null mutants for $S1P_1$ in neurons and in astrocytes. These conditional mutants exhibited the predicted pattern of stable disease progression, although clinical signs were attenuated when $S1P_1$ was deleted from astrocytes, indicating a non-neuronal and likely astrocyte locus for $S1P_1$ signaling [24]. Of note, the involvement of other non-CNS cell lineages expressing $S1P_1$, like microglia [49] and endothelial cells [50], which have roles in discrete phases of EAE or MS, have been little explored so far. A neuroprotective effect mediated by $S1P_1$ modulation in microglia cells was observed in mice challenged with ischemia and orally treated with AUY954, and in S1P1 knockout mice [51]. Furthermore, the same authors showed that in LPS-stimulated mouse primary microglia transfected with $S1P_1$ siRNA, gene expression of proinflammatory mediators such as TNF and IL-1β, but not IL-6, was suppressed [51]. In the present manuscript, we suggest the involvement of these immune cells in the anti-excitotoxic effect mediated by $S1P_1$ modulation. By immunofluorescence experiments we showed that bath application of ozanimod (the same condition of electrophysiology) has an anti-inflammatory effect and reduces the expression of the synaptopathic molecules IL-1β and, to a lesser extent, TNF in lesioned areas of EAE striatum, characterized by strong microglia/macrophage activation. In line with this, a global qPCR analysis of EAE striatal slices showed a significant reduction of IL-1β and TNF mRNA after ozanimod treatment together with an up-regulation of the mRNA of the M2-like marker FIZZ1. The anti-inflammatory effect of ozanimod was further supported by in vitro experiments showing that a direct and brief treatment of Th1-activated microglial BV2 cells with ozanimod was sufficient to downregulate the mRNAs of important pro-inflammatory cytokines.

Another potential mechanism responsible of ozanimod anti-inflammatory and neuroprotective activity involves T cells. Here, we showed that pretreatment of EAE T cells with ozanimod abrogates the excitotoxic effects of EAE T cells, likely through an anti-inflammatory action. These results highlight an additional way through which ozanimod could exert its beneficial effect also in MS patients, not only by reducing the infiltration of T cells, but also by modulating synaptotoxic T cells that circulate in the EAE/MS brain. It is worth noting that, by using an ex vivo MS chimeric model, we recently demonstrated that T cells derived from active MS patients exert a direct synaptotoxic effect, similarly to EAE lymphocytes [13]. Therefore, we expect that RRMS lymphocytes treated with ozanimod are less synaptotoxic in ex vivo MS chimeric experiments.

We believe that the central anti-excitotoxic effect exerted by $S1P_1$ modulation could have an impact also on clinical disability, as suggested by AUY954 icv delivery in EAE mice. To exclude a $S1P_1$-mediated peripheral restraint of T cells during treatment we verified the lack of lymphopenia in mice receiving central infusion of AUY954. These results are in accordance with the identification of non-immunological CNS mechanisms mediated by FTY720 and AUY954 [24].

In conclusion, in this study we show a selective and central action of $S1P_1$ modulators in ameliorating EAE glutamatergic synaptopathy and disease. Furthermore, we propose that ozanimod exerts a central anti-inflammatory activity by engaging the $S1P_1$ expressed by microglia cells and peripheral T cell infiltrating in the brain. Altogether these data strengthen the relevance of a neuro-immunomodulatory and -protective action of $S1P_1$ agonist in MS therapy.

Author Contributions: A.M., A.G., D.C., and G.M. conceived and designed the study. A.M., A.G., L.G., F.R.R., F.D.V., D.F., A.B., E.D., V.V., L.V., S.B. (Silvia Bullitta), K.S., S.C., S.B. (Sara Balletta), M.N., F.B. and M.S.B. acquired and analyzed the data. A.G., A.M., D.C., and G.M. drafted the manuscript. All the authors critically revised the article for important intellectual content and approved the version to be published. A.M. and A.G. contributed equally to this work, as first authors. D.C. and G.M. contributed equally to this work, as senior authors. All authors have read and agreed to the published version of the manuscript.

Funding: The study was supported by: a Celgene grant to D.C., a FISM grant (Fondazione Italiana Sclerosi Multipla-cod. 2019/S/1) to D.C. and F.R.R., national funding of the Italian Ministry of University and Research (MIUR-PRIN 2017-cod. 2017K55HLC) to D.C., and of the Italian Ministry of Health (GR-2016-02361163 to A.M.; GR-2016-02362380 to D.F.; GR-2018-12366154 to A.G.; RF-2018-12366144 to D.C. and G.M.; Ricerca corrente to IRCCS San Raffaele Pisana; Ricerca corrente and '5 per mille' public funding to IRCCS Neuromed). F.D.V. was supported by a research fellowship FISM (cod. 2018/B/2).

Acknowledgments: The authors thank Massimo Tolu for helpful technical assistance.

Conflicts of Interest: DC is the recipient of an Institutional grant from Celgene. No personal compensation was received. The founding sponsors had no role in the design of the study; in the collection, analyses, or interpretation of data; in the writing of the manuscript, and in the decision to publish the results. The other authors declare that they have no competing interests.

References

1. Frischer, J.M.; Bramow, S.; Dal-Bianco, A.; Lucchinetti, C.F.; Rauschka, H.; Schmidbauer, M.; Laursen, H.; Sorensen, P.S.; Lassmann, H. The relation between inflammation and neurodegeneration in multiple sclerosis brains. *Brain* **2009**, *132*, 1175–1189. [CrossRef]
2. Centonze, D.; Muzio, L.; Rossi, S.; Furlan, R.; Bernardi, G.; Martino, G. The link between inflammation, synaptic transmission and neurodegeneration in multiple sclerosis. *Cell Death Differ.* **2010**, *17*, 1083–1091. [CrossRef]
3. Rossi, S.; Furlan, R.; De Chiara, V.; Motta, C.; Studer, V.; Mori, F.; Musella, A.; Bergami, A.; Muzio, L.; Bernardi, G.; et al. Interleukin-1β causes synaptic hyperexcitability in multiple sclerosis. *Ann. Neurol.* **2012**, *71*, 76–83. [CrossRef]
4. Reich, D.S.; Lucchinetti, C.F.; Calabresi, P.A. Multiple Sclerosis. *N. Engl. J. Med.* **2018**, *378*, 169–180. [CrossRef]
5. Mandolesi, G.; Gentile, A.; Musella, A.; Fresegna, D.; De Vito, F.; Bullitta, S.; Sepman, H.; Marfia, G.A.; Centonze, D. Synaptopathy connects inflammation and neurodegeneration in multiple sclerosis. *Nat. Rev. Neurol.* **2015**, *11*, 711. [CrossRef]
6. Dendrou, C.A.; Fugger, L.; Friese, M.A. Immunopathology of multiple sclerosis. *Nat. Rev. Immunol.* **2015**, *15*, 545–558. [CrossRef]
7. Mandolesi, G.; Grasselli, G.; Musella, A.; Gentile, A.; Musumeci, G.; Sepman, H.; Haji, N.; Fresegna, D.; Bernardi, G.; Centonze, D. GABAergic signaling and connectivity on Purkinje cells are impaired in experimental autoimmune encephalomyelitis. *Neurobiol. Dis.* **2012**, *46*, 414–424. [CrossRef]
8. Mandolesi, G.; Musella, A.; Gentile, A.; Grasselli, G.; Haji, N.; Sepman, H.; Fresegna, D.; Bullitta, S.; De Vito, F.; Musumeci, G.; et al. Interleukin-1β alters glutamate transmission at purkinje cell synapses in a mouse model of multiple sclerosis. *J. Neurosci.* **2013**, *33*, 12105–12121. [CrossRef]
9. Rossi, S.; Muzio, L.; De Chiara, V.; Grasselli, G.; Musella, A.; Musumeci, G.; Mandolesi, G.; De Ceglia, R.; Maida, S.; Biffi, E.; et al. Impaired striatal GABA transmission in experimental autoimmune encephalomyelitis. *Brain Behav. Immun.* **2011**, *25*, 947–956. [CrossRef]
10. Azevedo, C.J.; Kornak, J.; Chu, P.; Sampat, M.; Okuda, D.T.; Cree, B.A.; Nelson, S.J.; Hauser, S.L.; Pelletier, D. In vivo evidence of glutamate toxicity in multiple sclerosis. *Ann. Neurol.* **2014**, *76*, 269–278. [CrossRef]
11. Gentile, A.; De Vito, F.; Fresegna, D.; Rizzo, F.R.; Bullitta, S.; Guadalupi, L.; Vanni, V.; Buttari, F.; Stampanoni Bassi, M.; Leuti, A.; et al. Peripheral T cells from multiple sclerosis patients trigger synaptotoxic alterations in central neurons. *Neuropathol. Appl. Neurobiol.* **2019**, *46*, 160–170. [CrossRef]
12. Rossi, S.; Lo Giudice, T.; De Chiara, V.; Musella, A.; Studer, V.; Motta, C.; Bernardi, G.; Martino, G.; Furlan, R.; Martorana, A.; et al. Oral fingolimod rescues the functional deficits of synapses in experimental autoimmune encephalomyelitis. *Br. J. Pharmacol.* **2012**, *165*, 861–869. [CrossRef]

13. Gentile, A.; Musella, A.; Bullitta, S.; Fresegna, D.; De Vito, F.; Fantozzi, R.; Piras, E.; Gargano, F.; Borsellino, G.; Battistini, L.; et al. Siponimod (BAF312) prevents synaptic neurodegeneration in experimental multiple sclerosis. *J. Neuroinflamm.* **2016**, *13*, 207. [CrossRef] [PubMed]
14. Parodi, B.; Rossi, S.; Morando, S.; Cordano, C.; Bragoni, A.; Motta, C.; Usai, C.; Wipke, B.T.; Scannevin, R.H.; Mancardi, G.L.; et al. Fumarates modulate microglia activation through a novel HCAR2 signaling pathway and rescue synaptic dysregulation in inflamed CNS. *Acta Neuropathol.* **2015**, *130*, 279–295. [CrossRef] [PubMed]
15. Smith, P.A.; Schmid, C.; Zurbruegg, S.; Jivkov, M.; Doelemeyer, A.; Theil, D.; Dubost, V.; Beckmann, N. Fingolimod inhibits brain atrophy and promotes brain-derived neurotrophic factor in an animal model of multiple sclerosis. *J. Neuroimmunol.* **2018**, *318*, 103–113. [CrossRef]
16. Derfuss, T.; Mehling, M.; Papadopoulou, A.; Bar-Or, A.; Cohen, J.A.; Kappos, L. Advances in oral immunomodulating therapies in relapsing multiple sclerosis. *Lancet Neurol.* **2020**, *19*, 336–347. [CrossRef]
17. Colombo, E.; Di Dario, M.; Capitolo, E.; Chaabane, L.; Newcombe, J.; Martino, G.; Farina, C. Fingolimod may support neuroprotection via blockade of astrocyte nitric oxide. *Ann. Neurol.* **2014**, *76*, 325–337. [CrossRef]
18. O'Sullivan, S.; Dev, K.K. Sphingosine-1-phosphate receptor therapies: Advances in clinical trials for CNS-related diseases. *Neuropharmacology* **2017**, *113*, 597–607. [CrossRef]
19. Subei, A.M.; Cohen, J.A. Sphingosine 1-phosphate receptor modulators in multiple sclerosis. *CNS Drugs* **2015**, *29*, 565–575. [CrossRef]
20. Brinkmann, V.; Billich, A.; Baumruker, T.; Heining, P.; Schmouder, R.; Francis, G.; Aradhye, S.; Burtin, P. Fingolimod (FTY720): Discovery and development of an oral drug to treat multiple sclerosis. *Nat. Rev. Drug Discov.* **2010**, *9*, 883–897. [CrossRef]
21. Kappos, L.; Bar-Or, A.; Cree, B.A.C.; Fox, R.J.; Giovannoni, G.; Gold, R.; Vermersch, P.; Arnold, D.L.; Arnould, S.; Scherz, T.; et al. Siponimod versus placebo in secondary progressive multiple sclerosis (EXPAND): A double-blind, randomised, phase 3 study. *Lancet* **2018**, *391*, 1263–1273. [CrossRef]
22. De Stefano, N.; Silva, D.G.; Barnett, M.H. Effect of Fingolimod on Brain Volume Loss in Patients with Multiple Sclerosis. *CNS Drugs* **2017**, *31*, 289–305. [CrossRef]
23. Noda, H.; Takeuchi, H.; Mizuno, T.; Suzumura, A. Fingolimod phosphate promotes the neuroprotective effects of microglia. *J. Neuroimmunol.* **2013**, *256*, 13–18. [CrossRef]
24. Choi, J.W.; Gardell, S.E.; Herr, D.R.; Rivera, R.; Lee, C.-W.; Noguchi, K.; Teo, S.T.; Yung, Y.C.; Lu, M.; Kennedy, G.; et al. FTY720 (fingolimod) efficacy in an animal model of multiple sclerosis requires astrocyte sphingosine 1-phosphate receptor 1 (S1P1) modulation. *Proc. Natl. Acad. Sci. USA* **2011**, *108*, 751–756. [CrossRef]
25. Di Menna, L.; Molinaro, G.; Di Nuzzo, L.; Riozzi, B.; Zappulla, C.; Pozzilli, C.; Turrini, R.; Caraci, F.; Copani, A.; Battaglia, G.; et al. Fingolimod protects cultured cortical neurons against excitotoxic death. *Pharmacol. Res.* **2013**, *67*, 1–9. [CrossRef]
26. Cipriani, R.; Chara, J.C.; Rodriguez-Antiguedad, A.; Matute, C. FTY720 attenuates excitotoxicity and neuroinflammation. *J. Neuroinflamm.* **2015**, *12*, 86. [CrossRef]
27. Luchtman, D.; Gollan, R.; Ellwardt, E.; Birkenstock, J.; Robohm, K.; Siffrin, V.; Zipp, F. In vivo and in vitro effects of multiple sclerosis immunomodulatory therapeutics on glutamatergic excitotoxicity. *J. Neurochem.* **2016**, *136*, 971–980. [CrossRef]
28. Centonze, D.; Muzio, L.; Rossi, S.; Cavasinni, F.; De Chiara, V.; Bergami, A.; Musella, A.; D'Amelio, M.; Cavallucci, V.; Martorana, A.; et al. Inflammation Triggers Synaptic Alteration and Degeneration in Experimental Autoimmune Encephalomyelitis. *J. Neurosci.* **2009**, *29*, 3442–3452. [CrossRef]
29. Scott, F.L.; Clemons, B.; Brooks, J.; Brahmachary, E.; Powell, R.; Dedman, H.; Desale, H.G.; Timony, G.A.; Martinborough, E.; Rosen, H.; et al. Ozanimod (RPC1063) is a potent sphingosine-1-phosphate receptor-1 (S1P1) and receptor-5 (S1P5) agonist with autoimmune disease-modifying activity. *Br. J. Pharmacol.* **2016**, *173*, 1778–1792. [CrossRef]
30. Cohen, J.A.; Comi, G.; Selmaj, K.W.; Bar-Or, A.; Arnold, D.L.; Steinman, L.; Hartung, H.P.; Montalban, X.; Kubala Havrdová, E.; Cree, B.A.C.; et al. Safety and efficacy of ozanimod versus interferon beta-1a in relapsing multiple sclerosis (RADIANCE): A multicentre, randomised, 24-month, phase 3 trial. *Lancet Neurol.* **2019**, *18*, 1021–1033. [CrossRef]

31. Hobson, A.D.; Harris, C.M.; van der Kam, E.L.; Turner, S.C.; Abibi, A.; Aguirre, A.L.; Bousquet, P.; Kebede, T.; Konopacki, D.B.; Gintant, G.; et al. Discovery of A-971432, An Orally Bioavailable Selective Sphingosine-1-Phosphate Receptor 5 (S1P5) Agonist for the Potential Treatment of Neurodegenerative Disorders. *J. Med. Chem.* **2015**, *58*, 9154–9170. [CrossRef]
32. Pan, S.; Mi, Y.; Pally, C.; Beerli, C.; Chen, A.; Guerini, D.; Hinterding, K.; Nuesslein-Hildesheim, B.; Tuntland, T.; Lefebvre, S.; et al. A monoselective sphingosine-1-phosphate receptor-1 agonist prevents allograft rejection in a stringent rat heart transplantation model. *Chem. Biol.* **2006**, *13*, 1227–1234. [CrossRef]
33. Mandolesi, G.; De Vito, F.; Musella, A.; Gentile, A.; Bullitta, S.; Fresegna, D.; Sepman, H.; Di Sanza, C.; Haji, N.; Mori, F.; et al. miR-142-3p Is a Key Regulator of IL-1beta-Dependent Synaptopathy in Neuroinflammation. *J. Neurosci.* **2017**, *37*, 546–561. [CrossRef]
34. Gentile, A.; Fresegna, D.; Federici, M.; Musella, A.; Rizzo, F.R.; Sepman, H.; Bullitta, S.; De Vito, F.; Haji, N.; Rossi, S.; et al. Dopaminergic dysfunction is associated with IL-1β-dependent mood alterations in experimental autoimmune encephalomyelitis. *Neurobiol. Dis.* **2015**, *74*, 347–358. [CrossRef]
35. Chu, F.; Shi, M.; Zheng, C.; Shen, D.; Zhu, J.; Zheng, X.; Cui, L. The roles of macrophages and microglia in multiple sclerosis and experimental autoimmune encephalomyelitis. *J. Neuroimmunol.* **2018**, *318*, 1–7. [CrossRef]
36. Fletcher, J.M.; Lalor, S.J.; Sweeney, C.M.; Tubridy, N.; Mills, K.H.G. T cells in multiple sclerosis and experimental autoimmune encephalomyelitis. *Clin. Exp. Immunol.* **2010**, *162*, 1–11. [CrossRef]
37. Macrez, R.; Stys, P.K.; Vivien, D.; Lipton, S.A.; Docagne, F. Mechanisms of glutamate toxicity in multiple sclerosis: Biomarker and therapeutic opportunities. *Lancet Neurol.* **2016**, *15*, 1089–1102. [CrossRef]
38. Pitt, D.; Nagelmeier, I.E.; Wilson, H.C.; Raine, C.S. Glutamate uptake by oligodendrocytes: Implications for excitotoxicity in multiple sclerosis. *Neurology* **2003**, *61*, 1113–1120. [CrossRef]
39. Chaudhry, B.Z.; Cohen, J.A.; Conway, D.S. Sphingosine 1-Phosphate Receptor Modulators for the Treatment of Multiple Sclerosis. *Neurotherapeutics* **2017**, *14*, 859–873. [CrossRef]
40. Landi, D.; Vollaro, S.; Pellegrino, G.; Mulas, D.; Ghazaryan, A.; Falato, E.; Pasqualetti, P.; Rossini, P.M.; Filippi, M.M. Oral fingolimod reduces glutamate-mediated intracortical excitability in relapsing-remitting multiple sclerosis. *Clin. Neurophysiol.* **2015**, *126*, 165–169. [CrossRef]
41. Behrangi, N.; Fischbach, F.; Kipp, M. Mechanism of Siponimod: Anti-Inflammatory and Neuroprotective Mode of Action. *Cells* **2019**, *8*, 24. [CrossRef] [PubMed]
42. Deogracias, R.; Yazdani, M.; Dekkers, M.P.J.; Guy, J.; Ionescu, M.C.S.; Vogt, K.E.; Barde, Y.-A. Fingolimod, a sphingosine-1 phosphate receptor modulator, increases BDNF levels and improves symptoms of a mouse model of Rett syndrome. *Proc. Natl. Acad. Sci. USA* **2012**, *109*, 14230–14235. [CrossRef] [PubMed]
43. Di Pardo, A.; Castaldo, S.; Amico, E.; Pepe, G.; Marracino, F.; Capocci, L.; Giovannelli, A.; Madonna, M.; van Bergeijk, J.; Buttari, F.; et al. Stimulation of S1PR5 with A-971432, a selective agonist, preserves blood-brain barrier integrity and exerts therapeutic effect in an animal model of Huntington's disease. *Hum. Mol. Genet.* **2018**, *27*, 2490–2501. [CrossRef] [PubMed]
44. Ren, M.; Han, M.; Wei, X.; Guo, Y.; Shi, H.; Zhang, X.; Perez, R.G.; Lou, H. FTY720 Attenuates 6-OHDA-Associated Dopaminergic Degeneration in Cellular and Mouse Parkinsonian Models. *Neurochem. Res.* **2017**, *42*, 686–696. [CrossRef]
45. Comi, G.; Kappos, L.; Selmaj, K.W.; Bar-Or, A.; Arnold, D.L.; Steinman, L.; Hartung, H.P.; Montalban, X.; Kubala Havrdová, E.; Cree, B.A.C.; et al. Safety and efficacy of ozanimod versus interferon beta-1a in relapsing multiple sclerosis (SUNBEAM): A multicentre, randomised, minimum 12-month, phase 3 trial. *Lancet Neurol.* **2019**, *18*, 1009–1020. [CrossRef]
46. Brana, C.; Frossard, M.J.; Pescini Gobert, R.; Martinier, N.; Boschert, U.; Seabrook, T.J. Immunohistochemical detection of sphingosine-1-phosphate receptor 1 and 5 in human multiple sclerosis lesions. *Neuropathol. Appl. Neurobiol.* **2014**, *40*, 564–578. [CrossRef]
47. Mullershausen, F.; Craveiro, L.M.; Shin, Y.; Cortes-Cros, M.; Bassilana, F.; Osinde, M.; Wishart, W.L.; Guerini, D.; Thallmair, M.; Schwab, M.E.; et al. Phosphorylated FTY720 promotes astrocyte migration through sphingosine-1-phosphate receptors. *J. Neurochem.* **2007**, *102*, 1151–1161. [CrossRef]
48. O'Sullivan, C.; Schubart, A.; Mir, A.K.; Dev, K.K. The dual S1PR1/S1PR5 drug BAF312 (Siponimod) attenuates demyelination in organotypic slice cultures. *J. Neuroinflamm.* **2016**, *13*, 31. [CrossRef]
49. Tham, C.-S.; Lin, F.-F.; Rao, T.S.; Yu, N.; Webb, M. Microglial activation state and lysophospholipid acid receptor expression. *Int. J. Dev. Neurosci.* **2003**, *21*, 431–443. [CrossRef]

50. Oo, M.L.; Thangada, S.; Wu, M.-T.; Liu, C.H.; Macdonald, T.L.; Lynch, K.R.; Lin, C.-Y.; Hla, T. Immunosuppressive and anti-angiogenic sphingosine 1-phosphate receptor-1 agonists induce ubiquitinylation and proteasomal degradation of the receptor. *J. Biol. Chem.* **2007**, *282*, 9082–9089. [CrossRef]
51. Gaire, B.P.; Lee, C.-H.; Sapkota, A.; Lee, S.Y.; Chun, J.; Cho, H.J.; Nam, T.-G.; Choi, J.W. Identification of Sphingosine 1-Phosphate Receptor Subtype 1 (S1P1) as a Pathogenic Factor in Transient Focal Cerebral Ischemia. *Mol. Neurobiol.* **2018**, *55*, 2320–2332. [CrossRef]

 © 2020 by the authors. Licensee MDPI, Basel, Switzerland. This article is an open access article distributed under the terms and conditions of the Creative Commons Attribution (CC BY) license (http://creativecommons.org/licenses/by/4.0/).

Article

The S100B Inhibitor Pentamidine Ameliorates Clinical Score and Neuropathology of Relapsing—Remitting Multiple Sclerosis Mouse Model

Gabriele Di Sante [1,2], Susanna Amadio [3,†], Beatrice Sampaolese [4,†], Maria Elisabetta Clementi [4], Mariagrazia Valentini [1], Cinzia Volonté [3,5], Patrizia Casalbore [5], Francesco Ria [1,2,*] and Fabrizio Michetti [6,7,*]

1. Department of Translational Medicine and Surgery, Section of General Pathology, Università Cattolica del Sacro Cuore, Largo Francesco Vito 1, 00168 Rome, Italy; gabriele.disante@unicatt.it (G.D.S.); mariagrazia.valentini@unicatt.it (M.V.)
2. Fondazione Policlinico Universitario A. Gemelli IRCCS, Largo Agostino Gemelli 1-8, 00168 Rome, Italy
3. Cellular Neurobiology Unit, Preclinical Neuroscience, IRCCS Santa Lucia Foundation, Via del Fosso di Fiorano 65, 00143 Rome, Italy; s.amadio@hsantalucia.it (S.A.); c.volonte@hsantalucia.it (C.V.)
4. Istituto di Scienze e Tecnologie Chimiche "Giulio Natta" SCITEC-CNR, Largo Francesco Vito 1, 00168 Rome, Italy; beatrice.sampaolese@scitec.cnr.it (B.S.); elisabetta.clementi@scitec.cnr.it (M.E.C.)
5. Institute for Systems Analysis and Computer Science, IASI-CNR, Largo Francesco Vito 1, 00168 Rome, Italy; patrizia.casalbore@cnr.it
6. Department of Neuroscience, Università Cattolica del Sacro Cuore, Largo Francesco Vito 1, 00168 Rome, Italy
7. IRCCS San Raffaele Scientific Institute, Università Vita-Salute San Raffaele, 20132 Milan, Italy
* Correspondence: francesco.ria@unicatt.it (F.R.); fabrizio.michetti@unicatt.it (F.M.); Tel.: +39-06-3015-4914 (F.R.); +39-06-3015-5848 (F.M.)
† Equally contributed as second authors.

Received: 24 February 2020; Accepted: 17 March 2020; Published: 18 March 2020

Abstract: S100B is an astrocytic protein acting either as an intracellular regulator or an extracellular signaling molecule. A direct correlation between increased amount of S100B and demyelination and inflammatory processes has been demonstrated. The aim of this study is to investigate the possible role of a small molecule able to bind and inhibit S100B, pentamidine, in the modulation of disease progression in the relapsing–remitting experimental autoimmune encephalomyelitis mouse model of multiple sclerosis. By the daily evaluation of clinical scores and neuropathologic-molecular analysis performed in the central nervous system, we observed that pentamidine is able to delay the acute phase of the disease and to inhibit remission, resulting in an amelioration of clinical score when compared with untreated relapsing–remitting experimental autoimmune encephalomyelitis mice. Moreover, we observed a significant reduction of proinflammatory cytokines expression levels in the brains of treated versus untreated mice, in addition to a reduction of nitric oxide synthase activity. Immunohistochemistry confirmed that the inhibition of S100B was able to modify the neuropathology of the disease, reducing immune infiltrates and partially protecting the brain from the damage. Overall, our results indicate that pentamidine targeting the S100B protein is a novel potential drug to be considered for multiple sclerosis treatment.

Keywords: S100B; multiple sclerosis; relapsing–remitting experimental autoimmune encephalomyelitis; pentamidine

1. Introduction

Multiple sclerosis (MS) is an autoimmune disease now recognized as a global disease, affecting more than 2.3 million persons worldwide, with an occurrence especially high in Western Europe and North America [1]. The pathologic hallmarks of the disease are demyelination and axonal loss, characterized by focal lesions/plaques that are scattered throughout the white and gray matter of the central nervous system (CNS). These plaques represent a combination of pathologic features, including edema, inflammation, gliosis, demyelination, and/or axonal loss. A variety of pathogenic processes have been implicated in plaque formation, including oxidative stress promoted by macrophages/microglia, neurotoxic factors secreted by activated T-cells, and autoantibodies directed at self-antigens. However, the key mechanisms in this disease are essentially unknown. Hence, the identification of such factors is still a necessary prerequisite to develop more efficacious therapeutic strategies [2].

S100B [3] is a small EF-related Ca^{2+}/and Zn^{2+}/binding protein, which is mainly synthesized by astrocytes, and, to a lesser extent, by oligodendrocytes in the CNS. It exerts both intracellular and extracellular actions. While a clearly defined intracellular function at present has not been delineated for this protein, secreted S100B is regarded to act as a paracrine and autocrine factor for astrocytes, with concentration-dependent effects. Under physiological conditions, astrocytes secrete S100B, which exerts a neurotrophic action in nanomolar concentration. Under stress conditions, including nervous tissue inflammation, astrocytes secrete S100B that has a neurotoxic effect at micromolar concentration, behaving as a danger/damage-associated molecular pattern (DAMP) molecule. S100B has been shown to interact with surrounding cell types mainly through the activation of the receptor for advanced glycation endproducts (RAGE), a ubiquitous, transmembrane immunoglobulin-like receptor known to act as both inflammatory intermediary and critical inducer of oxidative stress. S100B in biological fluids is regarded to be a reliable biomarker of active neural injury [4] but, more recently, evidence is accumulating that indicates that this protein plays a key role in the pathogenic processes of neural disorders for which it also acts as a biomarker, including acute brain injury, Alzheimer's disease (AD), Parkinson's disease (PD), and amyotrophic lateral sclerosis [5].

It is reasonable to hypothesize that high concentrations of S100B may play a promoting role in MS, based on correlative evidence. Elevated levels of S100B were detected in the cerebrospinal fluid (CSF) [6,7] and sera [7] of MS patients in the acute phase, being reduced in the stationary phase of the disease, and an increased expression of S100B has been detected in both active demyelinating and chronic active MS plaques [8]. In ex-vivo demyelinating models, a marked astrocytic elevation of S100B was observed upon demyelination, while inhibition of S100B action reduced demyelination and downregulated the expression of inflammatory molecules [7]. In addition, blockade of RAGE has been shown to suppress demyelination in a rodent demyelinating model of experimental autoimmune encephalomyelitis (EAE) [9], and, more recently, the S100B/RAGE axis has been shown to play a crucial role in oligodendrocyte myelination processes [10]. Overall, these data suggest a potential role of S100B in MS pathogenesis and, as a consequence, its suitability as a therapeutic target for the disease. However, the in vivo involvement of this molecule in MS processes has never been studied.

This work aims at investigating the effects of blocking S100B in a recognized experimental in vivo model of MS, such as the relapsing–remitting experimental autoimmune encephalomyelitis (RR–EAE) induced in Swiss Jim Lambert (SJL) mice. To this purpose, pentamidine isethionate (PTM) appears to be a suitable tool, since this antiprotozoal drug has been shown to inhibit S100B activity by blocking the interaction at the Ca^{2+}/p53 site of the protein [11–13]. Here, we will show a clinical amelioration of disease scores in RR–EAE SJL mice after PTM administration, accompanied by coherent variations of neuropathologic/biomolecular parameters.

2. Materials and Methods

2.1. Animal Procedures

RR–EAE in the SJL mouse model: to induce active EAE, we immunized female SJL (8–10 week-old) mice, purchased from Charles River, with an emulsion composed by a fragment of the proteo-lipid protein (PLP139–151, the immunodominant epitope) and complete Freund's adjuvant (CFA4X) containing 4 mg of heat-killed and dried Mycobacterium tuberculosis (strain H37Ra, ATTC 25177, and Bordetella Pertussis toxin (Sigma-Aldrich S.r.l., Milan, Italy). Immunization and treatments are described in the timetable in Figure S1a, according to the procedures described in our previous works [14–17]. Mice have been monitored daily for body weight, development of clinical signs and symptoms (CSS), and for disease remission/relapse. CSS have been scored using the scale described by Miller et al. [18]. In the light of the notion that blood–brain barrier (BBB) is damaged during the demyelinating disease, PTM was administered intraperitoneally, with a dosage of 4 mg/kg; SJL mice were randomly distributed into four different groups: untreated healthy controls (Ctrl), PBS–EAE group (vehicle), PTM-treated healthy controls, and PTM-treated EAE affected mice. Seven days after RR–EAE induction, the group of RR–EAE mice received a daily intraperitoneal administration of PTM (Sigma-Aldrich S.r.l., St. Louis, MO, USA) for 30 days (4 mg/kg).

Each group of mice was composed of 9 mice and the procedure was repeated in two experiments (with a total of 36 EAE affected mice, 18 treated with PTM and 18 with vehicle). Thirty-six additional SJL mice were not immunized but used as controls, 18 untreated and 18 treated with PTM). As expected from this RR–MS mouse model, 10% of mice were withdrawn from the protocol because of excessive reaction to immunization, unresponsiveness/anergy to immunization, or death for unknown causes. For ethical reasons according to the guidelines for animal wellness, mice with exaggerated symptoms were excluded and sacrificed before the foreseen timepoints (3 of the EAE group treated with vehicle and 1 of the EAE group treated with PTM). Hence, we analyzed 32 animals suffering from EAE (17 in the PTM-treated group and 15 in the vehicle-treated group). The animals were sacrificed at onset, remission, and relapse phases of the EAE (at least 5 mice for each timepoint). Specifically, mice euthanized during onset phase were chosen at the beginning of acute phase, when they reached a CSS between 2 and 3. Mice sacrificed during remission phase, were sacrificed when they presented at least 1 point of CSS below the peak reached during acute phase. On the contrary, relapse was analyzed sacrificing mice when they reached at least 0.5 point of CSS above their remission mean CSS. Ctrl, vehicle, and healthy groups were sacrificed accordingly. Mice were perfused with saline solution under deep anesthesia (87.5 mg/Kg ketamine and 12.5 mg/Kg xylazine; 0.1 mL/20 g mouse wt i.p.), the brain was removed and one hemisphere was used for morphological analysis, after fixation (48 h in 4% paraformaldehyde (PFA)), while the lysates extracted from the other hemisphere have been used for molecular biology assays and processed accordingly [19].

2.2. RT–qPCR Assay

Total RNA was isolated with a SV Total RNA Isolation System (Promega, Madison, WI, USA) and RNA concentration was evaluated by spectrophotometric reading at 280 and 260 nm. Total RNA was used for first strand cDNA synthesis with a High-Capacity cDNA Reverse Transcription Kit (Applied Biosystems). PowerUp™ SYBR® Green (Thermo Fisher Scientific, Waltham, MA, USA) Master Mix (2×) reagents were used according to the manufacturer's recommendations. The quantification of gene expression was obtained from Applied Biosystem 7900HT Fast Realtime PCR System. PowerUp™ SYBR® Green (Thermo Fisher Scientific, Waltham, MA, USA) Master Mix (2×) reagents were used according to the manufacturer's recommendations. Primers were bought from SIGMA-Aldrich (St. Louis, MO, USA): for *S100B* and inducible nitric oxide synthase (*iNOS*) after design by NCBI Primer-Blast program (https://www.ncbi.nlm.nih.gov/tools/primer-blast), while the other oligos were deduced from literature [20–24]. Each gene target quantification reaction was performed separately with the respective primer sets (Table S1). Conditions were as follows: 50 °C for 2 min, followed by 95

°C for 10 min, forty cycles at 95 °C for 15 s, followed by 60 °C for 1 min. The melt standard curve was at 95 °C for 15 s, followed by 60 °C for 1 min, 95 °C for 15 s, and finally, 60 °C for 15 s. Gene expression results were analyzed using Applied Biosystem software, SDS 2.4.1. Relative mRNA expression levels were calculated by normalizing examined genes against b actin using the $2^{-\Delta Ct}$ method.

2.3. Homogenate Preparation

Frozen brain (200 mg) was transferred to a mortar and the tissue crushed in liquid nitrogen with a pestle. During homogenization, the tissue was kept completely frozen to preserve functional and structural integrity of proteins. The samples were transferred to microcentrifuge tube with 1 mL of ProteoJET mammalian cell lysis reagent (Fermentas life Science, Waltham, MA, USA) plus a protease inhibitor cocktail, vortexed and then centrifuged at 13,000 rpm at 4 °C for 10 min. The protein concentration of supernatants was measured with Bio-Rad Protein Assay in 96-well microplates. All the samples were normalized to the same protein concentration and used according to specific experimental procedures.

2.4. ROS Detection

Radical oxygen species (ROS) levels were measured using the OxiSelect™ Intracellular ROS Assay Kit (Green Fluorescence; Cell Biolabs, Inc. San Diego, CA, USA) according to the manufacturer's instructions. Briefly: 50 µL (0.5 mg/mL of protein) of each homogenate were added to a 96-well plate suitable for fluorescence measurement. Then, 50 µL of catalyst and 100 µL of 2′, 7′-dichlorodihydrofluorescin diacetate (DCFH-DA) solution were added to each well. The plate was read with a fluorescence plate reader at 480 nm excitation/530 nm emission. A standard curve was made using 2′,7′-dichlorodihydrofluorescein standard solution (manufacture's solution).

2.5. NOS Activity

NOS enzyme activity was measured by colorimetric method using the NOS Activity Assay Kit (BioVision, Kampenhout, Belgium), according to the manufacturer's indications. Briefly, 30 µL (200 µg protein) of tissue homogenates were added in a 96-well plate. Successively, 40 µL of the reaction mix (NOS cofactor 1, NOS cofactor 2, NOS substrate, and nitrate reductase) were added in each well and, after 60 min of incubation, 50 µL of Griess reagent were added in each well, forming a purple azo dye by interaction with nitrites. The optical density of this dye was measured at 540 nm using a microplate reader. To generate the calibration curve, a standard solution, corresponding to 0, 250, 500, 750, 1000 pmol/well nitrites standard was used. Nitric Oxide Synthase Specific Activity was calculated = B/TxC = mU/mg of protein, where B is nitrite amount in sample well from standard curve, T is reaction time (60 min), and C is the amount of protein.

2.6. S100B Protein in Brain Homogenates

The quantitative measurement of S100B protein in brain homogenates was performed using S100B in vitro SimpleStep ELISA® (enzyme-linked immunosorbent assay) kit (Abcam, Cambridge, UK) according to the manufacturers' instructions. Briefly, 50 µL (1 mg/mL protein) of homogenates or standard samples were added to appropriate wells. After the addition of 50 µL of the antibody cocktail to each well, the plate was incubated 1 h at room temperature. Finally, after series of washes, 100 µL of TMB developing solution and 100 µL of stop solution were added to each well. The plate was read at 450 nm using a microplate reader and the quantitative measurement of S100B protein in mouse homogenates was reported as absorbance at 450 nm.

2.7. Tissue Processing and Histology

After fixation (4% PFA) and cryoprotection in sucrose 30%, 30 µm thick serial brain emisections taken from +1.34 to +4.66 bregma coordinates according to the atlas of Paxinos and Franklin [25], and spinal cord sections were cut with a cryostat (Leica CM1860 UV, Leica Biosystems, Nussloch, Germany). One out of every three sections was stained for immunofluorescence and/or immunohistochemistry analysis to detect the following antigens: CD68 (Bio-Rad/AbD Serotec, CA, US, 1:200), a marker of activated microglia, $CD4^+$ (Bio-Rad/AbD Serotec, Raleigh, NC, USA 1:50) for T-cell infiltrates, glial fibrillary acidic protein (GFAP; Cell Signaling Technology, MA, US, 1:500) and S100B (Novus biological, CO, USA, 1:1000) for astrocytes, myelin basic protein, MBP (Cell Signaling Technology Denver, MA, USA, 1:200) for myelin sheaths, and NeuN (Chemicon International, CA, USA, 1:100) for neurons.

2.8. Immunohistochemistry

After preincubation with 0.3% H_2O_2 in PBS, the sections were incubated at 4°C with primary antibodies in PBS-0.3% Triton X-100, 2% normal donkey serum (NDS). Following the use of biotinylated donkey anti-mouse or donkey anti-rat antibodies (Jackson ImmunoResearch Europe Ltd., Ely, UK), avidin-biotin-peroxidase reactions were performed (Vectastain, ABC kit, Vector, Burlingame, CA, USA), using 3,3'-diaminobenzidine (Sigma-Aldrich, St. Louis, MO, USA) as a chromogen. The sections were then analyzed using an Axioskop 2 optical microscope (Zeiss, Oberkochen, Germany), with Neurolucida software (MBF Bioscience, Williston, VT, USA) for image acquisition. The quantification of the tissue area positive for CD68/CD4/GFAP was performed with the NIH ImageJ software. The areas were calculated as CD68/CD4/GFAP positive area/total area analyzed, and indicated as percentages of stained area.

2.9. Immunofluorescence

The sections were blocked with 10% NDS in 0.3% Triton X-100 in PBS and incubated with primary antisera/antibodies in 0.3% Triton X-100 and 2% NDS in PBS, for 24–48 h at 4 °C and processed for immunofluorescence. The sections were washed thoroughly and incubated with appropriate fluorescent-conjugated secondary antibodies for 3 h at room temperature. The secondary antibodies (Jackson Immunoresearch, Philadelphia, PA, USA) in 0.3% Triton X-100 and 2% NDS in PBS were Alexa Fluor® 488-AffiniPure donkey anti-mouse IgG (1:200, green), Alexa Fluor® 488-AffiniPure donkey anti-rabbit IgG (1:200, green), Cy5-conjugated donkey anti-mouse IgG (1:100, blue), Cy3-conjugated donkey anti-rat IgG (1:100, red), and Cy3-conjugated donkey anti-rabbit IgG (1:100, red). After rinsing, the sections were stained with the nucleic acid blue dye Hoechst 33,342 (1:1000), mounted on slide glasses, covered with fluoromount medium (Sigma-Aldrich, St. Louis, MO, USA) and a coverslip, and analyzed by confocal microscopy. Immunofluorescence analysis was performed by confocal laser scanning microscope (Zeiss LSM 800) equipped with four laser lines: 405, 488, 561, and 639 nm. Brightness and contrast were adjusted with the Zen software 3.0 blue edition (Zeiss, Oberkochen, Germany). For lesion and cellular infiltrates detection, Luxol fast blue and cresyl fast violet combined staining (Kluver–Barrera) was performed.

2.10. Statistical Analysis

Student's *t*-test, one-way ANOVA or two-way ANOVA (with PTM-treatement as main factor) were performed to examine the effects and possible interaction of independent variables (GraphPad 6.0 software). When appropriate, post hoc comparisons were made using Tukey's HSD, with a significance level of $p < 0.05$. For statistical analysis of quantitative PCR data, the unpaired *t*-test was used to compare ΔCt values across the replicates, setting the *p*-value cut-off at 0.05.

All experimental work has been conducted in accordance with relevant national legislation on the use of animals for research, referring to the Code of Practice for the Housing and Care of Animals Used in Scientific Procedures and the protocol was approved by the Ethics Committee of animal welfare organization (OPBA) of the "Università Cattolica del Sacro Cuore" of Rome and by the Italian Ministry of Health (authorization number 321/2017-PR, protocol number 1F295.34/04-11-2016, date of approval 12th April 2017).

3. Results

3.1. PTM Treatment Ameliorates EAE

A total of 32 mice were examined for disease clinical course, immunohistochemistry for CD68 and Kluver–Barrera staining, in two independent experiments (Figure 1 and Figure S1b,c). As detected by immunohistochemistry for CD68 and Kluver–Barrera staining, significant increase in CD68 immunoreactivity with decrease in myelin content and parallel increase in cellularity could be seen in the CNS of immunized mice. Inflammatory lesions were observed with the typical periventricular infiltration and accumulation of mononuclear cells. In cerebellar sections, particularly evident demyelinating areas were highlighted by lack of MBP and NeuN immunofluorescent staining (Figure S1d). PTM-treated group showed a lower severity of symptoms at onset ($p = 0.02$ at day 9, $p = 0.05$ at day 10) and during the remission and relapse phases ($p = 0.04$ at day 13, $p = 0.002$ at day 21 and $p = 0.01$ at day 22), as evaluated by Student's t-test corrected for multiple comparisons using the Holm–Sidak method. This effect can be appreciated comparing also the sums and the means of disease scores of the two groups of mice respectively with a $p = 0.04$ and a $p = 0.01$ after nonparametric Mann–Whitney test. The disease scores of untreated and PTM-treated healthy controls are not shown, as both groups did not develop any sign of disease. Thus, these data indicated that PTM is able to delay the disease and to reduce its overall severity (Figure 1a–c)

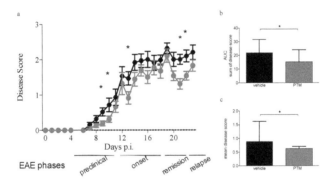

Figure 1. Clinical signs of neuroinflammation and demyelination of relapsing–remitting experimental autoimmune encephalomyelitis (RR–EAE) are ameliorated by pentamidine. (**a**) Clinical symptoms score (CSS) from day 1 to day 23 p.i. of 32 EAE-affected animals (4 mice were withdrawn from the experiments because of excessive reaction to immunization, unresponsiveness/anergy to immunization, or death for unknown causes), pentamidine isethionate (PTM)-treated (red) and vehicle treated (black). The circles and confidence bars represent the mean and the standard deviation of the CSS of the entire group of mice for each day p.i. Statistically significant differences of the CSS are observed at onset, during remission and at relapse of EAE (Student t-test: $p = 0.02$ at day 9 p.i.; $p = 0.05$ at day 10 p.i.; $p = 0.04$ at day 13 p.i.; $p = 0.002$ at day 21 p.i.; $p = 0.01$ at day 22 p.i.). At onset (days 12–15 p.i.), 10 mice (5 from each group, CSS comprised between 2 and 3 of each individual mouse) were sacrificed. Ten mice (5 from each group, at least 1 point of CSS below their individual peak CSS reached during acute phase) were sacrificed at remission (days 18–23 p.i). Finally, 12 mice (6 from each group, when they reached at least 0.5 points of CSS above their individual remission mean CSS) were sacrificed at relapse (day 30 p.i.). The average CSS values after 24 days p.i. are not shown due to the low number of mice (only 12 mice remained) and the heterogeneity among them. b and c: Mean of cumulative diseases (sum of CSS from day 1 to the day of sacrifice) of each mouse (**b**) and the mean of the diseases (sum of CSS from day 1 to the day of sacrifice divided by the number of days) of each animal (**c**). Both graphs show the two groups (with colors in concordance to figure a) displaying the overall significant impact of PTM on CSS and on the amelioration of symptoms. Statistical analysis has been performed using the Mann–Whitney test ($p = 0.04$ for b and $p = 0.01$ for c; * $p < 0.05$).

3.2. PTM Treatment Attenuates Neuroinflammation

To evaluate the impact of PTM on neuroinflammation, the expression of genes encoding for inflammatory cytokines during RR–EAE has been evaluated by qPCR performed on total mRNA extracted from the emi-brains of treated (EAE/PTM), untreated RR–EAE (EAE/vehicle), and CTRL mice samples (healthy PTM-treated mice).

A significant decrease of neuroinflammatory parameters in PTM-treated animals was revealed by the reduction of mRNA expression for Interferon γ, *IFNγ* (Figure 2a, vehicle vs. PTM $p = 0.03$, Mann–Whitney test) and for tumor necrosis factor α, *TNFα* (Figure 2b, vehicle vs. PTM $p = 0.003$, Mann–Whitney test). We could not observe a statistically valid difference in the levels of mRNA specific for interleukin β, *IL1β* (Figure 2c).

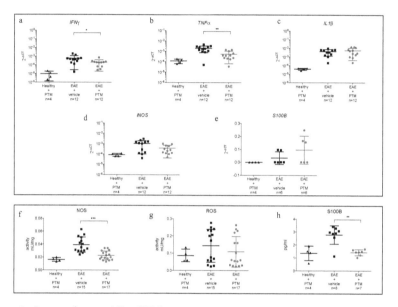

Figure 2. Impact of pentamidine (PTM) on gene expression and protein levels of inflammatory cytokines, S100B, nitric oxide synthase (NOS), and radical oxygen species (ROS) during EAE. (**a–c**) qPCR, performed on total mRNA extracted from the emi-brain of treated (PTM), untreated (vehicle) EAE, and CTRL (PTM treated healthy mice), shows a significant reduction in PTM-treated animals for the expression of *IFNγ* ((**a**), $p = 0.03$) and *TNFα* ((**b**), $p = 0.008$), despite no difference of *IL1β* (**c**). No change of inducible nitric oxide synthase *iNOS* and *S100B* mRNA is observed (**d,e**). Colorimetric activity assay for NOS shows a significant difference (**f**, vehicle vs. PTM $p < 0.0001$), while ROS activity does not reveal any change during PTM treatment (**g**). S100B has been measured by ELISA (**h**) resulting in a significant downregulation of the S100B protein ($p = 0.0085$). Statistical analysis has been performed using Mann–Whitney test. The qPCR analyses for cytokines and *iNOS* (**a–d**) have been performed on a total of 28 mice (12 EAE–PTM-treated, 12 EAE-vehicle-treated and 4 healthy-untreated), excluding EAE affected mice sacrificed during remission phase. The qPCR and the ELISA for S100B (**e,h**) were performed on mice of the second experiment (because the samples of first experiment were not usable in terms of quality and quantity) on a total of 19 mice (7 EAE–PTM-treated, 8 EAE-vehicle-treated and 4 healthy untreated); for S100B qPCR three results were missing (1 in EAE–PTM-treated and 2 EAE-vehicle-treated groups). Colorimetric activity assays for NOS and ROS (**f,g**) have been performed on 36 mice (17 EAE–PTM treated, 15 EAE–vehicle treated and 4 healthy untreated). For graphical reasons, y-axes are displayed in logarithmic for cytokines and iNOs mRNA expression (**a–d**) and linear y-axes for S100B mRNA expression (**e**) and colorimetric and ELISA assays (**f–h**). * $p < 0.05$; ** $p < 0.001$; *** $p < 0.0001$.

Finally, to confirm the predictive role of S100B levels on MS pathogenesis, we measured mRNA of *iNOS* and *S100B*, finding no differences in terms of gene expression levels (Figure 2d,e). NOS are a family of enzymes that catalyze the production of nitric NO from L-arginine. NO plays an important role in neurotransmission, vascular regulation, immune response, and apoptosis. Three isoforms of NOS have been identified: two constitutive enzymes, neuronal NOS (nNOs) and endothelial NOS (eNOS), and one inducible enzyme (iNOS). Colorimetric activity assay for NOS resulted in a significant difference in vehicle vs. PTM (Figure 2f, $p < 0.0001$, Mann–Whitney test), while no changes were observed in ROS activity (Figure 2g). The quantification of S100B by ELISA (Figure 2h) resulted in a significant downregulation of S100B protein after PTM treatment (vehicle vs. PTM, $p = 0.0085$, Mann–Whitney test). As found for iNOS, the variation of S100B protein expression did not correspond to a change in terms of mRNA. To further dissect whether PTM modulates inflammatory cytokines during the relapse phase, we correlated the gene expression levels of inflammatory cytokines with the disease course (Figure S2a,b). PTM reduced the expression levels of these genes, in particular during the relapse phase.

To support our observations about clinical scores, we examined the demyelination and immune infiltrates in the CNS of RR–EAE mice treated with/without PTM, as shown by immunohistochemistry and immunofluorescence analysis (Figures 3 and 4). As expected, in cerebral cortex–striatum, hippocampus, and cerebellum, the CD68 and GFAP labeling were apparently increased in EAE with respect to Ctrl mice (Figure 3 insets). Moreover, after PTM treatment, the CD68 and CD4 signals were in part decreased presenting low amount of cellular infiltrates, particularly in the cerebellum, where, however, the differences in quantification of the areas of CD68/CD4/GFAP/positive tissue, using NIH image software, did not reach statistical significance (Figure 3a–c and Figure S3). Moreover, the GFAP signal remained constant in all the analyzed brain areas (Figure 3). When examined by immunofluorescence, cerebellar sections also evidenced CD68-labelled infiltrates accompanied by demyelinating lesions in EAE mice, which appeared to be reduced in PTM-treated EAE mice. Also, S100B immunofluorescence appeared to be decreased in PTM-treated EAE mice, in accordance with S100B protein immunochemical measurements shown in Figure 2h (Figure 4).

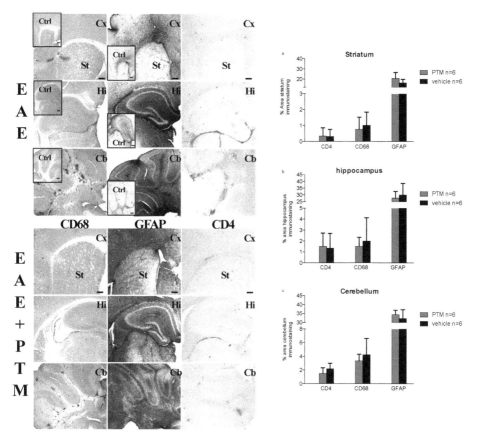

Figure 3. Effects of pentamidine (PTM) on immune infiltrates and glial fibrillary acidic protein (GFAP) in the central nervous system (CNS) of RR–EAE mice. EAE ($n = 6$) and EAE–PTM ($n = 6$) mice were sacrificed 30 days after immunization in the relapse/late onset of EAE (sacrificed when they reached at least 0.5 point of CSS above their individual remission mean CSS). Immunohistochemistry was performed with CD68, GFAP, and CD4 antibodies on serial brain emisections (30 µm) of control (Ctrl), EAE, and EAE–PTM mice. Cx = cortex; St = striatum; Hi = Hippocampus; Cb = cerebellum. Scale bars = 250 µm. The quantifications are displayed on the bar graph on the right of the figure (EAE–PTM-treated in red versus EAE-vehicle treated in black). Immunocytochemistry and DAB assays revealed that pentamidine seems to decrease the expression of CD68 in all the analyzed areas (**a–c**). CD4 infiltrates seem to be reduced by PTM in cerebellar area only (**c**), and GFAP percentages are increased in all the areas except in hippocampus (**b**). All these regional cell counts, although not significant, would indicate a different impact of the inhibition of S100B within each cerebral area.

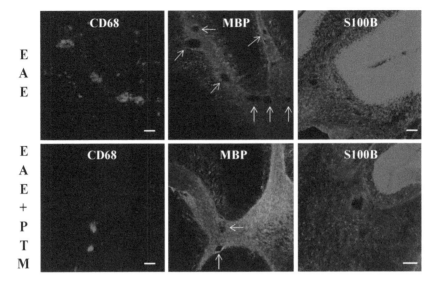

Figure 4. Immunofluorescence in representative cerebellar sections of EAE mice and PTM-treated EAE mice. CD68-labelled infiltrates accompanied by demyelinating lesions (arrows) are evident in EAE mice, which appear to be reduced in PTM-treated EAE mice, as shown by CD68 (red) and MBP (green) signals. Decreased S100B immunolabelling (blue) also appears in PTM-treated EAE mice. Scale bars = 100 µm.

4. Discussion

The present data indicate that PTM, which is regarded to block S100B protein, induces clinical disease scores amelioration accompanied by some improvement of neuropathologic and biomolecular parameters in a recognized experimental in vivo model of MS (RR–EAE in the SJL mice).

RR–EAE is a well-accepted preclinical model to study MS pathogenesis and therapy in the mouse: it shares with the human disease key immunological and pathological features, including patch inflammation, demyelination, axonal loss, and gliosis. Similarly to most common human MS manifestations (about 80%), RR–EAE in the SJL mouse is a CD4[+] T-cell-mediated autoimmune disease directed against protein components of CNS myelin, resulting in a relapsing–remitting clinical course of paralysis [26].

S100B is regarded as a DAMP, also sharing with these molecules some characteristics such as the interaction with RAGE, the ability to stimulate microglial migration, the non-canonical secretion modality that bypasses the Golgi route [5]. In particular, the protein has been shown to induce a RAGE-dependent autocrine loop in astrocytes, which results in a pro-inflammatory phenotype [27] also stimulating IL6 and TNFα secretion [28]. In addition, S100B, which has also been reported to be expressed and secreted by CD8[+] T- and NK-cells on stimulation [29], upregulates cyclooxygenase-2 [30], promotes migration/activation of microglia [29], induces iNOS, IL1β and TNFα expression, as well as the release of matrix metalloproteinase 9 and NO [31–35]. It is also noteworthy that, while S100B is currently regarded to be released from astrocytes with paracrine/autocrine effects, and a role for astrocytes in MS is now definitely recognized [36], its presence in oligodendrocytes [5] appears to deserve a special attention for a possible role in pathogenic MS processes.

In this study, we used PTM, an approved antiprotozoal drug [37,38] that is also known to inhibit S100B activity [11]. Because of this property, the PTM effects on RR–EAE SJL mice may be reasonably attributed to a block of S100B activity, although additional or even alternative mechanisms cannot be ruled out. It should also be noted that the inflammatory cytokines that we observed here to be affected by PTM administration are also known to be influenced by S100B [28,31,33]. Furthermore, S100B expression is reduced after PTM administration, as we have shown here in the present experimental MS animal model, also confirming what was observed in other pathological conditions where PTM exerts a protective role [39,40].

Our results thus propose that the inhibition of S100B has a significant impact on the extent of neuroinflammatory features, and this appears in line with the typical clinical and neuropathological heterogeneity of the model [41,42]. Remarkably, the expression levels of *IFNγ* and TNFα are strongly decreased by the PTM treatment, in line with the notion that these two cytokines play a relevant role in the CNS, in both physiological and pathological conditions [43]. Conversely, *IL-1β* expression levels are not modulated by PTM. This discrepancy is not surprising and might be ascribed to differences related to the RR–EAE model, to the specific brain areas analyzed, and, finally, to the local neuroinflammation level, parameters which are discriminative in the pathology, as also demonstrated in other EAE models [44].

All NOS isoforms are present in the CNS with a clear upregulation in reactive microglia/macrophages [45], resulting in a high generation of nitric oxide, a free radical found at higher than normal concentrations within inflammatory MS [46]. Confirming previous data, our results also show an increase in the activity of NOS in EAE mice and, furthermore, highlight that the treatment with PTM restores the activity of the enzyme to the control levels, although the effect of PTM on iNOS was already observed in vitro [47]. It still remains to be clarified the molecular mechanism(s) contributing to the reduction of NO production with the consequent anti-oxidative and anti-inflammatory potential.

As far as ROS are concerned, we have not found significant results in our experimental model. It has been widely ascertained that MS may be affected by oxidative stress. However, it was also reported that in RR–MS, the inflammatory process prevails, and oxidative stress is counteracted by antioxidant mechanisms [48]. Intriguingly, the levels of S100B protein, but not *S100B* mRNA, are reduced following PTM treatment. This might be due to a PTM action in protein synthesis steps following mRNA transcription. Alternatively, the binding of PTM to S100B might partially mask immunologically relevant epitopes of the protein.

PTM administration has also been shown to ameliorate clinical and/or neuropathologic/biomolecular parameters in other disorders involving the nervous system, such as AD, PD, sepsis-associated encephalopathy, and bowel inflammation, where its action has been reasonably attributed to the block of S100B activity [49–51]. In general, variations/manipulations of S100B concentration have been shown to directly correlate with clinical symptoms and/or biomolecular/pathological parameters of these disorders, which depend, as far it is known, on etiologic factors different from those hypothesized for MS, but that share with MS the occurrence of neuroinflammatory processes [5,52].

In summary, the present data identify PTM and its binding molecule S100B protein, respectively, as a novel potential drug and a therapeutic target for MS treatment. Of note, PTM is suitable for a rapid translation for clinical use, being an already-approved drug. Although PTM has limited capacity to permeate the BBB [53], and although a proportion is retained within the capillary endothelium [54], the BBB damage occurring during demyelinating diseases [55], as also shown by evidences of tracer leakage into the CNS [56], suggests that this drug may be easily available within the inflamed CNS, also when administered systemically, as in this work. Finally, the identification of S100B as a putative therapeutic target of MS pathogenetic processes might likely open novel perspectives for even more efficacious treatments.

Supplementary Materials: The following are available online at http://www.mdpi.com/2073-4409/9/3/748/s1. Figure S1: Experimental procedure of EAE (clinical outcome and neuropathology). Table S1: Primer sequences.

Author Contributions: "Conceptualization, C.V., F.R. and F.M.; methodology, G.D.S., S.A., M.E.C., and M.V.; software, G.D.S. and B.S.; validation, M.E.C.; formal analysis, G.D.S., S.A., M.E.C., and M.V.; investigation, S.A., B.S., and P.C.; data curation, G.D.S., M.E.C., and B.S.; writing—original draft preparation, G.D.S., S.A., B.S., M.E.C.,

C.V., F.R., and F.M.; writing—review and editing, G.D.S, S.A., B.S., M.E.C., C.V., P.C., F.R., and F.M.; visualization, F.R.; supervision, G.D.S., F.R., and F.M.; project administration, G.D.S.; funding acquisition, G.D.S., F.R., and F.M. All authors have read and agreed to the published version of the manuscript.".

Funding: This research was funded by FISM (Fondazione Italiana Sclerosi Multipla) cod 2018/R/11 (FM) and co-financed with 5-per-mille public funding.

Acknowledgments: We thank Ottavio Cremona and Luca Muzio (IRCSS Istituto Scientifico San Raffaele, Università Vita-Salute San Raffaele, Milan) for useful discussions concerning the experimental design. Residues, helix, and pdb drawings of graphical abstract are partially modified from [11] and adapted to the figure.

Conflicts of Interest: The authors declare no conflict of interest. The funders had no role in the design of the study; in the collection, analyses, or interpretation of data; in the writing of the manuscript, or in the decision to publish the results.

References

1. Magyari, M.; Sorensen, P.S. The changing course of multiple sclerosis: Rising incidence, change in geographic distribution, disease course, and prognosis. *Curr. Opin. Neurol.* **2019**, *32*, 320–326. [CrossRef]
2. Kaunzner, U.W.; Al-Kawaz, M.; Gauthier, S.A. Defining Disease Activity and Response to Therapy in MS. *Curr. Treat. Opt. Neurol.* **2017**, *19*, 20. [CrossRef]
3. Moore, B.W. A soluble protein characteristic of the nervous system. *Biochem. Biophys. Res. Commun.* **1965**, *19*, 739–744. [CrossRef]
4. Michetti, F.; Corvino, V.; Geloso, M.C.; Lattanzi, W.; Bernardini, C.; Serpero, L.; Gazzolo, D. The S100B protein in biological fluids: More than a lifelong biomarker of brain distress. *J. Neurochem.* **2012**, *120*, 644–659. [CrossRef] [PubMed]
5. Michetti, F.; D'Ambrosi, N.; Toesca, A.; Puglisi, M.A.; Serrano, A.; Marchese, E.; Corvino, V.; Geloso, M.C. The S100B story: From biomarker to active factor in neural injury. *J. Neurochem.* **2019**, *148*, 168–187. [CrossRef] [PubMed]
6. Michetti, F.; Massaro, A.; Murazio, M. The nervous system-specific S-100 antigen in cerebrospinal fluid of multiple sclerosis patients. *Neurosci. Lett.* **1979**, *11*, 171–175. [CrossRef]
7. Barateiro, A.; Afonso, V.; Santos, G.; Cerqueira, J.J.; Brites, D.; van Horssen, J.; Fernandes, A. S100B as a Potential Biomarker and Therapeutic Target in Multiple Sclerosis. *Mol. Neurobiol.* **2016**, *53*, 3976–3991. [CrossRef]
8. Petzold, A.; Eikelenboom, M.J.; Gveric, D.; Keir, G.; Chapman, M.; Lazeron, R.H.C.; Cuzner, M.L.; Polman, C.H.; Uitdehaag, B.M.J.; Thompson, E.J.; et al. Markers for different glial cell responses in multiple sclerosis: Clinical and pathological correlations. *Brain* **2002**, *125*, 1462–1473. [CrossRef]
9. Yan, H.; Zhou, H.-F.; Hu, Y.; Pham, C.T.N. Suppression of experimental arthritis through AMP-activated protein kinase activation and autophagy modulation. *J. Rheum. Dis. Treat.* **2015**, *1*, 5. [CrossRef]
10. Santos, G.; Barateiro, A.; Gomes, C.M.; Brites, D.; Fernandes, A. Impaired oligodendrogenesis and myelination by elevated S100B levels during neurodevelopment. *Neuropharmacology* **2018**, *129*, 69–83. [CrossRef]
11. Charpentier, T.H.; Wilder, P.T.; Liriano, M.A.; Varney, K.M.; Pozharski, E.; MacKerell, A.D.; Coop, A.; Toth, E.A.; Weber, D.J. Divalent metal ion complexes of S100B in the absence and presence of pentamidine. *J. Mol. Biol.* **2008**, *382*, 56–73. [CrossRef] [PubMed]
12. Pirolli, D.; Carelli Alinovi, C.; Capoluongo, E.; Satta, M.A.; Concolino, P.; Giardina, B.; De Rosa, M.C. Insight into a Novel p53 Single Point Mutation (G389E) by Molecular Dynamics Simulations. *Int. J. Mol. Sci.* **2010**, *12*, 128–140. [CrossRef] [PubMed]
13. Hartman, K.G.; McKnight, L.E.; Liriano, M.A.; Weber, D.J. The evolution of S100B inhibitors for the treatment of malignant melanoma. *Future Med. Chem.* **2013**, *5*, 97–109. [CrossRef] [PubMed]
14. Nicolò, C.; Di Sante, G.; Orsini, M.; Rolla, S.; Columba-Cabezas, S.; Romano Spica, V.; Ricciardi, G.; Chan, B.M.C.; Ria, F. Mycobacterium tuberculosis in the adjuvant modulates the balance of Th immune response to self-antigen of the CNS without influencing a "core" repertoire of specific T cells. *Int. Immunol.* **2006**, *18*, 363–374. [CrossRef]
15. Nicolò, C.; Sali, M.; Di Sante, G.; Geloso, M.C.; Signori, E.; Penitente, R.; Uniyal, S.; Rinaldi, M.; Ingrosso, L.; Fazio, V.M.; et al. Mycobacterium smegmatis expressing a chimeric protein MPT64-proteolipid protein (PLP) 139-151 reorganizes the PLP-specific T cell repertoire favoring a CD8-mediated response and induces a relapsing experimental autoimmune encephalomyelitis. *J. Immunol.* **2010**, *184*, 222–235. [CrossRef]

16. Nicolò, C.; Di Sante, G.; Procoli, A.; Migliara, G.; Piermattei, A.; Valentini, M.; Delogu, G.; Cittadini, A.; Constantin, G.; Ria, F. M tuberculosis in the adjuvant modulates time of appearance of CNS-specific effector T cells in the spleen through a polymorphic site of TLR2. *PLoS ONE* **2013**, *8*, e55819. [CrossRef]
17. Piermattei, A.; Migliara, G.; Di Sante, G.; Foti, M.; Hayrabedyan, S.B.; Papagna, A.; Geloso, M.C.; Corbi, M.; Valentini, M.; Sgambato, A.; et al. Toll-Like Receptor 2 Mediates In Vivo Pro- and Anti-inflammatory Effects of Mycobacterium Tuberculosis and Modulates Autoimmune Encephalomyelitis. *Front. Immunol.* **2016**, *7*, 191. [CrossRef]
18. Miller, S.D.; Karpus, W.J.; Davidson, T.S. Experimental Autoimmune Encephalomyelitis in the Mouse. In *Current Protocols in Immunology*; Coligan, J.E., Bierer, B.E., Margulies, D.H., Shevach, E.M., Strober, W., Eds.; John Wiley & Sons, Inc.: Hoboken, NJ, USA, 2010; ISBN 978-0-471-14273-7.
19. Penitente, R.; Nicolò, C.; Van den Elzen, P.; Di Sante, G.; Agrati, C.; Aloisi, F.; Sercarz, E.E.; Ria, F. Administration of PLP$_{139-151}$ Primes T Cells Distinct from Those Spontaneously Responsive In Vitro to This Antigen. *J. Immunol.* **2008**, *180*, 6611–6622. [CrossRef]
20. Xie, K.-Y.; Wang, Q.; Cao, D.-J.; Liu, J.; Xie, X.-F. Spinal astrocytic FGFR3 activation leads to mechanical hypersensitivity by increased TNF-α in spared nerve injury. *Int. J. Clin. Exp. Pathol.* **2019**, *12*, 2898–2908.
21. Fu, C.; Zhu, X.; Xu, P.; Li, Y. Pharmacological inhibition of USP7 promotes antitumor immunity and contributes to colon cancer therapy. *Onco Targets Ther.* **2019**, *12*, 609–617. [CrossRef]
22. Qiu, X.; Guo, Q.; Liu, X.; Luo, H.; Fan, D.; Deng, Y.; Cui, H.; Lu, C.; Zhang, G.; He, X.; et al. Pien Tze Huang Alleviates Relapsing-Remitting Experimental Autoimmune Encephalomyelitis Mice by Regulating Th1 and Th17 Cells. *Front. Pharmacol.* **2018**, *9*, 1237. [CrossRef]
23. Xu, Z.; Fouda, A.Y.; Lemtalsi, T.; Shosha, E.; Rojas, M.; Liu, F.; Patel, C.; Caldwell, R.W.; Narayanan, S.P.; Caldwell, R.B. Retinal Neuroprotection From Optic Nerve Trauma by Deletion of Arginase 2. *Front. Neurosci.* **2018**, *12*, 970. [CrossRef]
24. Cao, L.; Malon, J.T. Anti-nociceptive Role of CXCL1 in a Murine Model of Peripheral Nerve Injury-induced Neuropathic Pain. *Neuroscience* **2018**, *372*, 225–236. [CrossRef]
25. Franklin, K.B.J.; Paxinos, G. *Paxino's and Franklin's the Mouse Brain in Stereotaxic Coordinates: Compact*, 5th ed.; Elsevier Science & Technology: San Diego, CA, USA, 2019; ISBN 978-0-12-816160-9.
26. Lublin, F.D. Delayed, relapsing experimental allergic encephalomyelitis in mice. Role of adjuvants and pertussis vaccine. *J. Neurol. Sci.* **1982**, *57*, 105–110. [CrossRef]
27. Villarreal, A.; Aviles Reyes, R.X.; Angelo, M.F.; Reines, A.G.; Ramos, A.J. S100B alters neuronal survival and dendrite extension via RAGE-mediated NF-κB signaling. *J. Neurochem.* **2011**, *117*, 321–332. [CrossRef] [PubMed]
28. Ponath, G.; Schettler, C.; Kaestner, F.; Voigt, B.; Wentker, D.; Arolt, V.; Rothermundt, M. Autocrine S100B effects on astrocytes are mediated via RAGE. *J. Neuroimmunol.* **2007**, *184*, 214–222. [CrossRef] [PubMed]
29. Steiner, J.; Marquardt, N.; Pauls, I.; Schiltz, K.; Rahmoune, H.; Bahn, S.; Bogerts, B.; Schmidt, R.E.; Jacobs, R. Human CD8(+) T cells and NK cells express and secrete S100B upon stimulation. *Brain Behav. Immun.* **2011**, *25*, 1233–1241. [CrossRef] [PubMed]
30. Shanmugam, N.; Reddy, M.A.; Natarajan, R. Distinct roles of heterogeneous nuclear ribonuclear protein K and microRNA-16 in cyclooxygenase-2 RNA stability induced by S100b, a ligand of the receptor for advanced glycation end products. *J. Biol. Chem.* **2008**, *283*, 36221–36233. [CrossRef] [PubMed]
31. Adami, C.; Bianchi, R.; Pula, G.; Donato, R. S100B-stimulated NO production by BV-2 microglia is independent of RAGE transducing activity but dependent on RAGE extracellular domain. *Biochim. Biophys. Acta* **2004**, *1742*, 169–177. [CrossRef]
32. Bianchi, R.; Adami, C.; Giambanco, I.; Donato, R. S100B binding to RAGE in microglia stimulates COX-2 expression. *J. Leukoc. Biol.* **2007**, *81*, 108–118. [CrossRef]
33. Bianchi, R.; Giambanco, I.; Donato, R. S100B/RAGE-dependent activation of microglia via NF-kappaB and AP-1 Co-regulation of COX-2 expression by S100B, IL-1beta and TNF-alpha. *Neurobiol. Aging* **2010**, *31*, 665–677. [CrossRef] [PubMed]
34. Bianchi, R.; Kastrisianaki, E.; Giambanco, I.; Donato, R. S100B protein stimulates microglia migration via RAGE-dependent up-regulation of chemokine expression and release. *J. Biol. Chem.* **2011**, *286*, 7214–7226. [CrossRef] [PubMed]
35. Xu, J.; Wang, H.; Won, S.J.; Basu, J.; Kapfhamer, D.; Swanson, R.A. Microglial activation induced by the alarmin S100B is regulated by poly(ADP-ribose) polymerase-1. *Glia* **2016**, *64*, 1869–1878. [CrossRef] [PubMed]

36. Ponath, G.; Park, C.; Pitt, D. The Role of Astrocytes in Multiple Sclerosis. *Front. Immunol.* **2018**, *9*, 217. [CrossRef] [PubMed]
37. Drake, S.; Lampasona, V.; Nicks, H.L.; Schwarzmann, S.W. Pentamidine isethionate in the treatment of Pneumocystis carinii pneumonia. *Clin. Pharm.* **1985**, *4*, 507–516. [CrossRef]
38. Pearson, R.D.; Hewlett, E.L. Pentamidine for the treatment of Pneumocystis carinii pneumonia and other protozoal diseases. *Ann. Intern. Med.* **1985**, *103*, 782–786. [CrossRef]
39. Huang, L.; Zhang, L.; Liu, Z.; Zhao, S.; Xu, D.; Li, L.; Peng, Q.; Ai, Y. Pentamidine protects mice from cecal ligation and puncture-induced brain damage via inhibiting S100B/RAGE/NF-κB. *Biochem. Biophys. Res. Commun.* **2019**, *517*, 221–226. [CrossRef]
40. Costa, D.V.S.; Bon-Frauches, A.C.; Silva, A.M.H.P.; Lima-Júnior, R.C.P.; Martins, C.S.; Leitão, R.F.C.; Freitas, G.B.; Castelucci, P.; Bolick, D.T.; Guerrant, R.L.; et al. 5-Fluorouracil Induces Enteric Neuron Death and Glial Activation During Intestinal Mucositis via a S100B-RAGE-NFκB-Dependent Pathway. *Sci. Rep.* **2019**, *9*, 665. [CrossRef]
41. Simmons, S.B.; Pierson, E.R.; Lee, S.Y.; Goverman, J.M. Modeling the heterogeneity of multiple sclerosis in animals. *Trends Immunol.* **2013**, *34*, 410–422. [CrossRef]
42. Pandolfi, F.; Cianci, R.; Casciano, F.; Pagliari, D.; De Pasquale, T.; Landolfi, R.; Di Sante, G.; Kurnick, J.T.; Ria, F. Skewed T-cell receptor repertoire: More than a marker of malignancy, a tool to dissect the immunopathology of inflammatory diseases. *J. Biol. Regul. Homeost. Agents* **2011**, *25*, 153–161.
43. Hidaka, Y.; Inaba, Y.; Matsuda, K.; Itoh, M.; Kaneyama, T.; Nakazawa, Y.; Koh, C.-S.; Ichikawa, M. Cytokine production profiles in chronic relapsing-remitting experimental autoimmune encephalomyelitis: IFN-γ and TNF-α are important participants in the first attack but not in the relapse. *J. Neurol. Sci.* **2014**, *340*, 117–122. [CrossRef] [PubMed]
44. Kim, B.S.; Jin, Y.-H.; Meng, L.; Hou, W.; Kang, H.S.; Park, H.S.; Koh, C.-S. IL-1 signal affects both protection and pathogenesis of virus-induced chronic CNS demyelinating disease. *J. Neuroinflammation* **2012**, *9*, 217. [CrossRef] [PubMed]
45. Li, S.; Vana, A.C.; Ribeiro, R.; Zhang, Y. Distinct role of nitric oxide and peroxynitrite in mediating oligodendrocyte toxicity in culture and in experimental autoimmune encephalomyelitis. *Neuroscience* **2011**, *184*, 107–119. [CrossRef]
46. Abdel Naseer, M.; Rabah, A.M.; Rashed, L.A.; Hassan, A.; Fouad, A.M. Glutamate and Nitric Oxide as biomarkers for disease activity in patients with multiple sclerosis. *Mult. Scler. Relat. Disord.* **2019**, *38*, 101873. [CrossRef] [PubMed]
47. Kitamura, Y.; Arima, T.; Imaizumi, R.; Sato, T.; Nomura, Y. Inhibition of constitutive nitric oxide synthase in the brain by pentamidine, a calmodulin antagonist. *Eur. J. Pharmacol.* **1995**, *289*, 299–304. [CrossRef]
48. Adamczyk, B.; Adamczyk-Sowa, M. New Insights into the Role of Oxidative Stress Mechanisms in the Pathophysiology and Treatment of Multiple Sclerosis. *Oxid. Med. Cell. Longev.* **2016**, *2016*, 1973834. [CrossRef]
49. Esposito, G.; Capoccia, E.; Sarnelli, G.; Scuderi, C.; Cirillo, C.; Cuomo, R.; Steardo, L. The antiprotozoal drug pentamidine ameliorates experimentally induced acute colitis in mice. *J. Neuroinflammation* **2012**, *9*, 277. [CrossRef]
50. Cirillo, C.; Capoccia, E.; Iuvone, T.; Cuomo, R.; Sarnelli, G.; Steardo, L.; Esposito, G. S100B Inhibitor Pentamidine Attenuates Reactive Gliosis and Reduces Neuronal Loss in a Mouse Model of Alzheimer's Disease. *Biomed. Res. Int.* **2015**, *2015*, 508342. [CrossRef]
51. Rinaldi, F.; Seguella, L.; Gigli, S.; Hanieh, P.N.; Del Favero, E.; Cantù, L.; Pesce, M.; Sarnelli, G.; Marianecci, C.; Esposito, G.; et al. inPentasomes: An innovative nose-to-brain pentamidine delivery blunts MPTP parkinsonism in mice. *J. Control. Release* **2019**, *294*, 17–26. [CrossRef]
52. Ferraccioli, G.; Carbonella, A.; Gremese, E.; Alivernini, S. Rheumatoid arthritis and Alzheimer's disease: Genetic and epigenetic links in inflammatory regulation. *Discov. Med.* **2012**, *14*, 379–388.
53. Sekhar, G.N.; Watson, C.P.; Fidanboylu, M.; Sanderson, L.; Thomas, S.A. Delivery of antihuman African trypanosomiasis drugs across the blood-brain and blood-CSF barriers. *Adv. Pharmacol.* **2014**, *71*, 245–275. [PubMed]

54. Sanderson, L.; Dogruel, M.; Rodgers, J.; De Koning, H.P.; Thomas, S.A. Pentamidine movement across the murine blood-brain and blood-cerebrospinal fluid barriers: Effect of trypanosome infection, combination therapy, P-glycoprotein, and multidrug resistance-associated protein. *J. Pharmacol. Exp. Ther.* **2009**, *329*, 967–977. [CrossRef] [PubMed]
55. Osorio-Querejeta, I.; Alberro, A.; Muñoz-Culla, M.; Mäger, I.; Otaegui, D. Therapeutic Potential of Extracellular Vesicles for Demyelinating Diseases; Challenges and Opportunities. *Front. Mol. Neurosci.* **2018**, *11*, 434. [CrossRef] [PubMed]
56. Bell, L.; Koeniger, T.; Tacke, S.; Kuerten, S. Characterization of blood–brain barrier integrity in a B-cell-dependent mouse model of multiple sclerosis. *Histochem. Cell Biol.* **2019**, *151*, 489–499. [CrossRef]

© 2020 by the authors. Licensee MDPI, Basel, Switzerland. This article is an open access article distributed under the terms and conditions of the Creative Commons Attribution (CC BY) license (http://creativecommons.org/licenses/by/4.0/).

Article

A Clonal NG2-Glia Cell Response in a Mouse Model of Multiple Sclerosis

Sonsoles Barriola [1], Fernando Pérez-Cerdá [2,3], Carlos Matute [2,3], Ana Bribián [1] and Laura López-Mascaraque [1,*]

[1] Departamento de Neurobiología Molecular, Celular y del Desarrollo, Instituto Cajal-CSIC, 28002 Madrid, Spain; sonsolesbarriola@cajal.csic.es (S.B.); abribian@cajal.csic.es (A.B.)
[2] Centro de Investigación Biomédica en Red de Enfermedades Neurodegenerativas (CIBERNED), Departamento de Neurociencias, Universidad del País Vasco, 48940 Leioa, Spain; fernando.perez@ehu.eus (F.P.-C.); carlos.matute@ehu.eus (C.M.)
[3] Achucarro Basque Center for Neuroscience, 48940 Leioa, Spain
* Correspondence: mascaraque@cajal.csic.es

Received: 8 April 2020; Accepted: 19 May 2020; Published: 21 May 2020

Abstract: NG2-glia, also known as oligodendrocyte precursor cells (OPCs), have the potential to generate new mature oligodendrocytes and thus, to contribute to tissue repair in demyelinating diseases like multiple sclerosis (MS). Once activated in response to brain damage, NG2-glial cells proliferate, and they acquire a reactive phenotype and a heterogeneous appearance. Here, we set out to investigate the distribution and phenotypic diversity of NG2-glia relative to their ontogenic origin, and whether there is a clonal NG2-glial response to lesion in an experimental autoimmune encephalomyelitis (EAE) murine model of MS. As such, we performed in utero electroporation of the genomic lineage tracer, StarTrack, to follow the fate of NG2-glia derived from single progenitors and to evaluate their response to brain damage after EAE induction. We then analyzed the dispersion of the NG2-glia derived clonally from single pallial progenitors in the brain of EAE mice. In addition, we examined several morphological parameters to assess the degree of NG2-glia reactivity in clonally-related cells. Our results reveal the heterogeneity of these progenitors and their cell progeny in a scenario of autoimmune demyelination, revealing the ontogenic phenomena at play in these processes.

Keywords: NG2-glia; progenitors; multiple sclerosis; lineage; in utero electroporation; morphometric analyses; clonal analyses; lesioned brain

1. Introduction

Multiple sclerosis (MS) is a chronic, disabling autoimmune and neurodegenerative disorder targeting the white and gray matter of the central nervous system (CNS) [1–3]. The loss of myelin and oligodendrocytes, axonal loss, and glial scar formation in the brain are the hallmarks for MS. As such, much interest has been generated by NG2-glia (also referred to as oligodendrocyte progenitor cells (OPCs) due to their ability to form myelin and the fact that they undergo a reactive response to brain injury. These cells are the fourth-most distinct major glial cell population [4], and they constitute 8–9% of the total cells in the white matter and 2–3% of the total cells in the grey matter [5]. Moreover, they express the NG2 chondroitin sulfate proteoglycan (neural/glial antigen 2 -NG2), the alpha receptor for platelet-derived growth factor (PDGFRα), as well as other oligodendrocyte markers. NG2-glia also have several distinguishing physiological properties, expressing voltage-gated ion channels and participating in neural–glial interactions [6–8]. They are thought to fulfil diverse functions in the brain, and they are implicated in the development of the nervous system, in maintaining its homeostasis, in neuromodulation, and in sustaining (re-)myelination [9–12]. The importance of these cells for the proper functioning of the brain is also reflected in the fact that they are distributed throughout the entire brain and that a constant pool of NG2-glia is maintained in the brain [13,14].

In animal models used to probe CNS lesions, NG2 cells proliferate and migrate to the site of injury, where they differentiate into oligodendrocytes [15–17]. In response to brain injury, some NG2-glia change their morphology and adopt a reactive phenotype, like astrocytes [18–20]. This morphology is characterized by thickened and highly ramified processes, as opposed to the relatively thin and unbranched prolongations of non-reactive NG2-glia [17]. Furthermore, NG2-glia display both molecular and behavioral heterogeneity following brain injury [17,21,22], producing both beneficial and deleterious effects [8,21,23–27]. However, little is known about the existence of certain subsets of NG2-glia that produce a specific response to brain damage and whether their fate may already be determined early in embryonic development.

A population of NG2-glia are known to arise from embryonic neural progenitor cells (NPCs) located in the subpallium [28]. Moreover, NG2-glia that come from dorsal NPCs (pallium) migrate following radial glia processes and group in clones throughout the cortex [13,29,30]. Thus, it is essential to track individual NPCs, and to achieve heritable and stable labelling of their progeny. Here we used the StarTrack approach, a reliable and proven method for clonal analysis based on the random genomic integration and expression of twelve fluorescent proteins [18]. In this study, we focused on the progeny of NG2-glia derived from individual embryonic glial fibrillary acidic protein-expressing progenitors (GFAP$^+$) in the brain of mice in which experimental autoimmune encephalomyelitis (EAE) was induced. Employing StarTrack in in these EAE-lesioned mice allowed us to analyze the clonal distribution and morphological cell response of NG2-glia in EAE lesions. Accordingly, we performed a morphometric analysis on the progeny of these cells, evaluating different parameters to assess the morphological changes to NG2-glia in the EAE mouse model. This analysis enabled us to assess the intensity of NG2-glia activation and to identify the heterogeneity in the phenotypes, clonal size, and cell fate of the individual progenitors singled out by the StarTrack approach.

2. Materials and Methods

2.1. Animals

C57BL/6 mice employed in this study were the same as those previously used to analyze the clonal response of astrocytes in a murine model of MS [19]. In brief, mice from Janvier Labs were housed in standard cages at the Universidad del País Vasco (UPV)-EHU animal facility, maintained on a 12 h controlled light–dark cycle with food and water available ad libitum. The study was carried out in accordance with the European Union recommendations on the use and welfare of experimental animals (2010/63/EU), and those of the Spanish Ministry of Agriculture (RD 1201/2005 and L 32/2007). The Bioethical Committee at the UPV-EHU approved the protocol. Nine adult StarTrack-electroporated mice (three sham and six EAE mice) were used. Clones were analyzed in all EAE mice. Animals that displayed NG2-glial clones—three of the EAE-lesioned mice (two males and one female)—were selected for the clonal analysis of NG2-glia. For the morphometric analysis, a selection of ten cells from EAE-lesioned mice and six cells from sham was made.

2.2. StarTrack DNA Vectors

Clonal analysis was accomplished using the StarTrack approach. StarTrack DNA vectors were produced as described previously [13,19,31] and the StarTrack mixture consisted of twelve PiggyBac constructs containing the six fluorescent nuclear and cytoplasmic proteins driven by the GFAP promoter (XFP), along with the hyperactive PiggyBac transposase (CMW-hyPBase) construct. The fluorescent proteins produced by the DNA vectors were the yellow fluorescent protein (YFP), monomeric Kusabira Orange (mKO), mCerulean, mCherry, mT-Sapphire and enhanced green fluorescent protein (EGFP).

2.3. In Utero Electroporation (IUE)

IUE was performed as previously described [19]. Embryonic day (E)14 pregnant mice were anesthetized with 4% isoflurane (2 mL/L: Isova vet, Centauro), their uterine horns were exposed by midline laparotomy and the embryos they contained were visualized by trans-illumination with cold light. The plasmid mixture was injected into the lateral ventricle (LV) of each embryo, which then received one or two trains of five square pulses (35 V, 50 ms duration, 950 ms intervals). Finally, the uterine horns were placed back into the abdominal cavity and the electroporated embryos were allowed to continue their normal development until birth.

2.4. EAE Induction

Chronic, relapsing EAE was induced in C57BL/6-electroporated mice during the eighth post-natal week [19,32,33]. Each animal was first immunized with a subcutaneous injection (300 µL) of a 1:1 ratio of myelin oligodendrocyte glycoprotein antigen solution (MOG35-55 peptide, 200 ng/mouse: Sigma, Barcelona, Spain) and complete Freund's adjuvant (CFA), followed by an intraperitoneal injection of pertussis toxin (500 ng: Sigma) in 100 µL of phosphate-buffered saline (PBS) on the day of immunization and two days later. CFA is a solution that contains *Mycobacterium tuberculosis* H37RA (8 mg/mL) in incomplete Freund's adjuvant. EAE was scored double-blind each day: 0, no noticeable signs of EAE; 1, flaccid tail; 2, paralyzed tail; 3, impairment or loss of muscle tone in hindlimbs; 4, unilateral partial hindlimb paralysis; 5, total bilateral hindlimb paralysis; 6, complete hindlimb paralysis and loss of muscle tone in the forelimbs; 7, complete paralysis of the forelimbs and hindlimbs; and 8, moribund. In our experiments, the motor symptoms in mice with EAE initiated around 10 days' post-immunization and progressively aggravated until reaching a peak typically at day 21, and declined slightly thereafter during the chronic phase [32]. EAE was successfully induced in all mice used in this study, and the scores representing the symptoms of the three EAE mice were 1.75 (nearly paralyzed tail), 3, and 4.5 (see Figure 2C from Bribián et al., 2018 [19]). Since tissue damage and demyelination parallels the symptoms, we assumed that the NG2-glial clonal response was maximal at that peak of the symptoms and accordingly, analyzed brain tissues at that stage. Results between animals were homogeneous.

2.5. Immunohistochemistry

Mice were perfused 21 days' post-induction (dpi) with 4% paraformaldehyde (PF) in a phosphate buffer (PB). They were then post-fixed for over 2 h in the same solution and stored at 4 °C in PBS. Coronal vibratome sections (50 µm) were washed and permeabilized three times with 0.5% Triton X-100 (PBS-T), washed three times in 0.1% PBS-T, and blocked for 30 min at room temperature (RT) with 5% normal goat serum (NGS, S26-100ML: Merck-Millipore). Brain sections were incubated overnight at 4 °C with the following antibodies in 5% NGS and 0.1% PBS-T: rabbit anti-PDGFRα (1:300, 3174S: Cell Signaling) and biotinylated tomato lectin (TL, 1:50, L0651: Sigma-Aldrich). After washing the brain slices three times with 0.1% PBS-T, they were incubated for 2 h at RT with a secondary antibody coupled to Alexa 633 (1:1000, Invitrogen) or a Streptavidin–Alexa Fluor 633 conjugate (1:1000, S21375: Invitrogen Life Technologies (Carlsbad,. CA, USA). Prior to visualization, they were washed 6 times in 0.1% PBS-T and then 1× PBS.

2.6. Imaging Acquisition and Data Analysis

The expression of the different fluorescent proteins was first checked under an epifluorescence microscope (Nikon, Eclipse F1) equipped with filters (Semrock) optimized for the following fluorophores: YFP (FF01-520/15), mKO (FF01-540/15), Cerulean (FF01-405/10), mCherry (FF01-590/20), Cy5 (FF02-628/40-25), GFP (FF01-473/10), and UV-2A (FF01-334/40-25). Consequently, images were acquired on a confocal microscope (Leica, TCS-SP5) and the emission for each fluorescent protein was obtained in separated channels using different excitation (Ex) and emission (Em) wavelengths (in nanometers, nm): mT-Sapphire (Ex: 405; Em: 520–535), mCerulean (Ex: 458; Em: 468–480), EGFP (Ex: 488; Em: 498–510), YFP (Ex: 514; Em: 525–535), mKO (Ex: 514; Em: 560–580), mCherry (Ex: 561; Em: 601–620), and Alexa 633 (Ex: 633; Em: 650–760). Laser lines were situated between 25% and 40%, and maximum projections were obtained using the confocal (LASAF Leica) and NIH-ImageJ software. Affected or lesioned areas were localized by TL staining and the perimeters of the lesion site were defined using the "enlarge" tool of NIH-ImageJ software, with a distance of 50 µm between the concentric perimeters. The Simple Neurite Tracer (SNT) plugin (NIH-ImageJ) [34] and a Scholl analysis [35,36] were used for the morphological analysis. The statistical analysis of the data and the graphical representations were performed using the R statistical software package (version 3.5: R Core Team, 2018), and the Prism 5 (GraphPad) software. Statistical significance was evaluated using either a two-tailed unpaired Student's t test for 2-group comparisons or a one-way ANOVA followed by Dunnett's post hoc test for multiple group comparisons. Values with a confidence interval of 95% ($p < 0.05$) were considered statistically-significant and significant differences between the groups are indicated in the graphs with asterisks: * $p < 0.05$, ** $p < 0.01$, *** $p < 0.001$.

3. Results

3.1. Spatial Distribution of the Cortical Progeny of NG2-Glia Derived from Single Embryonic Progenitors in EAE-Lesioned Mouse Brain

To evaluate how the NG2-glia progenitors responded to EAE lesions in the phase of symptom improvement, we combined the genomic StarTrack tool with the induction of EAE in mice. We first targeted individual E14 dorsal (pallial) progenitors in the LV (Figure 1A) [13]) through IUE of the StarTrack mixture (driven by the GFAP promoter) with the hyperactive PiggyBac transposase (CMW-hyPBase: Figure 1A). Since up to 4096 different color-code combinations can be achieved with this method, it facilitates a precise clonal analysis of the progeny of single labelled cells with an active GFAP promoter [37]. Eight weeks after birth, the MOG peptide was administered to analyze the clonal NG2-glia response to EAE lesions 21 dpi (Figure 1A).

To perform the clonal analysis of NG2-glia in EAE mice, cortical regions containing StarTrack-labelled cells close to affected areas were selected (Figure 1B). These areas were characterized by the presence of perivascular inflammatory infiltrates and enlarged perivascular spaces revealed by TL staining (Figure 1B). This TL staining allowed a perimeter to be drawn around the core of the lesions in both the cortex and striatum. To analyze labelled cells in the affected area, the perimeter surrounding the lesion core was amplified four times using the "enlarge" tool (NIH-ImageJ), obtaining borders at 50 µm, 100 µm, 150 µm, and 200 µm (Figure 1B). Clonal dispersion of the sibling astrocytes and sibling NG2-glia close to the lesion sites (Figure 1B,E) was identified in the cortical layers (Figure 1B,D,E, and Figure 2A), the corpus callosum (Figures 1C and 2B), and in part of the striatum (Figure 2B). The identity of the NG2-glia was confirmed through the expression of PDGFRα (Figure 1C) and sibling NG2-glia were identified by their uniform fluorescent color-code (i.e., the expression and localization of the XFPs). While some NG2-glia clones were distributed within or in close proximity to the lesion site (Figure 1D), others were located far from the affected areas (Figure 1E). Thus, combining EAE induction and StarTrack we were able to evaluate the clonal response of NG2-glia in an inflammatory and lesioned mouse brain.

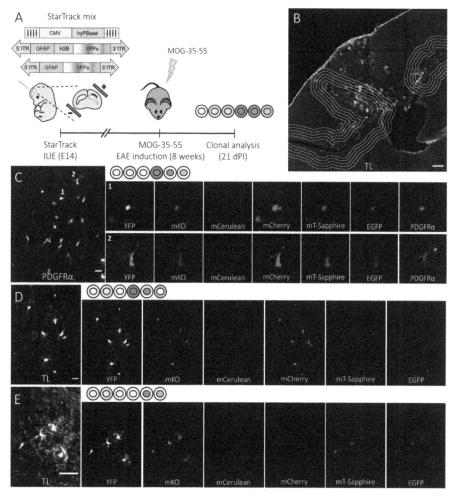

Figure 1. StarTrack-labelled NG2-glia progeny of individual dorsal subventricular zone (SVZ) progenitors after the induction of EAE. (**A**) Time-line showing the experimental approach: in utero electroporation (IUE) of the StarTrack mix into E14 dorsal neural progenitor cells (NPCs) in the lateral ventricle (LV), induction of experimental autoimmune encephalomyelitis (EAE) in mice eight weeks after electroporation, and clonal analysis 21 days' post-induction (dpi). The StarTrack mixture contained a hyperactive PiggyBac transposase (hyPBase) and 12 plasmids that encoded six different fluorescent proteins (XFPs) expressed in either the nucleus or cytoplasm. A glial fibrillary acidic protein (GFAP) promoter drove XFP expression. (**B**) Coronal view of the brain showing the location of EAE lesions in the cerebral cortex, corpus callosum, and striatum, as identified by tomato lectin (TL) labelling. A line surrounding the lesion core and four concentric compartments separated by 50 µm was used to measure the distance of the NG2-glia sibling cells from the lesion. StarTrack-labelled both astrocytes and NG2-glia around the different lesion sites. (**C–E**) Detail of the coronal views of sibling NG2-glia dispersed throughout the corpus callosum (**C**) and cerebral cortex (**D,E**). The color-codes and the corresponding XFPs expressed are detailed at the side of the merged images. NG2-glia were identified by platelet-derived growth factor (PDGFRα$^+$) staining (**C**: 1 and 2), TL was used to identify the EAE affected and lesion areas (**E**), and the non-affected zones (**D**). Scale bars, 200 µm (**B**) and 20 µm (**C–E**).

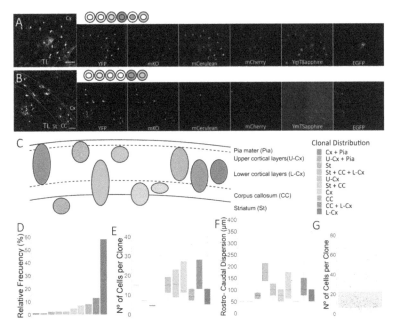

Figure 2. Clonal distribution of NG2-glia in coronal sections. (**A**,**B**) Detail of sibling NG2-glia close to the lesion in the cerebral cortex (**A**) and striatum and corpus callosum (**B**). Affected areas were identified by tomato lectin (TL) staining and clonally-related cells showed equivalent color combinations, detailed at the side of the merged images. (**C**) Scheme showing the diverse patterns of clonal NG2-glia dispersion. Dorsal E14 progenitors from the lateral ventricle produced different patterns of clonal NG2-glia progeny in the EAE-lesioned brain, with a heterogeneous clonal dispersion within the striatum (St), corpus callosum (CC), the upper- (U-Cx) and lower-cortical layers (L-Cx]), and the pia mater (Pia). (**D**) The relative frequency of the different clonal distributions estimates the frequency of each clone pattern in the EAE mice, indicating that NG2-glia clones derived from pallial progenitors expanded preferentially within the lower cortical layers. (**E**) The number of cells per clone varied depending on the cell fates of the clone. The largest clones were those distributed within the CC. (**F**) The largest clones exhibited a more extended rostro-caudal axis dispersion with sibling cells occupying a maximum average dispersion of 330 μm. (**G**) The clones varied in size from 3 to 80 cells, with an average of 17 cells per clone. Scale bars, 50 μm (**A**,**B**).

3.2. Clonal Distribution of the Cell Progeny from Individual Embryonic Pallial NG2-Glia Progenitors in EAE-Lesioned Brains

We evaluated the clonal fate of NG2-glia generated from dorsal embryonic NPCs in the LV, and of clonally-related NG2-glia distributed within the cerebral cortex, corpus callosum, and striatum (Figures 1B and 2). Sibling NG2-glia form clusters of up to 80 cells in lesioned brains, with clones having an average size of 17 cells (n = 84 analyzed clones, 1393 NG2-glial cells: Figure 2G). NG2-glia clones were distributed sparsely in specific regions within these three areas (Figure 2C), with different patterns of dispersion relative to the astrocyte clones [31]. Thus, we identified different patterns of NG2-glia clonal cell dispersion (Figure 2C), with clonally-related cells limited to the striatum, corpus callosum, or the upper or lower layers of the cortical cortex (Figure 2A). To obtain the relative frequency of each clone pattern, we calculated the number of times that each clone pattern appeared, divided by the total number of NG2-glia clones in the three EAE mice. Other clones expanded and colonized multiple regions of the brain (Figure 2B) but interestingly, none of the NG2-glia clones exclusively populated the pia matter (Figure 2D). As mentioned above, some NG2-glia clones were located in the striatum, with clonally-related cells situated in both the corpus callosum and lower cortical layers (Figure 2B,C). In addition, while 8% of the total clones were confined to the corpus callosum, 13% of NG2-glia clones populated both the corpus callosum and the lower cortical layers (Figure 2B,C,D). Nevertheless, most of the clones generated from NPCs in this EAE-lesion scenario (58% of the NG2-glia clones) were distributed in clusters found in the lower cortical layers (Figure 2A,D).

Interestingly, the average of NG2-glia clone size was larger when the clones colonized more than one region (Figure 2E). The distribution of clones with more cells corresponded to those clones in which some of the sibling cells were located in the corpus callosum. These larger clones populated both the striatum, the corpus callosum, and the lower cortical layers (Figure 2E). In addition, clones with cells in the corpus callosum showed the highest average spread in the rostro-caudal axis. NG2-glia clones were distributed in the striatum, the cortex, and pia, or just in the cortex where clones displayed the lower rostro-caudal dispersion (Figure 2F). This indicates that there was a correlation between rostro-caudal axis extension and the number of siblings NG2-glia cells per clone (Figure 2E,F).

Thus, NG2-glia clones derived from dorsal progenitors of the embryonic SVZ were distributed in ten different patterns of clonal dispersion, showing heterogeneity in relation to both their size and rostro-caudal distribution.

3.3. Relationship Between Sibling NG2-Glia and EAE Brain Lesions

The clonal distribution of NG2-glia was evaluated by performing StarTrack clonal analysis in mice in which EAE lesions were induced. The NG2-glia clonal analysis revealed the distribution of different NG2-glia clones relative to the lesion. We quantified the number of sibling cells within the lesion core or in the four concentric perimeters (Figures 1B and 3A), as well as the clonally-related cells outside the defining borders of the lesion (Figure 3B), analyzing 642 sibling cells from 48 different clones (Figure 3). Clones with cells inside the 200 µm perimeter around the lesion (Figure 3A) were considered as inner clones, whereas the rest were considered as outer clones (Figure 3B). We measured the distance from the lesion core of all sibling NG2-glia cells, showing a very heterogeneous dispersion (Figure 3A,B). Inner clones (Figure 3A) displayed multiple distribution patterns, the most characteristic was that in which there was an even dispersion of clonally-related cells around the lesion core (Figure 3A: clones 01–04, 06). Nevertheless, some inner clones were more widely dispersed, with sibling cells in and far from the lesion core (Figure 3A: clone 05). The average number of cells per inner clone was significantly higher than that of the outer clones ($p < 0.001$: Figure 3C). In addition, 60% of the analyzed clones analyzed were in close proximity to the lesion (Figure 3D).

3.4. Morphometric Analyses of NG2-Glia in EAE Lesions

To analyze the cell heterogeneity in the morphological response of NG2-glia to EAE, we examined the StarTrack-labelled NG2-glia close to the lesion. Since NG2-glia react in response to brain damage, by thickening and branching their processes, we applied morphometric parameters to identify and quantify these changes. First, StarTrack-labelled NG2-glia within or far from the lesion core were divided into the two main subtypes, type I and type II cells, according to the primary branch thickness. The morphology of type I ($n = 6$ cells) and type II ($n = 4$ cells) was compared to that of sham NG2-glia ($n = 6$ cells) through both conventional morphological analysis and Scholl analysis with the SNT plugin (NIH-ImageJ: Figure 4D). Different cell parameters were analyzed to evaluate the progression of NG2-glia arborization in response to lesion (adapted from Yamada et al., 2013 [36]: Figure 4A–C,E–G): cell body area and volume, domain area and volume, circularity, solidity, average branch thickness, total thickness, number of branches, total number of intersections, ending Scholl radius, and total branch length. The cell body area and the domain area are 2D parameters (μm^2) to measure the surface of the cell and the convex hull area from the projection image, respectively. The region enclosed by a polygon that connects the end-points of NG2-glia processes is considered as the convex hull area. The domain and cell body volume are 3D parameters (μm^3) and consider the width of the cell, measuring the convex hull area and the cell body area rendered from z-series data sets, respectively.

Both, domain area and volume, are parameters that indicate the tridimensional non-overlapping domain that NG2-glia occupy [5,14,38]. The circularity (a 2D parameter) is the ratio between the NG2-glia cell body area and its perimeter, indicating how round the cell is, a perfect circle attributed a value of one. By contrast, the solidity, also a 2D parameter, is the ratio between the body cell area and the convex hull area, showing how filled the convex hull area is. Solidity is a parameter that indirectly estimates how thick the cell processes are, or how branched the cell is., and Wwhen the cell body area coincides with the convex hull area, the solidity is one. A significant increase in both the cell body area and the domain area of the two morphological types of NG2-glia was detected relative to the sham cells (Figure 4H,I), although the significance was higher in the case of type II NG2-glia. When the circularity, the cell body volume, the domain volume, and the solidity were compared, these parameters were only significantly higher in type II NG2-glia. Nevertheless, the type I NG2-glia exhibited a tendency towards an increase in those parameters compared to the sham cells (Figure 4J–M).

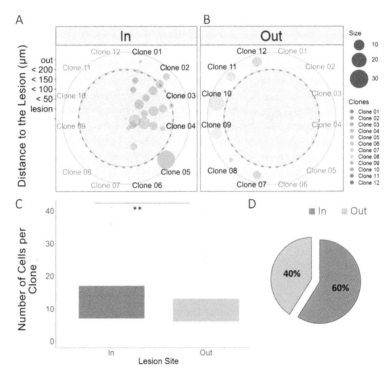

Figure 3. Dispersion of NG2-glia clones relative to the lesion. (**A,B**) Radial plots of sibling NG2-glia dispersion according to their distance from the EAE lesion and the number of clonally-related cells. The radii represent the lesion core area and the different consecutive perimeters (up to 200 μm) surrounding that area at an interval of 50 μm. All cells further than 200 μm away from the lesion core were considered to be outside of the lesion area (green area), and cells within 200 μm were considered to be inside the lesion area (pink area). Quadrants separate sibling cells of inner clones (01–06) (**A**) and from the outer clones (07–12) (**B**). Inner clones were described as clones with one or more clonally-related cells within at least 200 μm from the core. Sibling cells of six outer clones and six inner clones were represented, the sibling cells sharing the same color-code. (**C**) There were more clonally-related cells that formed clones within the lesion site (inner clones) than those that formed outer clones. Two-tailed unpaired Student's t test was used to evaluate the statistically-significant differences between the two groups: ** $p < 0.001$. (**D**) From the NG2-glia clones analyzed ($n = 48$), 60% had at least one sibling cell within the 200 μm perimeter around the lesion. More inner clones than outer clones were found.

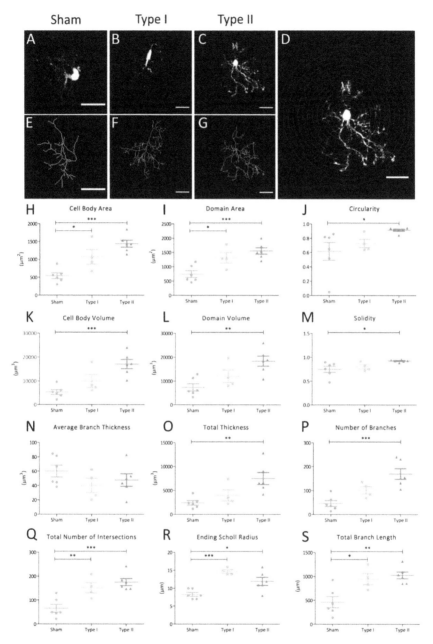

Figure 4. Morphometric parameters showing the shift of NG2-glia towards hypertrophy in the EAE brain. (**A–G**) Representative z-stack projection of StarTrack-labelled NG2-glia. Morphological analysis comparing sham NG2-glia (**A,E**) with both cell types identified in EAE mice: type I (**B,F**) and type II (**C,G**). Fill out images obtained once all paths were drawn using the Simple Neurite Tracer (SNT) plugin (NIH-ImageJ (**A–C**). Measurements of cell body area and volume, domain area and volume, circularity, solidity and thickness were acquired using this image. The respective rendered paths of the NG2-glia in filled out images (**E–G**). The number of branches were automatically measured from these

rendered images. Representation of the Scholl analysis on a type II NG2-glia cell (**D**). The interval between each consecutive Scholl circle radius was 4 μm. The total number of intersections, total branch length and ending Scholl radius were obtained in this analysis. (**H–S**) Graphs of the parameters that describe the change in NG2-glia towards a more severe morphology. The morphology of type I and II NG2-glia was compared with that of sham NG2-glia. Statistically-significant differences across the groups were evaluated using a one-way ANOVA followed by Dunnett's post hoc test for multiple group comparisons: * $p < 0.05$, ** $p < 0.01$, *** $p < 0.001$. Scale bars = 20 μm.

Total thickness (μm^3) is a parameter that refers to the volume of all the branches with a threshold of 0.05. In this study, the number of branches included both the major and primary processes, and all the minor branches that derived from them (the secondary, tertiary, and higher order processes) [39]. Interestingly, the average branch thickness (μm^3) appeared to be the same for type I, type II, and sham NG2-glia morphologies, this being a measure of the relationship between the total thickness of a cell and its number of branches (Figure 4N). Nevertheless, the total thickness and the number of branches was significantly higher in type II NG2-glia and it tended to be higher in type I NG2-glia compared to sham NG2-glia (Figure 4O,P). Thus, NG2-glia not only increase in size and the domain they occupy in response to EAE-induction but also, in the arborization of their processes and their volume, with some thicker branches in type II NG2-glia.

In addition, the Scholl 3D analysis compared the ending Scholl radius (μm), the number of total intersections and the total branch length (μm). The ending Scholl radius is the distance from the cell nucleus to the last concentric circle that surrounds the NG2-glia soma, with a process intersection. The concentric circles of the Scholl analysis have 4 μm intervals in-between them, being the cell nucleus, the center of those circumferences that surround the analyzed cell (Figure 4D) and the number of total intersections measures how many branches cross the concentric Scholl circles. There was a significant increment in cell complexity of both the type I and type II NG2-glia, in both the total number of intersections and total branch length (Figure 4Q,S). This difference with regard the sham cells was more evident in type II NG2-glia. Nevertheless, even if the ending Scholl radius was higher in the EAE groups, there was no tendency for this parameter to increase when the cell changed towards hypertrophy. Our data revealed that in a lesioned scenario, NG2-glia acquire a complex morphology, based on an increment in the total number of intersections (Figure 4Q) in approximately the same Scholl radius (Figure 4R).

Together, our morphological analysis revealed NG2-glia cells in the brain become more complex in response to EAE, also experiencing an increment in size and domain. There was a progressive change in the morphology of NG2-glia in response to lesion, adopting their two morphology types close to the lesion site (type I and II NG2-glia). Type I NG2-glia morphology tended to have a less reactive morphology than type II NG2-glia.

3.5. Clonal NG2-Glia Response to EAE Lesion

Once the morphological parameters of both NG2-glia types had been determined in EAE brains, we analyzed the clonal response in 137 cells from twelve different clones (Figures 5A and 2B) using the perimeters delimiting the infiltrates and lesions (Figure 1B). Activation of these clonally-related cells generated NG2-glia clones of type I, type II, or mixed sibling NG2-glia clones. Those clones formed exclusively by type II NG2-glia were restricted to the lesion core (Figure 5A,B, clones 01–03, 10), whereas type I NG2-glia clones were mainly distributed at a distance from the lesion core (Figure 5A,B: clones 05–07), although some were located close to the lesion core (Figure 5A,B: clones 11, 12). Finally, mixed clones, formed by type I and II NG2-glia were evident at both locations (Figure 5A,B: clones 04, 08 and 09). An analysis of the inner clones showed that sibling cells of type II clones concentrated within the perimeter 150 μm from the infiltrate (Figure 5C), whereas type I NG2-glia sibling cells and mixed clones were dispersed homogeneously from the lesion core (Figure 5C). Thus, while type II cells tended to group around the lesion core, type I cells were distributed more sparsely across the inner to outer regions. Hence, there was a heterogeneous response of clonally-related NG2-glia to lesion.

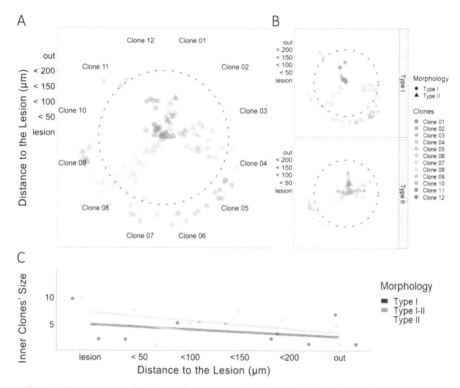

Figure 5. Heterogeneous clonal NG2-glia response to EAE lesions. (**A**) Radial plot of sibling NG2-glia dispersion according to their distance from the EAE-lesion site and the morphology of each cell ($n = 34$). The concentric radii define the lesion core area, with the different perimeters (up to 200 µm) surrounding that area drawn at an interval of 50 µm. Lateral quadrants separate type I (dots) and type II (triangles) NG2-glia into two radial plots. Sibling cells have the same color-code. (**B**) Clonally-related type II NG2-glia inner clones (yellow) accumulate close to the lesion core, within the 150 µm perimeter, and at a higher rate when closer to the lesion core. Type I sibling cells (purple) extended away from the lesion beyond the final 200 µm perimeter and there were a number of sibling cells distributed evenly at all distances from the lesion. Mixed clones of type I and type II sibling cells (green) tend to accumulate closer to the lesion core, although some clones were located outside of the 200 µm perimeter. (**C**) Taken together, type II NG2-glia in inner clones are preferentially located within the 150 µm perimeter and concentrate at a higher rate at the lesion, and closer to it. By contrast, type I cells are dispersed evenly inside and beyond the lesion perimeters established.

4. Discussion

This study reveals the response of individual NG2-glia progenitors to the damage induced in the EAE mouse model of MS. Using the StarTrack approach in EAE mice, we tracked the progeny of individual embryonic NG2-glia progenitors from the dorsal SVZ in vivo. Labelled sibling NG2-glia dispersed throughout the white and grey matter in a clonal manner. Moreover, a morphometric analysis identified changes in these cells relative to the EAE lesion. Thus, our data revealed that the heterogeneous distribution of NG2-glia in response to EAE lesions is at least in part related to the nature of their progenitors.

4.1. Heterogeneity in the Clonal Distribution of NG2-Glia in a Mouse Model of MS

Astrocytes derived from dorsal NPCs in the LV are distributed clonally in either the pia mater, corpus callosum, or cortex [13]. Moreover, sibling astrocytes in the cortex are restricted to become vascular or protoplasmic astrocytes, meaning that their distribution and hence, their function, is determined by the embryonic NPCs [13]. NG2-glia have different temporal and spatial origins during development [28,40]. In particular, the progeny of pallial progenitors migrate to the corpus callosum and cerebral cortex, revealing the relationship between their heterogeneity and their ontogenic origin [31]. Here, we assessed the ten different distributions of NG2-glia clones, showing that specific pallial embryonic progenitors are already committed to a particular clonal distribution in the cortex, pia, corpus callosum, and striatum. As such, clones of NG2-glia derived from embryonic pallial progenitors are more frequently distributed within the lower cortical layers. Furthermore, some sibling NG2-glia may be widely dispersed throughout the striatum, corpus callosum, cerebral cortex, and/or the pia, unlike the astroglial clones [31]. This might imply that NG2-glia heterogeneity is determined early in development. Likewise, this widespread cell dispersion may indicate that they are not limited to act in a small area or that their functions are not so specific.

Clones close to the EAE lesion appear to be larger than those far from it, which could be explained by proliferation and migration towards the lesion. Indeed, NG2-glial cells proliferate in a normal brain [5,14,38] and after brain damage replacing those NG2-glia that die or differentiate into oligodendrocytes [21,41–43]. NG2-glia also became hypertrophic and migrate towards the lesion to participate in the formation of the glial scar [21,44]. Moreover, type II NG2-glia clones, corresponding to reactive and hypertrophic cells, tend to accumulate within the lesion core, producing clones with more sibling cells close to the affected area. This indicates that there is a relationship between the distance from the lesion and NG2-glia activation [17,43], and that there is a clonal response. Moreover, while those clones close to the lesion were reactive, some others with sibling cells within the lesion core remained with a non-hypertrophic morphology. Thus, NG2-glial clones are heterogeneous in both their distribution and in their shift towards a reactive cell phenotype after EAE-induction. These different responses of NG2-glia clones not only depend on the molecular environment but also on the progenitor identity, as occurs in astroglial clones in response to EAE injury [19]. NG2-glia could be preferentially connected to each other in a clonal way, similar to the preferential gap junctions between sibling astrocytes [45], which might explain the clonal response of NG2-glia after brain damage. Indeed, NG2-glia share synaptic connections with neurons [7,46] and some connectivity between NG2-glia may exist due to their homeostatic role, and in the control of NG2-glia cell density through active growth and self-retraction of their processes [22,42,43].

Our data reveals that the number of clonally-related NG2-glia cells per clone in the EAE-model is similar to that reported in the same age control mice, although the variance appeared to be higher [13]. Hence, some clones may substitute others, as reported with NG2-glia during development [28].

4.2. Heterogenic and Clonal NG2-Glia Responses to EAE Lesions

Central features of the connection between the NG2-glia lineage and their response to EAE lesions were studied here. The different clonal responses to brain damage observed have marked implications in terms of the functional heterogeneity of NG2-glia following brain damage [21,23–27,44], implying that this heterogeneity is established early in embryonic development. Our results reveal that groups of NG2-glia clones display a heterogeneous response to brain damage and perivascular infiltration.

NG2-glia in a 3-nitropropionic acid-lesioned striatum tend to proliferate and accumulate within the perimeter of the lesion [17]. In addition, sibling NG2-glia cells, derived from dorsal progenitors, distribute through the grey and white matter around the lesions, with more clones of NG2-glia close to the lesion than far from it. This suggests that these progenitors may be determined to react upon brain injury and indeed, specific NG2-glia populations proliferate and react upon acute brain injury in both the grey and white matter [46]. Interestingly, endogenous NG2-glia can react to brain demyelination in order to protect axons, maintaining the velocity of axonal transmission with newly-formed but defective myelin [6]. It is crucial to understand the mechanisms regulating the NG2-glia response to demyelination in order to improve remyelination and functional axon recovery. Our results reveal that the NG2-glia response is already determined in their progenitors and therefore, our findings will be of interest for future therapeutic remyelination strategies.

4.3. NG2-Glia Activate in Response to EAE Lesions, Undergoing a Progressive Morphological Change to Reactivity Characterized by an Increase in Their Size, Domain, and Arborization

For the first time, we have analyzed morphometric parameters to define the reactive phenotype of NG2-glia in response to EAE lesion. Although some morphological changes of reactive NG2-glia are known [17,21,44], it is unclear how NG2-glia evolves in the face of the severity and hypertrophy in EAE. We found that NG2-glia increase their size, tridimensional non-overlapping domains, or complexity in response to different brain insults [17,44]. By contrast, NG2-glia complexity and size decrease in the first 72 h after ischemia, indicating cell death rather than hypertrophy [45]. These parameters are restored one to two weeks later [45], suggesting that neighboring NG2-glia might maintain cell density and homeostasis [21,47,48].

After injury, NG2-glia undergo a graded continuum of morphological alterations over time and they adopt, at least, two morphologically distinct states as they become reactive: a less severe or non-reactive morphology (type I cells), and a hypertrophic morphology (type II cells). All these changes occur gradually, passing through intermediate stages in between the non-reactive and the hypertrophic NG2-glia. Reactive NG2-glia acquire a rounder shape, thickening their branches to fill their spatial domains and enlarging their individual domain. Interestingly, NG2-glia migrate to the lesion after brain injury and lose their individual domain [21]. Together with their shift to hypertrophy, this change may imply that these cells are involved in the formation of the glial scar after brain damage to restrict the lesion and to promote tissue repair. Moreover, these cells control reactive gliosis [21,26,43]. Reactive NG2-glial cells increment NG2 expression [17,45,49], which is implicated in cell adhesion [50] and associated with a decrease in the differentiation to oligodendrocytes, as well as with an increment in synaptic contacts [51]. Scholl parameters revealed growing complexity in the processes of NG2-glia following activation and hypertrophy, showing that hypertrophied NG2-glia are more branched and complex [17].

Our findings regarding the different clonal response to brain damage have marked implications in understanding the functional heterogeneity of NG2-glia in response to brain damage [21,23–27,44], implying that this heterogeneity is established early in embryonic development. Defining the heterogeneity of NG2 cells in relation to their ontogenic origin is important to address the in vivo reprogramming of NG2-glia to neurons in CNS injury mouse models [52–54].

In conclusion, we present relevant features of the NG2-glia lineage in terms of the response to EAE lesions. We demonstrate that NG2-glia from individual progenitors differ in both their spatial fate and in their functional response to EAE lesions.

Author Contributions: Conceptualization, L.L.-M., A.B., and C.M.; methodology, A.B. and F.P.-C.; formal analysis, S.B.; resources, L.L.-M.; C.M.; writing—original draft preparation, S.B. and L.L.-M.; writing—review and editing, S.B. and L.L.-M.; supervision, L.L.-M.; funding acquisition, L.L.-M. and C.M. All authors have read and agreed to the published version of the manuscript.

Funding: This research was funded by research Grants from the Fundación Ramón Areces (Ref. CIVP9A5928), MINECO (BFU2016-75207-R), SAF2016-75292-R, CIBERNED and Gobierno Vasco (IT1203-19).

Acknowledgments: Sonsoles Barriola is affiliated to the UAM PhD Program in Neuroscience. We are grateful to the UPV-EHU animal facility, the Imaging and Microscopy facilities at the Instituto Cajal-CSIC and the Achucarro Basque Center for their excellent technical assistance and Mark Sefton for editorial assistance. We are also thankful to Lina María Delgado García, Rebeca Sánchez González, Ana Cristina Ojalvo Sanz and Nieves Salvador Cabos for their help with the morphometric and statistical analysis.

Conflicts of Interest: The authors have no conflicts of interest to declare.

References

1. Lassmann, H. Cortical lesions in multiple sclerosis: Inflammation versus neurodegeneration. *Brain* **2012**, *135*, 2904–2905. [CrossRef] [PubMed]
2. Prins, M.; Schul, E.; Geurts, J.; van der Valk, P.; Drukarch, B.; van Dam, A.M. Pathological differences between white and grey matter multiple sclerosis lesions. *Ann. N. Y. Acad. Sci.* **2015**, *1351*, 99–113. [CrossRef] [PubMed]
3. Van Munster, C.E.P.; Jonkman, L.E.; Weinstein, H.C.; Uitdehaag, B.M.; Geurts, J.J. Gray matter damage in multiple sclerosis: Impact on clinical symptoms. *Neuroscience* **2015**, *303*, 446–461. [CrossRef] [PubMed]
4. Nishiyama, A.; Chang, A.; Trapp, B.D. NG2+ glial cells: A novel glial cell population in the adult brain. *J. Neuropathol. Exp. Neurol.* **1999**, *58*, 1113–1124. [CrossRef]
5. Hill, R.A.; Nishiyama, A. NG2 cells (polydendrocytes): Listeners to the neural network with diverse properties. *Glia* **2014**, *62*, 1195–1210. [CrossRef]
6. Dimou, L.; Gallo, V. NG2-glia and their functions in the central nervous system. *Glia* **2015**, *63*, 1429–1451. [CrossRef]
7. Orduz, D.; Maldonado, P.P.; Balia, M.; Vélez-Fort, M.; de Sars, V.; Yanagawa, Y.; Emiliani, V.; Angulo, M.C. Interneurons and oligodendrocyte progenitors form a structured synaptic network in the developing neocortex. *Elife* **2015**, *2015*, 1–53. [CrossRef]
8. Nakano, M.; Tamura, Y.; Yamato, M.; Kume, S.; Eguchi, A.; Takata, K.; Watanabe, Y.; Kataoka, Y. NG2 glial cells regulate neuroimmunological responses to maintain neuronal function and survival. *Sci. Rep.* **2017**, *7*, 42041. [CrossRef]
9. Xiao, L.; Ohayon, D.; McKenzie, I.A.; Sinclair-Wilson, A.; Wright, J.L.; Fudge, A.D.; Emery, B.; Li, H.; Richardson, W.D. Rapid production of new oligodendrocytes is required in the earliest stages of motor-skill learning. *Nat. Neurosci.* **2016**, *19*, 1210–1217. [CrossRef]
10. Hill, R.A.; Li, A.M.; Grutzendler, J. Lifelong cortical myelin plasticity and age-related degeneration in the live mammalian brain. *Nat. Neurosci.* **2018**, *21*, 683–695. [CrossRef]
11. Hughes, E.G.; Orthmann-Murphy, J.L.; Langseth, A.J.; Bergles, D.E. Myelin remodeling through experience-dependent oligodendrogenesis in the adult somatosensory cortex. *Nat. Neurosci.* **2018**, *21*, 696–706. [CrossRef] [PubMed]
12. Van Tilborg, E.; de Theije, C.G.M.; van Hal, M.; Wagenaar, N.; de Vries, L.S.; Benders, M.J.; Rowitch, D.H.; Nijboer, C.H. Origin and dynamics of oligodendrocytes in the developing brain: Implications for perinatal white matter injury. *Glia* **2018**, *66*, 221–238. [CrossRef] [PubMed]
13. García-Marqués, J.; Núñez-Llaves, R.; López-Mascaraque, L. NG2-glia from pallial progenitors produce the largest clonal clusters of the brain: Time frame of clonal generation in cortex and olfactory bulb. *J. Neurosci.* **2014**, *34*, 13–2305. [CrossRef] [PubMed]
14. Boda, E.; Di Maria, S.; Rosa, P.; Taylor, V.; Abbracchio, M.P.; Buffo, A. Early phenotypic asymmetry of sister oligodendrocyte progenitor cells after mitosis and its modulation by aging and extrinsic factors. *Glia* **2015**, *63*, 271–286. [CrossRef] [PubMed]
15. Keirstead, H.S.; Levine, J.M.; Blakemore, W.F. Response of the oligodendrocyte progenitor cell population (Defined by NG2 labelling) to demyelination of the adult spinal cord. *Glia* **1998**, *22*, 161–170. [CrossRef]
16. McTigue, D.M.; Wei, P.; Stokes, B.T. Proliferation of NG2-positive cells and altered oligodendrocyte numbers in the contused rat spinal cord. *J. Neurosci.* **2001**, *21*, 3392–3400. [CrossRef]
17. Jin, X.; Riew, T.-R.; Kim, H.L.; Choi, J.-H.; Lee, M.-Y. Morphological characterization of NG2 glia and their association with neuroglial cells in the 3-nitropropionic acid-lesioned striatum of rat. *Sci. Rep.* **2018**, *8*, 5942. [CrossRef]
18. Martín-López, E.; García-Marques, J.; Núñez-Llaves, R.; López-Mascaraque, L. Clonal astrocytic response to cortical injury. *PLoS ONE* **2013**, *8*, e74039. [CrossRef]
19. Bribian, A.; Pérez-Cerdá, F.; Matute, C.; López-Mascaraque, L. Clonal Glial Response in a Multiple Sclerosis Mouse Model. *Front. Cell. Neurosci.* **2018**, *12*, 375. [CrossRef]
20. Okada, S.; Hara, M.; Kobayakawa, K.; Matsumoto, Y.; Nakashima, Y. Astrocyte reactivity and astrogliosis after spinal cord injury. *Neurosci. Res.* **2018**, *126*, 39–43. [CrossRef]

21. Hughes, E.G.; Kang, S.H.; Fukaya, M.; Bergles, D.E. Oligodendrocyte progenitors balance growth with self-repulsion to achieve homeostasis in the adult brain. *Nat. Neurosci.* **2013**, *16*, 668–676. [CrossRef] [PubMed]
22. Valny, M.; Honsa, P.; Waloschkova, E.; Matuskova, H.; Kriska, J.; Kirdajova, D.; Androvic, P.; Valihrach, L.; Kubista, M.; Anderova, M. A single-cell analysis reveals multiple roles of oligodendroglial lineage cells during post-ischemic regeneration. *Glia* **2018**, *66*, 1068–1081. [CrossRef] [PubMed]
23. Tanaka, Y.; Tozuka, Y.; Takata, T.; Shimazu, N.; Matsumura, N.; Ohta, A.; Hisatsune, T. Excitatory GABAergic activation of cortical dividing glial cells. *Cereb. Cortex* **2009**, *19*, 2181–2195. [CrossRef] [PubMed]
24. Kang, Z.; Wang, C.; Zepp, J.; Wu, L.; Sun, K.; Zhao, J.; Chandrasekharan, U.; DiCorleto, P.E.; Trapp, B.D.; Ransohoff, R.M.; et al. Act1 mediates IL-17-induced EAE pathogenesis selectively in NG2+ glial cells. *Nat. Neurosci.* **2013**, *16*, 8–1401. [CrossRef] [PubMed]
25. Seo, J.H.; Miyamoto, N.; Hayakawa, K.; Pham, L.D.D.; Maki, T.; Ayata, C.; Kim, K.W.; Lo, E.H.; Arai, K. Oligodendrocyte precursors induce early blood-brain barrier opening after white matter injury. *J. Clin. Invest.* **2013**, *123*, 782–786. [CrossRef]
26. Rodriguez, J.P.; Coulter, M.; Miotke, J.; Meyer, R.L.; Takemaru, K.I.; Levine, J.M. Abrogation of β-Catenin signaling in oligodendrocyte precursor cells reduces glial scarring and promotes axon regeneration after CNS injury. *J. Neurosci.* **2014**, *34*, 10285–10297. [CrossRef]
27. Liu, J.; Dietz, K.; Hodes, G.E.; Russo, S.J.; Casaccia, P. Widespread transcriptional alternations in oligodendrocytes in the adult mouse brain following chronic stress. *Dev. Neurobiol.* **2018**, *78*, 152–162. [CrossRef]
28. Kessaris, N.; Fogarty, M.; Iannarelli, P.; Grist, M.; Wegner, M.; Richardson, W.D. Competing waves of oligodendrocytes in the forebrain and postnatal elimination of an embryonic lineage. *Nat. Neurosci.* **2006**, *9*, 173–179. [CrossRef]
29. Winkler, C.C.; Yabut, O.R.; Fregoso, S.P.; Gomez, H.G.; Dwyer, B.E.; Pleasure, S.J.; Franco, S.J. The Dorsal Wave of Neocortical Oligodendrogenesis Begins Embryonically and Requires Multiple Sources of Sonic Hedgehog. *J. Neurosci.* **2018**, *38*, 5237–5250. [CrossRef]
30. Tong, C.K.; Fuentealba, L.C.; Shah, J.K.; Lindquist, R.A.; Ihrie, R.A.; Guinto, C.D.; Rodas-Rodriguez, J.L.; Alvarez-Buylla, A. A dorsal SHH-dependent domain in the V-SVZ produces large numbers of oligodendroglial lineage cells in the postnatal brain. *Stem Cell Rep.* **2015**, *5*, 461–470. [CrossRef]
31. García-Marqués, J.; López-Mascaraque, L. Clonal Identity Determines Astrocyte Cortical Heterogeneity. *Cereb. Cortex* **2013**, *23*(6), 1463–1475.
32. Matute, C.; Torre, I.; Pérez-Cerdá, F.; Pérez-Samartín, A.; Alberdi, E.; Etxebarria, E.; Arranz, A.M.; Ravid, R.; Rodríguez-Antigüedad, A.; Sánchez-Gómez, M.; et al. P2X(7) receptor blockade prevents ATP excitotoxicity in oligodendrocytes and ameliorates experimental autoimmune encephalomyelitis. *J. Neurosci.* **2007**, *27*, 33–9525. [CrossRef] [PubMed]
33. Pampliega, O.; Domercq, M.; Soria, F.N.; Villoslada, P.; Rodríguez-Antigüedad, A.; Matute, C. Increased expression of cystine/glutamate antiporter in multiple sclerosis. *J. Neuroinflammation* **2011**, *8*, 63. [CrossRef] [PubMed]
34. Tavares, G.; Martins, M.; Correia, J.S.; Sardinha, V.M.; Guerra-Gomes, S.; das Neves, S.P.; Marques, F.; Sousa, N.; Oliveira, J.F. Employing an open-source tool to assess astrocyte tridimensional structure. *Brain Struct. Funct.* **2017**, *222*, 1989–1999. [CrossRef]
35. Scholl, D.A. Dendritic organization in the neurons of the visual and motor cortices of the cat. *J. Anat.* **1953**, *87*, 387–406.
36. Yamada, J.; Jinno, S. Novel objective classification of reactive microglia following hypoglossal axotomy using hierarchical cluster analysis. *J. Comp. Neurol.* **2013**, *521*, 1184–1201. [CrossRef]
37. Figueres-Oñate, M.; Garciá-Marqués, J.; López-Mascaraque, L. UbC-StarTrack, a clonal method to target the entire progeny of individual progenitors. *Sci. Rep.* **2016**, *6*, 33896. [CrossRef]
38. Butt, A.M.; Hamilton, N.; Hubbard, P.; Pugh, M.; Ibrahim, M. Synantocytes: The fifth element. *J. Anat.* **2005**, *207*, 695–706. [CrossRef]
39. Khakh, B.S.; Sofroniew, M.V. Diversity of astrocyte functions and phenotypes in neural circuits. *Nat. Neurosci.* **2015**, *18*, 942–952. [CrossRef]
40. Malatesta, P.; Hack, M.A.; Hartfuss, E.; Kettenmann, H.; Klinkert, W.; Kirchhoff, F.; Götz, M. Neuronal or glial progeny: Regional differences in radial glia fate. *Neuron* **2003**, *37*, 751–764. [CrossRef]

41. Simon, C.; Götz, M.; Dimou, L. Progenitors in the adult cerebral cortex: Cell cycle properties and regulation by physiological stimuli and injury. *Glia* **2011**, *59*, 869–881. [CrossRef] [PubMed]
42. Dimou, L.; Götz, M. Glial cells as progenitors and stem cells: New roles in the healthy and diseased brain. *Physiol. Rev.* **2014**, *94*, 709–737. [CrossRef] [PubMed]
43. Hesp, Z.C.; Yoseph, R.Y.; Suzuki, R.; Jukkola, P.; Wilson, C.; Nishiyama, A.; McTigue, D.M. Proliferating NG2-Cell-Dependent Angiogenesis and Scar Formation Alter Axon Growth and Functional Recovery After Spinal Cord Injury in Mice. *J. Neurosci.* **2018**, *38*, 1366–1382. [CrossRef] [PubMed]
44. Jäkel, S.; Dimou, L. Glial cells and their function in the adult brain: A journey through the history of their ablation. *Front. Cell. Neurosci.* **2017**, *11*, 24. [CrossRef]
45. Bonfanti, E.; Gelosa, P.; Fumagalli, M.; Dimou, L.; Viganò, F.; Tremoli, E.; Cimino, M.; Sironi, L.; Abbracchio, M.P. The role of oligodendrocyte precursor cells expressing the GPR17 receptor in brain remodeling after stroke. *Cell Death Dis.* **2017**, *8*, e2871. [CrossRef]
46. Boda, E.; Viganò, F.; Rosa, P.; Fumagalli, M.; Labat-Gest, V.; Tempia, F.; Abbracchio, M.P.; Dimou, L.; Buffo, A. The GPR17 receptor in NG2 expressing cells: Focus on in vivocell maturation and participation in acute trauma and chronic damage. *Glia* **2011**, *59*, 1958–1973. [CrossRef]
47. Kirby, B.B.; Takada, N.; Latimer, A.J.; Shin, J.; Carney, T.J.; Kelsh, R.N.; Appel, B. In vivo time-lapse imaging shows dynamic oligodendrocyte progenitor behavior during zebrafish development. *Nat. Neurosci.* **2006**, *9*, 1506–1511. [CrossRef]
48. Parolisi, R.; Boda, E. NG2 Glia: Novel Roles beyond Re-/Myelination. *Neuroglia* **2018**, *1*, 151–175. [CrossRef]
49. Levine, J.M. Increased expression of the NG2 chondroitin-sulfate proteoglycan after brain injury. *J. Neurosci.* **1994**, *14*, 4716–4730. [CrossRef]
50. Wight, T.N.; Kinsella, M.G.; Qwarnström, E.E. The role of proteoglycans in cell adhesion, migration and proliferation. *Curr. Opin. Cell Biol.* **1992**, *4*, 793–801. [CrossRef]
51. Schäfer, M.K.E.; Tegeder, I. NG2/CSPG4 and progranulin in the posttraumatic glial scar. *Matrix Biol.* **2018**, *68-69*, 571–588.
52. Heinrich, C.; Bergami, M.; Gascón, S.; Lepier, A.; Viganò, F.; Dimou, L.; Sutor, B.; Berninger, B.; Götz, M. Sox2-mediated conversion of NG2 glia into induced neurons in the injured adult cerebral cortex. *Stem Cell Rep.* **2014**, *3*, 1000–1014. [CrossRef] [PubMed]
53. Pereira, M.; Birtele, M.; Shrigley, S.; Benitez, J.A.; Hedlund, E.; Parmar, M.; Ottosson, D.R. Direct Reprogramming of Resident NG2 Glia into Neurons with Properties of Fast-Spiking Parvalbumin-Containing Interneurons. *Stem Cell Rep.* **2017**, *9*, 742–751. [CrossRef] [PubMed]
54. Yavarpour-Bali, H.; Ghasemi-Kasman, M.; Shojaei, A. Direct reprogramming of terminally differentiated cells into neurons: A novel and promising strategy for Alzheimer's disease treatment. *Prog. Neuro-Psychopharmacol. Biol. Psychiatry* **2020**, *98*, 109820. [CrossRef] [PubMed]

© 2020 by the authors. Licensee MDPI, Basel, Switzerland. This article is an open access article distributed under the terms and conditions of the Creative Commons Attribution (CC BY) license (http://creativecommons.org/licenses/by/4.0/).

Article

Suppression of the Peripheral Immune System Limits the Central Immune Response Following Cuprizone-Feeding: Relevance to Modelling Multiple Sclerosis

Monokesh K. Sen [1], Mohammed S. M. Almuslehi [1,2], Erika Gyengesi [1], Simon J. Myers [3], Peter J. Shortland [3], David A. Mahns [1,*] and Jens R. Coorssen [4,*]

[1] School of Medicine, Western Sydney University, Locked Bag 1797, Penrith, NSW 2751, Australia; monokesh.sen@westernsydney.edu.au (M.K.S.); m.almuslehi@westernsydney.edu.au (M.S.M.A.); e.gyengesi@westernsydney.edu.au (E.G.)
[2] Department of Physiology, College of Veterinary Medicine, Diyala University, Diyala, Iraq
[3] School of Science and Health, Western Sydney University, Locked Bag 1797, Penrith, NSW 2751, Australia; s.myers@westernsydney.edu.au (S.J.M.); p.shortland@westernsydney.edu.au (P.J.S.)
[4] Department of Health Sciences, Faculty of Applied Health Sciences, and Department of Biological Sciences, Faculty of Mathematics and Science, Brock University, St. Catharines, Ontario, ON L2S 3A1, Canada
* Correspondence: d.mahns@westernsydney.edu.au (D.A.M.); jcoorssen@brocku.ca (J.R.C.); Tel.: +02-4620-3784 (D.A.M.); +905-688-5550 (J.R.C.)

Received: 6 September 2019; Accepted: 18 October 2019; Published: 24 October 2019

Abstract: Cuprizone (CPZ) preferentially affects oligodendrocytes (OLG), resulting in demyelination. To investigate whether central oligodendrocytosis and gliosis triggered an adaptive immune response, the impact of combining a standard (0.2%) or low (0.1%) dose of ingested CPZ with disruption of the blood brain barrier (BBB), using pertussis toxin (PT), was assessed in mice. 0.2% CPZ(±PT) for 5 weeks produced oligodendrocytosis, demyelination and gliosis *plus* marked splenic atrophy (37%) and reduced levels of CD4 (44%) and CD8 (61%). Conversely, 0.1% CPZ(±PT) produced a similar oligodendrocytosis, demyelination and gliosis but a smaller reduction in splenic CD4 (11%) and CD8 (14%) levels and *no* splenic atrophy. Long-term feeding of 0.1% CPZ(±PT) for 12 weeks produced similar reductions in CD4 (27%) and CD8 (43%), as well as splenic atrophy (33%), as seen with 0.2% CPZ(±PT) for 5 weeks. Collectively, these results suggest that 0.1% CPZ for 5 weeks may be a more promising model to study the 'inside-out' theory of Multiple Sclerosis (MS). However, neither CD4 nor CD8 were detected in the brain in CPZ±PT groups, indicating that CPZ-mediated suppression of peripheral immune organs is a major impediment to studying the 'inside-out' role of the adaptive immune system in this model over long time periods. Notably, CPZ(±PT)-feeding induced changes in the brain proteome related to the suppression of immune function, cellular metabolism, synaptic function and cellular structure/organization, indicating that demyelinating conditions, such as MS, can be initiated in the absence of adaptive immune system involvement.

Keywords: inside-out; outside-in; oligodendrocytosis; demyelination; gliosis; histology; top-down proteomics; bioinformatics; mitochondria

1. Introduction

Currently, there are two competing theories regarding the pathophysiology underlying the initiation of Multiple Sclerosis (MS): 'outside-in' and 'inside-out' [1–4]. The former proposes that a dysregulated peripheral immune system leads to an autoimmune response against myelin components of the central nervous system (CNS). The central concept of this theory has been built mainly on the basis of studies using the experimental autoimmune encephalomyelitis (EAE) animal model [5–7] and correlation of the end stage of treatment (e.g., paralysis and demyelination) with clinical tests and post-mortem samples from MS patients. In EAE, animals are injected with exogenous antigens such as myelin basic protein (MBP), proteolipid protein (PLP) or myelin oligodendrocyte glycoprotein (MOG) and complete Freund's adjuvant (CFA), activating peripheral immune cells, including T- and B-cells. When this immune response is combined with breach of the blood brain barrier (BBB) by injection of pertussis toxin (PT), autoreactive adaptive immune cells from the periphery migrate into the CNS leading to degeneration of oligodendrocytes (OLG), demyelination and gliosis [5,6,8–11]. It is argued that a similar process results in autoimmune cell migration into the CNS of MS patients [7,12–15]. The EAE animal model is thus the favourite choice of many researchers investigating the autoimmune aspects of MS [16].

There are, however, key differences between the EAE model and clinical MS. First, EAE relies on the use of exogenous antigens (MBP/PLP/MOG), whereas the autoimmune response in humans occurs spontaneously and is only detected following repeated episodes of clinical symptoms [17,18]. Second, the immune reaction in EAE is driven mainly by $CD4^+$ T-cells [19,20], whereas in MS, $CD8^+$ T-cells predominate [21–23]. Moreover, MS is a disease of the human cerebral and cerebellar cortices, whereas the effects of EAE are generally localized to the spinal cord [6,10,24–27], with largely non-overlapping changes in the brain proteomes being reported in EAE and MS patients [28,29]. Although therapies developed in EAE improve outcomes in animals, these therapies generally have more limited success in clinical MS in terms of halting disease initiation and progression [17,30]. In contrast to this 'outside-in' theory, the 'inside-out' hypothesis suggests that MS is initiated by an underlying degeneration of OLG and consequent demyelination that leads to the production of endogenous myelin antigens (e.g., peptidyl arginine deiminase, MBP, MOG and PLP) that then trigger an immune response in the CNS [3,4,31]. Histological evidence indicates that the loss of OLG and glial activation can occur in the absence of, or with only a limited number of, peripheral immune cells [32–35] and myelin injury [36]. The possibility of OLG triggering a secondary adaptive immune response has been reported following long-term diphtheria toxin exposure [37] or following a peripheral immune challenge after short-term CPZ-feeding (termed 'cuprizone autoimmune encephalitis' [31]). Cuprizone (CPZ) is synthesised by combining cyclohexanone and oxaldihydrazone [38]. While the mechanism of its toxic actions remain ill-defined, copper chelation [39] and dis-homeostasis of iron, zinc, sodium and manganese have been reported [40–44]. Such ion imbalance leads to endoplasmic reticulum stress, reduced mitochondrial ATP synthesis, and increased production of reactive oxygen and nitrogen species (reviewed in [2]). OLG appear to be preferentially susceptible to CPZ toxicity and likely degenerate due to their high energy demands and lower levels of anti-oxidant enzymes [45–47].

Intriguingly, longer-term CPZ-feeding did not evoke a peripheral immune response in the CNS [31,37,48]. Whether the failure of numerous CPZ studies [31,37,48] to observe an immune response associated with long duration feeding (>5 weeks) is due to a toxic effect of CPZ on the peripheral immune system remains unclear [31,37], but CPZ has been shown to have deleterious effects on immune organs like the spleen [49] and thymus [50].

To address this issue, this study compared whether a low (0.1%) or standard (0.2%) dose of CPZ when combined with disruption of the BBB by PT recruited peripheral immune cells into the CNS in mice. At first, a low and a standard dose of CPZ were used for 5 weeks and the amount of oligodendrocytosis (i.e., degeneration or loss of OLG), gliosis and demyelination was quantified; the low dose produced significant demyelination of the corpus callosum (CC), with limited suppression of splenic CD4/8 and no change in overall splenic mass. Having observed that 0.1% CPZ produced comparable oligodendrocytosis, but had less severe effects on the peripheral immune system, in the second study, 0.1% CPZ-feeding was extended to 12 weeks (±PT) to test whether a slower, progressive demyelination (i.e., more reminiscent of MS) and less severe effects on the peripheral immune organs could trigger an adaptive immune response in the CNS. Histological analyses were used to assess oligodendrocytosis, demyelination and gliosis in the CNS, as well as the levels of adaptive immune cells (CD4 and CD8) in brain and spleen.

In a subset of the mice in both experiments, the whole brain proteome was assessed using a well-established 'top-down' approach (i.e., two-dimensional gel electrophoresis coupled with liquid chromatography and tandem mass spectrometry) to identify changes in protein abundance correlated with key changes in molecular pathways [49,51–53]. To best understand the underlying molecular/cellular processes, a 'top-down' proteomic analysis was critical to identifying key protein species or *proteoforms*, the biologically active entities [54–56]. Thus, while much is assumed in the literature regarding the actions of CPZ at the molecular level by extrapolation of effects in vitro or at the cellular level, only such quantitative analyses can help to directly understand the underlying effects of CPZ(±PT).

2. Materials and Methods

2.1. Animals, Feeding, Injection and Monitoring

Seven-week-old male C57Bl/6 mice (n = 108) were purchased from the Animal Resources Centre, Murdoch, WA, Australia (www.arc.wa.gov.au) and co-housed (2 mice) in individual ventilated GM500 cages (Tecniplast, Buguggiate, VA, Italy) in the local animal care facility (School of Medicine, Western Sydney University). Animals were allowed to acclimatise for one week to the new environment prior to initiation of CPZ-feeding. Mice were maintained in a controlled environment (12-hour (h) light/dark cycle: 8am–8pm light, 8pm–8am dark, 50–60% humidity and at 21–23 °C, room temperature (RT)) throughout the entire period. Standard rodent powder chow (Gordon's specialty stockfeeds, Yanderra, NSW, Australia) and water were available *ad libitum*. Oral feeding of CPZ ([Bis(cyclohexanone)oxaldihydrazone, Sigma-Aldrich, St. Louis, MO, USA], 0.1–0.2% *w/w* freshly mixed with rodent chow was used to induce oligodendrocytosis as previously described [2,57–59]. To breach the BBB, the same methods as previously established for EAE were used i.e., 2–3 intraperitoneal (IP) injections of PT [8,60–63]—but adapted so that the breach of the BBB was timed (i.e., 400 ng on days 14, 16, and 23) to coincide with the reported onset of CPZ-induced oligodendrocytosis, demyelination and gliosis [2,57–59]. The efficacy of BBB breach has been shown using immunoglobulin G staining in the CPZ-fed mice [64]. CPZ groups (0.1% and 0.2%) were fed freshly prepared (daily) CPZ in rodent chow for either 5 (n = 10/group) or 12 (n = 12/group) weeks. Age-matched, naïve control (Ctrl, 5-week study n = 10 and 12-week study n = 12) and PT only (5-week study n = 10 and 12-week study n = 12) groups were used. Mice were weighed at the beginning of the studies, weekly throughout, and prior to culling, and the data from both groups (5 and 12 weeks) were combined (Figure 1). Research and animal care procedures were approved by the Western Sydney University Animal Ethics Committee (ethics code: A10394) in accordance with the Australian Code of Practice for the Care and Use of Animals for Scientific Purposes as laid out by the National Health and Medical Research Council of Australia.

2.2. Histology and Immunohistochemistry

2.2.1. Tissue Preparation

At the end of each feeding period (5 or 12 weeks), all mice were terminally overdosed with sodium pentabarbitone (250 mg/kg, Lethobarb™, Tory laboratories, Glendenning, NSW, Australia) and perfused with 30 mL of 0.9% saline followed by 50 mL of cold 4% paraformaldehyde (PFA, Sigma-Aldrich) for ~5 minutes (min). Brain and spleen samples were collected and post fixed with 4% PFA at 4 °C for one week and stored in 0.01 M phosphate buffered saline (PBS, Sigma-Aldrich) solution containing 0.02% sodium azide (Amresco, Solon, OH, USA) at 4 °C for ≤1 month or until sectioned. Spleen weights were measured (n = 3 and 5 from 5- and 12-week studies, respectively) and expressed as a function of body weight (Supplementary Figure S5c). Prior to sectioning, whole brain and spleen were immersed in 30% sucrose for 48 h at RT for cryo-protection followed by embedding in 4% gelatine (Chem-Supply, Gillman, SA, Australia) at −20 °C. Brains (5 weeks: 50 µm and 12 weeks: 40 µm) and spleen (20 µm) were sectioned coronally on a Leica cryostat (Leica, Wetzlar, HE, Germany). Sections were transferred to either 6-well plate (Sakura, Torrance, CA, USA) containing cold (5–6 °C)

0.01 M PBS (free floating) or mounted onto 0.5% gelatine-coated slides (Knittel Glass, Braunschweig, NI, Germany) as described previously [65].

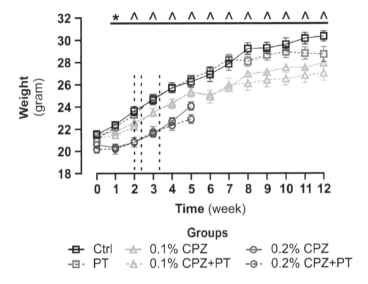

Figure 1. Body weight changes induced by CPZ-feeding. All mice gained weight over time. Groups fed the highest dose of CPZ(±PT) in each experiment gained weight significantly more slowly compared to other groups. Vertical dash lines indicate the timing of individual PT injections (i.e., days 14, 16, and 23). Data are expressed as mean ± SEM. Two-way ANOVA and Tukey post hoc analysis were used to determine differences among groups (* $p < 0.05$, ˆ $p < 0.0001$, 5-week study $n = 22$ animals/group of which $n = 12$ animals/group continued feeding for 12 weeks).

2.2.2. Silver Myelin Staining and Analysis

Silver staining of myelin was performed at RT as previously described [66,67]. Briefly, tissue sections were mounted onto 0.5% gelatine-coated slides and air dried for 48 h before immersion in 10% formalin (Sigma-Aldrich) for 2 weeks to increase the contrast of the staining. Sections were stained in a large glass container in parallel to maintain the consistency of staining. Slides were washed with distilled water and pre-treated with lipid-solvent pyridine and acetic anhydride solution (ratio 2:1) for 30 min. Sections were then rehydrated with serial dilutions of ethanol 80, 60, 40 and 20% for 20 seconds (sec) in each step followed by two washes with distilled water. Slides were then immersed in ammonical silver nitrate (Chem-Supply) containing developing solution (0.2% ammonium nitrate, 0.2% silver nitrate and 5% sodium carbonate) for 45 min. Sections were then dehydrated by rinsing sequentially using 20, 40, 60, 80, and 100% ethanol for 20 sec in each step. Sections were cleared by immersing the slides in xylene for 5 min and then sealed using cover slips (Knittel Glass) and ~1 mL mounting medium (Merck, Darmstadt, HE, Germany), and air dried for 72 h. The sections were viewed with an Olympus Carl Zeiss Bright Field Microscope (Zeiss, Jena, TH, Germany) and all images were captured at the same microscope settings (i.e., a fixed exposure time, magnification and illumination intensity). In ImageJ (https://imagej.nih.gov/) software, the region of interest (ROI) was contoured, and the mean optical density quantified (sum of each pixel intensity [range black (0) to white (256)] divided by the number of pixels in the ROI). To quantify the amount of myelin present (which stains black), data were expressed as the reciprocal of the light intensity (i.e., the smaller the value, the lower the myelin content) and normalised to the Ctrl groups (n = 3–5 animals/group and 5–9 sections/animal). The effectiveness of CPZ-feeding is frequently determined by loss of myelin from the midline corpus callosum (MCC), in this study the effectiveness of CPZ-induced demyelination was confirmed in the MCC and lateral corpus callosum (LCC). Anatomical landmarks were identified as described previously [68,69].

2.2.3. Immunofluorescence Staining and Analysis

All staining was performed at RT. Free floating brain (40–50 µm) and slide-mounted spleen (20 µm) coronal tissue sections were washed thrice with warm (40–50 °C) 0.01 M PBS to remove the gelatine and then immersed in 5–10% goat serum (Sigma-Aldrich) for 2 h with agitation at 50 rpm on a shaker table to block non-specific antibody (Ab) binding sites. Sections were then incubated (12 h) with primary monoclonal Ab to either neurite outgrowth inhibitor A (rabbit anti-Nogo A, 1:500, Merck-Millipore, Burlington, MA, USA), glial fibrillary acidic protein (mouse anti-Gfap-Alexa 488, 1:1000, Merck-Millipore), ionized calcium-binding adapter molecule 1 (rabbit anti-Iba 1, 1:1000, Wako, Chuo-Ku, OSA, Japan), anti-cluster of differentiation (rabbit anti-CD4, 1:200, Abcam, Cambridge, UK), or anti-CD8 (mouse anti-CD8, 1:100, Santa-Cruz Biotechnology, Dallas, TX, USA) diluted in 0.01 M PBS containing 0.1% Triton X100 (Tx100). Sections were then washed thrice with 0.01 M PBS and incubated with corresponding Alexa Fluor (either 488 or 555, dilution: 1:500) secondary Ab (diluted in 0.01 M PBS/0.1% Tx100 solution for 2 h while shaking at 50 rpm). Sections were then rinsed thrice with 0.01 M PBS and 1.5 µg/mL Vectasheild™ plus 4′,6-diamidino-2-phenylindole (DAPI, Vector Laboratories, Burlingame, CA, USA) to counterstain nuclei. Slides were sealed with cover slips and stored in the dark at 4 °C until analysis. Images were captured as before (using an Olympus Carl Zeiss Fluorescence Microscope) using the same fixed parameters (exposure and magnification). Quantification was performed using ImageJ, measuring the fluorescence intensity of Gfap, Iba 1, CD4 and CD8 from each ROI as described in silver myelin staining and analysis (n = 3–5 animals/group and 5–10 sections/animal). Cells positively stained for Gfap, Iba 1 and Nogo A (and co-stained with DAPI) were counted using the unbiased stereo investigator optical fractionator workflow software [65,70]. To obtain the cell density, total cell number was divided by total measured volume and the data expressed as 10^4 cells/mm^3 (n = 3–5 animals/group and 5–9 sections/animal).

2.3. Two-Dimensional Gel Electrophoresis (2DE) and Analysis

2.3.1. Sample Collection, Homogenisation and Protein Estimation

At the end of each experiment (i.e., 5 or 12 weeks) period, mice (n = 5 animals/group) were euthanized by overdose of isoflurane (Cenvet, Blacktown, NSW, Australia) exposure. Whole brains were collected following decapitation and immediately rinsed with ice cold 0.01 M PBS containing a cocktail of protease, kinase, and phosphatase inhibitors (Sigma-Aldrich, [51,52]) to remove any traces of blood. Tissue homogenisation was accomplished by automated frozen disruption using a Mikro-Dismembrator (40 Hz for 1 min, Sartorius, Göttingen, NI, Germany) to facilitate optimal protein extraction [52,71]. Powdered tissue samples were mixed with cold 20 mM HEPES hypotonic lysis buffer (Amresco) containing the cocktail of protease, kinase, and phosphatase inhibitors and vortexed for 90 sec followed by the restoration of isotonicity using the addition of an equivalent volume of ice cold 0.02 M PBS and incubated for 5 min on ice. The samples were then centrifuged (Beckman Coulter, Indianapolis, IN, USA) at 125,000× g (using a SW 55 Ti rotor) at 4 °C for 2 h. The resulting first supernatant (SP1) was collected as total cytosolic soluble protein (SP). The pellet was washed with ice cold 0.01 M PBS to extract any remaining cytosolic proteins (SP2) and centrifuged 8 h at 125,000× g. The resulting supernatant (SP2) was pooled with the SP1 fraction, and this combined soluble protein fraction was concentrated using an Amicon Ultra-4 centrifugal 3 KD cut-off filter column (Merck-Millipore). To prevent salt interference during the 1st dimension isoelectric focussing step, the resulting SP samples were washed three times using cold 4 M urea (Amresco) buffer supplemented with the inhibitor cocktail. The pellet containing membrane proteins (MP) was resuspended with cold 2DE solubilisation buffer (8 M urea, 2 M thiourea and 4% CHAPS) containing the inhibitor cocktail [71,72]. Total protein concentrations in SP and MP fractions were measured using the EZQ™

Protein Quantitation Kit (Life Technologies, Eugene, OR, USA) according to the manufacturer's instructions, using bovine serum albumin (Amresco) as the standard.

2.3.2. Protein Separation

At first, proteins were separated based on their isoelectric point (1st dimension) as follows: 100 µg of proteins were loaded onto an immobilised pH gradient (IPG, 7 cm, non-linear, Bio-Rad, Hercules, CA, USA) strips and passively rehydrated for 16 h at RT. Rehydrated IPG strips were placed in the Protean isoelectric focusing (IEF) tray (Bio-Rad) and IEF was carried out in a PROTEAN IEF system (Bio-Rad) for high-throughput protein resolution, initially at 250 V for 15 min which then increased linearly to 4000 V at 50 µA/gel for 2 h, with multiple electrode wick changes during voltage ramping to facilitate desalting. The following parameters were used during IEF: focus temperature: 17 °C, desalting: 15 min, linear gradient: 2 h, holding voltage: 500 V. Following IEF, IPG strips were either stored at −20 °C or immediately resolved in the second dimension using sodium dodecyl sulphate-polyacrylamide gel electrophoresis (SDS-PAGE). Second dimension (2D) was carried out using 1 × 84 × 70 mm 12.5% acrylamide SDS-PAGE gels. Prior to 2D, IPG strips were incubated for 10 min with 130 mM DTT in equilibration buffer (6 M urea, 20% glycerol, 2% SDS and 375 mM tris) followed by 10 min alkylation in equilibration buffer containing 350 mM acrylamide. IPG strips were then inserted on top of the SDS-PAGE gels and covered with warm (40–50 °C) ~300 µL agarose solution (0.5% low melting agarose, Bio-Rad) containing 2% bromophenol blue (Bio-Rad). Electrophoresis was carried out at 4 °C by applying 150 V for 5 min followed by 90 V for 3 h or until the bromophenol dye reached the bottom of the gels as described previously [51,52,71,73].

2.3.3. Protein Fixation and Staining

Upon completion of electrophoresis, gels (5-week study n = 180 gels and 12-week study n = 120 gels) were fixed in 10% methanol and 7% acetic acid solution for 1 h at RT (on a shaker at 50 rpm). Gels were rinsed with distilled water for 3 × 20 min to remove residual methanol and acetic acid followed by staining with 50 mL of high sensitive colloidal Coomassie Brilliant Blue (G-250, Amresco) and continuous shaking, as previously described [52,74–78]. After 20 h, the solution was discarded and stained gels were washed using 50 mL of 0.5 M sodium chloride solution for 3 × 15 min to remove excess Coomassie dye. Scanning of gels was carried out at 100 µm resolution using a Typhoon^{TM} FLA-9000 gel imager (GE Healthcare, Chicago, IL, USA). Excitation/emission wave lengths were 685/>750 nm and the photomultiplier tube was set to 600 V. Gels were preserved in 20% ammonium sulphate (50 mL/gel) and stored at 4 °C until spot excision [52].

2.3.4. Protein Resolution, Detection and Image Analysis

Quantitative analysis of gels was carried out using Delta 2D image analysis software (www.decodon.com/delta2d-version4.0.8, DECODON, Greifswald, MV, Germany) as described previously [49,52,53,74,78]. Briefly, total spot numbers were calculated from the raw images using the Delta 2D automated spot detection system, while gel edges and the protein ladder were excluded [52,79]. Gel images were warped and fused to generate a master image ensuring consistent spot matching. The fluorescent volumes of individual spots (i.e., protein abundance) were expressed as a function of all spot volumes detected and were measured using Delta 2D to assess changes across the different experimental groups. Four spot inclusion criteria were applied: 1) any changes in spot volumes were detectable in all biological replicates (n = 5 animals/group) and their associated technical replicate gels (n = 3 gels/fraction/animal, i.e., 15 gels/experimental group); 2) the relative standard deviation for technical replicates did not exceed 30% within individual animals; 3) values had to differ significantly ($p < 0.05$, t-test) between the naïve Ctrl group and at least one test group; and 4) have a fold change of ≥1.5 (increased/decreased fluorescence), to be considered genuine changes, and thus candidates for analysis by LC/MS/MS. These criteria allowed for reliable and reproducible identification of CPZ, CPZ+PT or PT associated proteoform changes. The fold change ($p < 0.05$, one-way ANOVA) of differentially abundant protein spots was calculated by dividing the average grey value (i.e., fluorescence intensity of the protein spot of each experimental group compared to the naïve Ctrl group) and presented in $\log_2$ scale as fold change relative to Ctrl. Precision Plus Protein Kaleidoscope molecular weight (MW) and 2DE isoelectric point *(pI)* standard (Bio-Rad) calibration gels ($n = 3$) were used to calculate the experimental *pI* and MW of resolved proteoforms. The coefficient of variation (standard deviation/mean) for the MW and *pI* migration was 2.6% and 1.4% for 2DE standards, respectively ($n = 3$), and 4% for MW ladders for experimental gels ($n = 20$). To quantify the gel shift indicative of protein post-translational modification (PTM), experimental MW and *pI* values were plotted relative to theoretical values; significant changes were indicated when the experimental measure fell above or below the 95% confidence intervals of the MW and *pI* calibration curves. The results are plotted as the average for both 5 and 12 weeks combined.

2.3.5. In-Gel Protein Spot Digestion

Protein spots of interest were excised manually and de-stained for 2 × 15 min with 50 mM ammonium bicarbonate (Sigma-Aldrich) solution containing 50% acetonitrile (Sigma-Aldrich). After complete removal of Coomassie dye, gel pieces were dehydrated using 100% acetonitrile. In-gel digestion was carried out by adding 20 μL of freshly prepared trypsin (12.5 ng/μL, Promega Corporation, Madison, WI, USA) to a solution of 50 mM ammonium bicarbonate for 8 h at 4 °C. Digested peptides were removed from the gel by 30 min sonication. The solution was then acidified by the addition of 2 μL of 2% formic acid (Merck-Millipore). The resulting peptide solution was concentrated to 10 μL using a Speed Vac™ vacuum concentrator (1400 rpm for 10–15 min, John Morris Scientific, Chatswood, NSW, Australia) and stored at -80 °C for future use or immediately subjected to LC/MS/MS [52].

2.3.6. Liquid Chromatography Tandem Mass Spectrometry (LC/MS/MS)

The concentrated peptide solutions were analysed by LC/MS/MS using a nanoAcquity ultra performance liquid chromatography system linked to a Xevo QToF mass spectrometer as previously described [49,52,80,81]. In brief, peptide sample solutions (3 μL) were loaded onto a C18 symmetry trapping column (20 mm × 180 μm), and desalted for 3 min at 5 μL/min flow rate using 1% solvent B (LC/MS grade 1% acetonitrile and 0.1% formic acid) in solvent A (Mili-Q water + 0.1% formic acid). The peptides were washed off the trapping column at 400 nL/min onto a C18 BEH analytical column (75 μm × 100 mm), packed with 1.7 μm particles of pore size 130 Å using a 60 min ramped LC protocol. The initial solvent composition was held at 1% B for 1 min followed by linear ramping to 50% B over 30 min. A further linear ramp to 85% B commenced at 31 min and held until 37 min before the solvent composition was returned to 1% B. Separated peptides were analysed using tandem mass spectrometry with a constant cone voltage of 25 V and source temperature of 100 °C, implementing an emitter tip that tapered to 10 μm at 2.3 kV. A data-directed acquisition (DDA) approach was performed, which continuously scanned across the *m/z* range 350–1500 for peptides of charge state 2^+–4^+ with an intensity of more than 50 counts/sec, with a maximum of three ions in any given 3 sec scan. Selected peptides were de-isotoped, fragmented and the masses measured. The ramped collision energy profile was set to 15–35 V at low mass and 30–40 V at high mass. The mass of the precursor peptide was then excluded for 30 sec. The DDA was via Masslynx software (version 4.1, Micromass, Manchester, UK) and converted to a peak list file (PKL) format using the ProteinLynx Global Server (Waters, Milford, CT, USA). Data were analysed using MASCOT Daemon (www.matrixscience.com) and queried against the SwissProt and MSPnr100 databases (see www.wehi.edu.au) using delimited and species-specific searches to identify the protein species using the following MASCOT parameter settings: the enzyme trypsin and taxonomy *Mus musculus* (mouse) were fixed. Moreover, no fixed modification was selected whereas variable modifications were carbamidomethyl (C), deamidated (NQ), oxidation (M) and propionamide (C). Only two missed cleavages of lysine or arginine residues were allowed; mass tolerance of parent and MS/MS ions was set to ±0.05 Da and peptide charge state was 2^+–4^+. The results of the search were filtered by excluding peptide hits with a *p*-value greater than 0.05. While the SwissProt and MSPnr100 databases were both used to best ensure confirmation of a protein identity, the higher of the two scores was documented in Table 1. When multiple proteins were detected from the same spot, the following criteria were applied to identify the most abundant and thus the most likely to have contributed to the originally detected change in spot volume: 1) The highest MASCOT score (>100) with a sequence coverage ≥5%; and 2) ≥4 unique matched peptides.

2.3.7. Literature Mining and Bioinformatics

A PubMed (www.ncbi.nlm.nih.gov/pubmed/) literature search was carried out for papers published in the English language using the identified canonical protein name with either CPZ, EAE or MS to find literature relevant to molecular/cellular functions. The UniProt (www.uniprot.org) database was used to obtain the gene and UniProt accession number (ID) of the identified protein species and analysis of subcellular localization [82]. A mapping of genes according to their classification and molecular functions was derived from protein analysis through evolutionary relationships (PANTHER, www.pantherdb.org) database using gene IDs of each identified protein [53]. Cellular components, biological processes and physiological pathways of the identified protein species were categorised using the database for annotation, visualization and integrated discovery (DAVID, version 6.8, david.ncifcrf.gov) database. UniProt accession IDs were used in the DAVID database to categorise proteins according to their GO (gene ontology, www.geneontology.org) cellular components, and biological processes. DAVID also characterised the physiological pathways associated with the identified protein species according to the Kyoto encyclopaedia of genes and genomes (KEGG, www.genome.jp/kegg) category [83]. Protein species were further characterised and grouped using the search tool for retrieval of interacting genes/proteins (STRING; version 10, string-db.org) to identify potential protein–protein interactions (PPI, [84]). Using the STRING database, a PPI map was generated in which each node represents a protein and connecting lines represent evidence of association (with line thickness indicating the strength of the potential interaction). Such associations are based on text mining, co-expression, co-occurrence, databases, experiments, neighbourhood and gene fusion of the identified proteins [85,86].

Table 1. 2DE LC/MS/MS analyses identified 33 proteins from 5- and 12-week studies.

Spot ID/Tissue Fraction	UniProt ID	Protein Name	Gene ID	Score/ Coverage %	Unique Peptides	MW/pI Theoretical	MW/pI Experimental	Highest Fold Change 5 W	Highest Fold Change 12 W	Data Base	Reference CPZ	Reference EAE	Reference MS
A4/MP	G3UVV4	Hexokinase 1	Hk1	270/10	10	101.8/6.2	220/6.4	0.66(0.2)↓	0.48(0.1PT)↓	M	-	-	[87]
5N3/SP	P05063	Fructose-bisphosphate aldolase C	Aldoc	510/33	12	39.3/6.6	73.4/5.9	0.32(0PT)↓	×	S	-	[88]	[89]
D2/MP	Q99KI0	Aconitate hydratase	Aco2	407/25	18	85.4/8.1	182.9/8.2	2.37(0.2PT)↑	0.23(0PT)↓	M	-	-	[90]
A10/MP	Q8K2B3	Succinate dehydrogenase flavoprotein subunit	Sdha	230/12	8	72.5/7	96/6	0.57(0.1PT)↓	0.61(0.1PT)↓	M	[91]	-	-
F2/MP	P08249	Malate dehydrogenase	Mdh2	983/57	25	36/8.9	70.2/6.9	2.30(0.1PT)↑	×	S	-	[88]	[92]
5O5R/MP	Q91WD5	NADH dehydrogenase iron-sulfur protein 2	Ndufs2	173/10	5	52.5/6.5	52.2/5.5	0.59(0PT)↓	1.77(0.1PT)↑	M	[91]	-	-
5R5/SP	Q03265	ATP synthase subunit-α	Atp5a1	177/26	15	59.7/9.2	117.4/8.2	2.76(0PT)↑	0.37(0.1)↓	S	[91]	-	-
A37/MP	Q60930	Voltage-dependent anion-selective channel protein	Vdac2	155/14	5	31.6/7.4	18.7/6.6	0.42(0.2PT)↓	0.50.1PT↓	M	-	-	[88]
A34/MP	P05201	Aspartate aminotransferase	Got1	350/26	12	46.2/6.6	29.5/7.4	0.69(0.2PT)↓	0.38(0.1PT)↓	M	-	-	[89]
A1/MP	B2RXT3	Ogdhl protein	Ogdhl	291/10	10	114.5/6.4	243.3/5.9	0.69(0.2)↓	0.65(0.1PT)↓	M	-	-	-
B23/MP	Q8VDQ8	NAD-dependent protein deacetylase sirtuin-2	Sirt2	529/36	13	39.4/6.35	27.8/7	0.69(0.1PT)↓	0.33(0.1PT)↓	M	-	[93]	[93]
5C1/SP	Q91HI5	Isovaleryl-CoA dehydrogenase	Ivd	650/39	17	46.3/8.5	79.4/6.2	0.42(0.2PT)↓	1.80(0PT)↑	S	[49]	-	-
5C5/SP	Q91WQ3	Tyrosine-tRNA ligase	Yars	333/28	17	59/6.57	124.9/6.4	0.32(0.2PT)↓	1.97(0PT)↑	S	-	[94]	[95]
A22/MP	P26443	Glutamate dehydrogenase 1	Glud1	324/12	7	61.3/8	106.8/6.8	2.24(0.2PT)↑	×	M	-	-	-
A16/MP	Q99MN9	Propionyl co-enzyme A carboxylase-β	Pccb	277/16	6	50.4/5.7	60.5/6.3	×	0.43(0.1PT)↓	M	-	-	-
B20/MP	P30275	Creatine kinase U-type	Ckmt1	137/22	11	46.9/8.4	83.2/7.5	0.44(0.2PT)↓	0.35(0.1)↓	M	-	[96]	-
A14/MP	P50396	Rab GDP dissociation inhibitor-α	Gdi1	227/20	8	50.5/4.9	77/5.9	0.6(0.2)↓	0.56(0.1PT)↓	M	-	-	[96]
5O4/MP	Q61598	Rab GDP dissociation inhibitor-β	Gdi2	148/10	5	50.5/5.9	55.6/5.7	0.38(0PT)↓	×	M	-	-	-
5L6/MP	P46460	Vesicle-fusing ATPase	Nsf	466/36	33	82.6/6.5	116/6.3	0.52(0.2PT)↓	0.71(0.1PT)↓	S	-	-	-
B17/MP	P11798	Calcium/calmodulin-dependent protein kinase type II subunit-α	Camk2a	225/16	7	54/6.6	53.2/7	×	0.42(0PT)↓	M	[97]	[98]	-
A11/MP	O08599	Syntaxin-binding protein 1	Stxbp1	244/12	5	68.7/6.3	90.6/6.3	0.71(0.2)↓	1.61(0PT)↑	M	-	[94]	-
A5/MP	A0A0J9YUE9	Dynamin 1	Dnm1	172/9	8	93.9/6.2	200/6.2	1.61(0.2PT)↑	0.46(0.1PT)↓	M	[49]	-	-
E14/MP	Q90 8B3	Charged multivesicular body protein	Chmp4b	176/15	4	24.9/4.6	23.7/5	×	1.65(0.1PT)↑	M	-	-	-
5H2K/MP	P03995	Glial fibrillary acidic protein	Gfap	1802/72	45	49.8/5.2	83.7/5.0	1.86(0.2PT)↑	1.58(0PT)↑	S	[99]	[100]	[101]
E2/MP	P08551	Neurofilament light polypeptide	Nefl	1594/57	92	61.4/4.6	53/4.9	0.71(0.2)↓	2.08(0.1PT)↑	M	-	[102]	[103]
5G3/SP	Q90898	Actin-related protein 2/3 complex subunit 5	Arpc5l	152/26	4	16.9/6.3	26/5.9	0.62(0.2PT)↓	×	S	-	-	[104]
5I1/SP	P42208	Septin-2	Sept2	180/21	7	41.5/6.1	81/5.7	0.30(0PT)↓	1.53(0PT)↑	S	-	[93]	-
5I2/SP	Q9Z2Q6	Septin-5	Sept5	159/22	9	42.7/6.2	80.4/5.9	0.36(0.2PT)↓	0.46(0.1PT)↓	S	-	[100]	-
4L1/MP	P18872	Guanine nucleotide-binding protein G(o) subunit-α	Gnao1	186/15	5	40.6/5.3	68.5/4.7	7.35(0.2)↑	1.48(0.1PT)↑	M	-	[105]	-
4O3/MP	Q9D7N9	Adipocyte plasma membrane-associated protein	Apmap	148/8	5	46.4/5.9	85/5.6	4.09(0.1)↑	1.46(0.1PT)↑	M	-	-	-
5R1/MP	P14211	Calreticulin	Calr	187/13	7	47.9/4.3	155/4.4	2.50(0.2PT)↑	1.83(0.1PT)↑	M	-	[105]	[106]
5G2/SP	P62259	14-3-3 protein epsilon	Ywhae	110/23	6	29.1/4.6	49.9/4.5	×	×	S	-	[100]	[107]
5E1/SP	Q90154	Leukocyte elastase inhibitor A	Serpinb1a	191/36	18	42.5/5.8	76.7/5.5	0.28(0.2PT)↓	×	S	[49]	-	-

Key: MP, membrane protein; SP, soluble protein; MW, molecular weight; pI, isoelectric point; S, Swiss-Prot; M, MSPnr100; PT, pertussis toxin; 0.1, 0.1% CPZ; 0.1PT, 0.1% CPZ+PT; 0.2, 0.2% CPZ; 0.2PT, 0.2% CPZ+PT; W, week; ×, unchanged; ↑, increase; ↓, decrease; -, not found or investigated (details are shown in Figure 5). Some of the spots contained more than one clearly identifiable protein; presented here are the hits with the highest score, coverage and peptide count. UniProt and gene IDs were derived from the UniProt database. MASCOT score, sequence coverage, theoretical (MW/pI), and unique peptides number were acquired from the MASCOT database search. Experimental (MW/pI) was derived from 2D gels of identified protein spot. References are from the published literature in PubMed on CPZ, EAE and MS and used to compare currently identified proteins with the existing literature.

2.4. Western Blot (WB)

2.4.1. Sample Preparation

Stored brain, spinal cord and spleen samples ($n = 3$/group) were homogenised in the deep-frozen state as described earlier. Equal ratios (~1 µL/1 µg tissue) of sample and pre-chilled lysis buffer (25 mM Tris, 1 mM EDTA and 1 mM EGTA) containing the inhibitor cocktail were used to solubilize the powdered samples and protein was recovered using centrifugation (at 125,000× g, 4 °C, for 1 h). Protein quantification was then carried out as previously described using the EZQ protein quantitation kit (see above).

2.4.2. Procedure

Total protein extract (brain, spinal cord and spleen) and CD4/8 recombinant proteins were resolved by 10% SDS-PAGE (100 V for 2 h at 4 °C) and transferred (100 V for 2 h at 4 °C) onto 0.22 µm pore size polyvinylidene difluoride (PVDF, Merck-Millipore) membrane using transfer buffer containing 25 mM tris, 192 mM glycine and 20% methanol. The membranes were incubated in blocking buffer containing non-fat dry skimmed milk (5% w/v, Coles, Hawthorn East, VIC, Australia) and polyvinylpyrrolidone (1% w/v, Sigma-Aldrich), in 0.05% Tris buffered saline-Tween 20 (TBST) for 1 h at RT on an orbital shaker (at 50 rpm). Primary Abs for CD4 (rabbit anti-CD4, 1:500, Abcam) and CD8 (mouse anti-CD8, 1:75, Santa-Cruz Biotechnology) were incubated for 1 h at RT. Blots were then washed thrice with 0.05% TBST at 10 min intervals and horseradish peroxidase-conjugated (HRP) secondary Ab (goat anti-rabbit- or goat anti-mouse-HRP: CD8 1:500, Santa-Cruz Biotechnology and CD4 1:2000, Abcam) was added and incubated for 1 h at RT. Chemiluminescent visualization of the transferred proteins was carried out using an enhanced chemiluminescence detection reagent (500 µL/cm^2 membrane, Merck-Millipore). Blots were scanned for 2 sec on the ImageQuant™ FUJI LAS-4000 biomolecular imager (GE Healthcare). ImageJ software was used to quantify the density of a band of interest on a blot by using a rectangular box to define the band. This band intensity was expressed as a raw value ($n = 3$ bands/animal, $n = 3$ animals/group) and presented relative to Ctrl.

2.4.3. Transfer Efficiency

Replicate 1D SDS-PAGE gels were resolved in parallel and one stained with Coomassie Brilliant Blue prior to, and the other after, transfer onto PVDF membrane to determine the transfer efficiency of proteins. Imaging was carried out using a Typhoon™ FLA-9000 gel imager. The density of the bands ($n = 3$ bands/gel) with the molecular weights corresponding to the known molecular weights of CD4/8 (37 and 50 KD, respectively) were quantified using Multigauge image analysis software-version 3.0 (Fujifilm, Minato-Ku, TYO, Japan).

2.4.4. T-Cell Detection Limits in Peripheral and CNS Tissues

To measure the detection limit of CD4 and CD8 antibody signals in WBs, spleen samples from naïve mice and commercial CD4/8 recombinant proteins (Sino Biological, Wayne, PA, USA) were used. For CD4, the lowest detectable signal was achieved using 5 µg of spleen protein, whereas 10 µg of total spleen protein was needed to detect a CD8 band. The minimal detectable concentrations of the commercial recombinant protein standards were 5 ng and 5 µg for CD4 and CD8, respectively (Figure 3a). These lowest detectable concentrations for both groups (spleen and commercial samples) were used as positive (spike) controls to establish the expected minimal detection of CD4 and CD8 in WBs of total brain protein from the different experimental groups.

2.5. Statistical Analysis and Graphing

Statistical analyses were performed using GraphPad Prism-version 7.03 (www.graphpad.com, San Diego, CA, USA) software. Data were analysed using either one or two-way analysis of variance

(ANOVA) or an unpaired two-tailed t-test and, where appropriate, Newman-Keuls or Tukey post hoc analyses to determine specific differences among groups. Data are presented as means ± standard error of the mean (SEM), otherwise indicated in the text. Statistical significance was accepted when $p < 0.05$. Figures were assembled using CorelDRAW-version 2018 (www.coreldraw.com, Ottawa, ON, Canada) and Photoshop CS6 (Adobe, San Jose, CA, USA) image processing software. All Nogo A images were adjusted only for colour contrast (Supplementary Figures S1 and S2).

3. Results

3.1. Body Weight

Mice in all groups gained weight over the duration of feeding, but this was significantly slower in CPZ(±PT)-fed animals ($p < 0.05$, Figure 1). Significant reductions in weight gain started at week 1 after 0.2% CPZ±PT and week 2 after 0.1% CPZ±PT-feeding and continued until week 11. At week 12, 0.1% CPZ±PT groups were significantly ($p < 0.05$) different compared to the Ctrl group but not to the PT group. No direct or combined effect ($p > 0.05$) of PT was found in any group at any time point.

3.2. Marked Demyelination, Oligodendrocytosis and Gliosis

In both the 5- and 12-week studies, the Ctrl and PT groups exhibited intense silver staining of myelin in the midline corpus callosum (MCC) and lateral corpus callosum (LCC), whereas the 0.1% and 0.2% CPZ-fed(±PT) groups displayed a marked, concentration-dependent loss of silver staining ($p < 0.05$, Figure 2a; Supplementary Figures S1a, S2a, S3a). No differences in the extent of demyelination were found between the MCC and LCC regions in any of the groups with either duration of CPZ-feeding [($p > 0.05$, Figure 2a, (MCC); Supplementary Figure S3a, (LCC)]. Importantly, prolonged 0.1% CPZ (±PT)-feeding for 12 weeks produced a similar amount of demyelination to that seen at 5 weeks with 0.2% CPZ ($p < 0.05$, Figure 2a; Supplementary Figures S1a, S2a, and S3a). Consistent with the significant reduction in silver staining, 5 or 12 weeks of CPZ-feeding produced a significant loss (>90%) of mature, Nogo A positive OLG in the MCC and LCC; feeding with 0.1% was as effective as 0.2% CPZ at inducing OLG loss ($p < 0.05$, Figure 2b; Supplementary Figures S1b, S2b, S3b). Importantly, there were no differences in OLG loss when using 0.1% or 0.2% CPZ for 5 weeks, and 0.1% CPZ-feeding for 12 weeks produced a comparable loss of OLG to that seen at 5 weeks ($p > 0.05$, Figure 2b; Supplementary Figures S1b, S2b, S3b). Similarly, there were marked dose-dependent increases ($p < 0.05$) in the number and intensity of Gfap and Iba 1 staining (Figure 2c,d; Supplementary Figures S1c,d, S2c,d, S3c,d) in the CPZ-fed(±PT) groups compared to Ctrl or PT only animals (which were not different from each other, $p > 0.05$). Taken together, these results indicate that low dose CPZ-feeding for 5 weeks produced an almost complete loss of OLG but a more limited (i.e., slower) demyelination and gliosis response. When this feeding regime was prolonged for 12 weeks, it produced comparable changes to those seen at 5 weeks using 0.2% CPZ. PT had no direct or synergistic effects on any of the histological parameters studied at either time point.

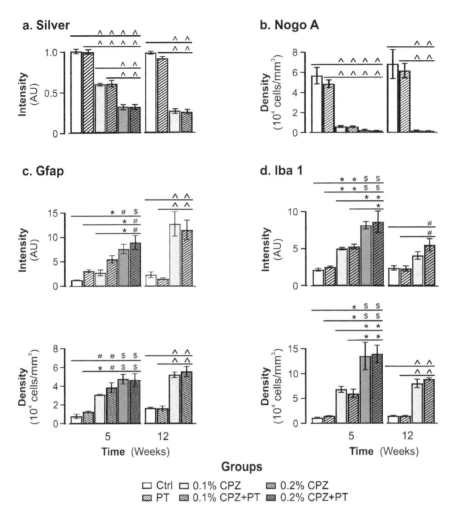

Figure 2. Quantification of demyelination, cell death and gliosis in the midline corpus callosum. (**a**) Silver staining. CPZ-feeding(±PT) led to significant demyelination and reduced silver staining intensity at 5 and 12 weeks. PT alone had no effect. Feeding 0.1% CPZ for longer (12 weeks) produced a comparable demyelination to that seen with 0.2% for 5 weeks. (**b**) Nogo A. CPZ-feeding(±PT) led to significant oligodendrocytosis with 0.1% CPZ was as effective as 0.2% CPZ at either time point. PT alone had no effect. (**c**) Gfap. Staining intensity increased in a dose dependent manner, in the CPZ(±PT) treated groups and this was associated with an increase the number of Gfap positive astrocytes at both time points. PT only did not evoke a Gfap response. (**d**) Iba 1. Increased Iba 1 fluorescence intensity and number of Iba 1 positive microglia were seen in both 5- and 12-week groups. PT alone produced no microglial response. Data are presented as mean ± SEM. One-way ANOVA and Tukey post hoc analysis was used to determine differences among groups (* $p < 0.05$, # $p < 0.01$, $ $p < 0.001$ and ^ $p < 0.0001$). Quantitation based on analysis of 5–9 sections/animal, 3–5 animals/group.

3.3. Detection and Localisation of CD4 and CD8 T Cells

Immunofluorescence staining of brain sections failed to detect CD4$^+$ and CD8$^+$ positive cells in the CC in CPZ(±PT) groups (Supplementary Figure S4a, even when using high antibody titres and long incubation times). This was not due to the lack of antibody sensitivity, as immunofluorescent CD4 and CD8 positive cells were seen in histological sections of spleen (Supplementary Figure S6a). Likewise, CD4/8 signals in whole brain protein samples were undetectable by western blot (WB) analysis (Figure 3c), even at high protein loads (up to 120 µg). This was not due to the lack of protein transfer from SDS-PAGE gel to PVDF membrane, as transfer efficacy at the molecular weights corresponding to CD4/8 (37 and 50 KD, respectively) were 93 ± 1.6% and 98 ± 0.7%, respectively (Supplementary Figure S4b). Furthermore, CD4/8 signals were also detected by WB analysis when spleen (Figure 3a,c) and EAE spinal cord were used as positive controls (Supplementary Figure S4c; EAE was induced using an established method [8]). The capacity and sensitivity of WB to detect CD4/8 signal was also confirmed by other control experiments in which brain homogenates were spiked with spleen homogenate or commercially available CD4/8 recombinant proteins (Figure 3b). Notably, measurement of splenic weight showed a significant ($p < 0.05$) decrease in splenic mass (37 ± 0.1%) in 0.2% CPZ-fed(±PT) groups whereas no change was observed in 0.1% CPZ(±PT) groups at 5 weeks (Supplementary Figure S5a,c). In contrast, a significant ($p < 0.05$) reduction of splenic mass was observed in 0.1% CPZ(±PT) groups at 12 weeks (Supplementary Figure S5b,c). Moreover, 5 weeks of CPZ-feeding also resulted in a significant dose-dependent reduction in CD4/8 in spleen compared to Ctrl (Figure 3d, 0.2% > 0.1%, $p < 0.05$). Following 5 weeks of 0.1% CPZ-feeding(±PT), the reduction of CD4 (11 ± 0.03%) and CD8 (14 ± 0.03%) signal intensity was less marked than that seen with 0.2% CPZ(±PT) groups (CD4 44 ± 0.03% and CD8 61 ± 0.04%). Following 12 weeks of 0.1% CPZ(±PT), further reductions in spleen CD4 (27 ± 0.01%) and CD8 (43 ± 0.02%) signal intensity and splenic atrophy (33 ± 0.1%) were observed. In addition, immunofluorescence staining of spleen sections (n = 10 sections/animal and n = 3 animals/group) indicated a significant ($p < 0.05$) reduction of CD4 and CD8 in the 0.2% CPZ-fed group (Supplementary Figure S6a,b).

3.4. Brain Proteome Changes

All samples yielded well-resolved proteomes encompassing the full MW/pI range of the gels. Representative images of soluble (SP) and membrane (MP) proteomes from Ctrl and 0.2% CPZ-fed mice are shown in Figure 4a. A total of ~1650 consensus spots (i.e., protein spots that resolved consistently and were analysed across all gels) were detected from the combined analyses of whole brain soluble and membrane proteomes (Figure 4a; Supplementary Table S1). The spot quantification from the different groups from both time points is summarized in Supplementary Table S1. The different groups yielded comparable numbers of resolved protein spots at 5 weeks ($p > 0.05$), and this was also true for the 12-week samples ($p > 0.05$, Supplementary Table S1). In total, 845 ± 8 and 793 ± 11 spots were resolved from the soluble and membrane fractions, respectively, in the 5-week study, whereas 824 ± 11 and 717 ± 3 spots were resolved from the soluble and membrane fractions, respectively, in the 12-week study. Database hits of high-quality and confidence were returned following LC/MS/MS analysis and identified 33 different proteoforms from these spots including 73% in the membrane proteome and 27% in the soluble proteome (Table 1).

Table 1 summarizes the best identified proteoforms within each spot that displayed a 100% reproducible change across technical (n = 3 gels/fraction) and biological (n = 5 animals/group) replicates. All identified proteins had a MASCOT score exceeding 100, with 46% between 100–200; 21% between 201–300; 9% between 301–400; 6% between 401–500; and 18% exceeding 500. Moreover, each identification was based on at least 4 unique peptides: 61% based on 4–10 peptides; 27% based on 11–20 peptides; and 9% based on over 30 peptides. Similarly, sequence coverage was always ≥5% with 5–10% for 6 proteins; 11–20% for 11 proteins; 21–30% for 8 proteins; 31–40% for 5 proteins; and >50% for 3 proteins (Table 1). The combination of high MASCOT scores and the presence of ≥4 unique peptides with high coverage highlight the high confidence of the protein identifications.

As also shown in Table 1 and Figure 4b, several of the identified proteoforms displayed a mismatch between their theoretical and experimental MW and *pI*, indicative of post-translational modifications (e.g., phosphorylation and glycosylation). Only, 1 proteoform showed an increase whereas 19 (58%) showed a decrease and 13 (39%) showed the same experimental *pI* relative to the theoretical *pI*. In contrast, twenty-five (75%) proteoforms increased, 4 (12%) decreased, while another 4 remained unchanged in their experimental MW relative to theoretical MW. Interestingly, a subset of proteoforms was found showing an approximate doubling of the experimental MW relative to theoretical—hexokinase 1, aconitate hydratase, ATP synthase subunit-α, ogdhl protein, tyrosine-tRNA ligase and dynamin 1—potentially indicative of dimerization, whereas an approximate tripling of MW may indicate calreticulin trimers (Table 1) suggesting a possible increase in oligomerization/self-association of proteoforms due to CPZ-feeding(±PT).

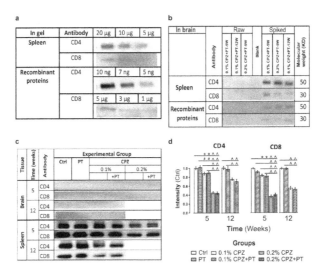

Figure 3. Western blot analysis. (**a**) Measurement of the detection limit of CD4/8 signal using naïve Ctrl spleen homogenates and CD4/8 recombinant proteins in gels. (**b**) Confirmation of CD4/8 signal detection using brain tissue samples (60 µg) spiked with either 5 µg or 10 µg of spleen homogenate or 5 ng and 5 µg commercial CD4 and CD8 recombinant proteins. (**c**) No CD4/8 signal was detected in CPZ(±PT) brain samples whereas a reduced CD4/8 signal intensity was found in spleen. (**d**) Quantification of splenic CD4/8 blots showed a significant reduction of CD4 and CD8 signal intensity in spleens of specific groups. Cropped CD4/8 western blots are presented unaltered and shown in their entirety in Supplementary Figure S7. Data are presented as mean (±SEM) relative to the Ctrl mice. One-way ANOVA and Newman-Keuls Multiple Comparison post hoc analysis were used to determine differences among groups (* $p < 0.05$, # $p < 0.01$ and ^ $p < 0.0001$).

Statistical comparisons of the spot intensities between all groups revealed significant and reproducible changes ($p < 0.05$) in 33 spots across Ctrls and experimental groups in the 5- and 12-week studies (Figure 5). Among the identified proteoforms, not all changes were sustained, or shared, between the 5- and 12-week studies (Supplementary Figure S8). In total, 23 shared proteoform changes were found in both the 5- and 12- week studies (i.e., ≥1.5-fold change in both studies), whereas 7 proteoforms that changed in the 5-week study (≥1.5-fold) did not in the 12-week study (<1.5 fold) and 3 other proteoforms were changed by 12 weeks (≥1.5-fold) but not at 5 weeks (<1.5 fold). The presence of changes at 5 weeks that were not evident at 12 weeks indicates that, relative to their age matched controls, such changes are time-dependent and resolved by 12 weeks.

Whether this return to control levels during prolonged feeding with 0.1% CPZ is due to compensatory mechanisms, aging or other processes remains unknown.

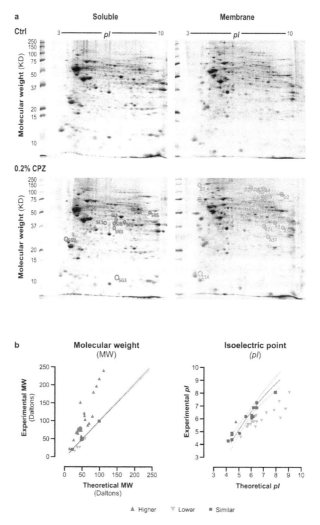

Figure 4. Top-down proteomic analysis. (**a**) Representative two-dimensional gel images of soluble (SP) and membrane (MP) brain proteomes from naïve Ctrl and 0.2% CPZ-fed groups used to detect proteoform changes. Proteoforms were resolved on the basis of their isoelectric point (*pI*) and molecular weight (MW). The total number of spots across different groups at both 5 and 12 weeks is given in Supplementary Table S1. Delta 2D software analysis revealed 33 unique spots for which the spot volume changed by at least 1.5-fold in at least one experimental group; soluble proteoforms are indicated by red circles and membrane proteoforms by green circles. The identities of the protein species are shown in Table 1 (n = 15 gels/fraction, n = 5 animals/fraction, 5-week study n = 180 gels and 12-week study n = 120 gels). (**b**) Comparison between theoretical and experimental MW (left) and *pI* (right) of identified proteoforms. Red represents increase, green decrease and blue indicates no statistically significant difference between the experimental and theoretical values. Purple dashed lines indicate 95% confidence intervals and the solid line represents full agreement between experimental and theoretical values.

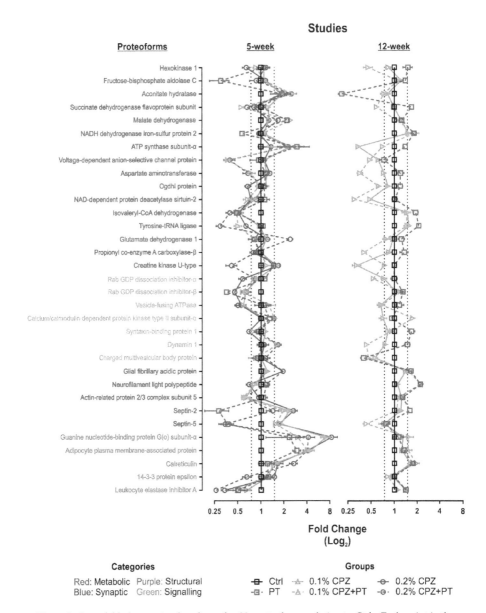

Figure 5. Log$_2$ fold changes in abundance for 33 proteoforms relative to Ctrl. Each point is the average change in spot volume ratio from all treated groups (PT, 0.1% CPZ, 0.1% CPZ+PT, 0.2% CPZ and 0.2% CPZ+PT) relative to Ctrls from the triplicate gels resolved for each of the 5- and 12-week samples. Only significant changes in abundance ($p < 0.05$) that exceeded the established 1.5-fold criteria (dash lines) in at least 1 experimental group were selected for excision, processing, and protein identification. Solid and dashed lines connect the protein changes within each experimental group with and without PT injection, respectively. Proteoforms (top left) indicate the different functional categories including metabolic (red), synaptic (blue), structural (purple) and signalling (green). Analysis was based on n = 180 gels (5-week study) and, n = 120 gels (12-week study); n = 5 animals/group.

Key: MP, membrane protein; SP, soluble protein; MW, molecular weight; *pI*, isoelectric point; S, Swiss-Prot; M, MSPnr100; PT, pertussis toxin; 0.1, 0.1% CPZ; 0.1PT, 0.1% CPZ+PT; 0.2, 0.2% CPZ; 0.2PT, 0.2% CPZ+PT; W, week; ×, unchanged; ↑, increase; ↓, decrease; -, not found or investigated (details are shown in Figure 5). Some of the spots contained more than one clearly identifiable protein; presented here are the hits with the highest score, coverage and peptide count. UniProt and gene IDs were derived from the UniProt database. MASCOT score, sequence coverage, theoretical (MW/*pI*), and unique peptides number were acquired from the MASCOT database search. Experimental (MW/*pI*) was derived from 2D gels of identified protein spot. References are from the published literature in PubMed on CPZ, EAE and MS and used to compare currently identified proteins with the existing literature.

3.5. Literature Mining

Literature mining via PubMed was used to assess the likely function(s) of identified proteoforms. This confirmed 8 proteoforms (creatine kinase U-type, glutamate dehydrogenase 1, vesicle-fusing ATPase, propionyl co-enzyme A carboxylase-β, tyrosine-tRNA ligase, actin-related protein 2/3 complex subunit 5, charged multi-vesicular body protein and adipocyte plasma membrane-associated protein) not previously related to CPZ, EAE or MS, and 25 proteins (but not specific proteoforms) previously associated with CPZ (8), EAE (13) and MS (16) studies (Table 1).

3.6. Biological Processes and Pathways

The 33 differentially expressed proteoforms were subjected to bioinformatic analysis using PANTHER, DAVID, UniProt and STRING for association with protein classes, molecular functions, physiological pathways, biological processes, cellular components, subcellular localizations and protein–protein interactions. Analysis using PANTHER indicated that the main protein classes were enzyme modulator (15%), nucleic acid binding (10%), oxidoreductase (10%) with 19% unclassified (Supplementary Figure S9a). Molecular function analysis using PANTHER indicated catalytic activity (40%) and binding (26%) roles with 18% unclassified (Figure 6a). However, the potential functions of all the proteoforms were inferred by literature mining for information on the canonical proteins (see *Discussion*). Physiological pathway analysis associated 26% of the proteins with KEGG metabolic pathways categories using DAVID (Figure 6b). GO biological process analysis showed the main categories were oxidation-reduction (13%) and transportation (13%), with 15% unclassified (Supplementary Figure S9b). Moreover, GO cellular component analysis (Figure 6c) indicated that most of the proteins were either cytoplasmic (25%), extracellular exosome (21%), mitochondrial (20%) or related to the myelin sheath (20%). Furthermore, subcellular localization analysis using UniProt revealed the main categories to be mitochondrial (31%) and cytoplasmic (26%; Supplementary Figure S9c). The identified proteoforms were also analysed using STRING software, providing PPI maps; this indicated that 15 of the 33 proteoforms (hexokinase 1, fructose-bisphosphate aldolase C, aconitate hydratase, succinate dehydrogenase flavoprotein subunit, malate dehydrogenase, NADH dehydrogenase iron-sulfur protein 2, ATP synthase subunit-α, voltage-dependent anion-selective channel protein, aspartate aminotransferase, ogdhl protein, isovaleryl-CoA dehydrogenase, tyrosine-tRNA ligase, glutamate dehydrogenase 1, propionyl co-enzyme A carboxylase-β and creatine kinase U-type) were potentially involved in major 'functional interactions' particularly with regard to the metabolic proteins (Figure 6d). Strong one-to-one connections were also indicated between synaptic (rab GDP dissociation inhibitor-α, rab GDP dissociation inhibitor-β, vesicle-fusing ATPase, calcium/calmodulin-dependent protein kinase type II subunit-α and dynamin 1) and structural (glial fibrillary acidic protein and neurofilament light polypeptide) proteoforms. This therefore, also revealed a second cluster of structural and signalling proteoforms. Among the connected proteoforms, 38% showed the highest connecting value (0.9), and 29% and 33% were high (0.7) and medium (0.4), respectively; no low value (0.15) connections were found.

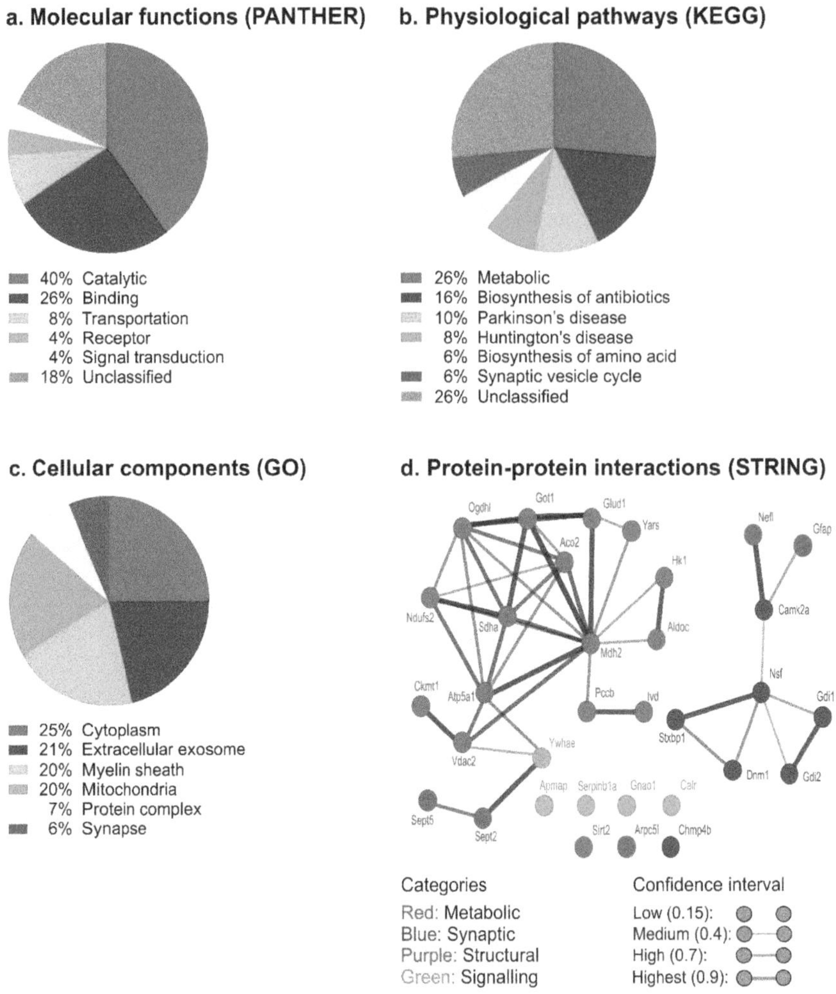

a. Molecular functions (PANTHER)

- 40% Catalytic
- 26% Binding
- 8% Transportation
- 4% Receptor
- 4% Signal transduction
- 18% Unclassified

b. Physiological pathways (KEGG)

- 26% Metabolic
- 16% Biosynthesis of antibiotics
- 10% Parkinson's disease
- 8% Huntington's disease
- 6% Biosynthesis of amino acid
- 6% Synaptic vesicle cycle
- 26% Unclassified

c. Cellular components (GO)

- 25% Cytoplasm
- 21% Extracellular exosome
- 20% Myelin sheath
- 20% Mitochondria
- 7% Protein complex
- 6% Synapse

d. Protein-protein interactions (STRING)

Categories
Red: Metabolic
Blue: Synaptic
Purple: Structural
Green: Signalling

Confidence interval
Low (0.15):
Medium (0.4):
High (0.7):
Highest (0.9):

Figure 6. Functional clustering and protein–protein interactions. Pie charts show the distribution of proteins according to (**a**) Molecular functions (characterized using PANTHER), (**b**) Physiological pathways (categorised using KEGG), (**c**) GO cellular components. (**d**) Protein–protein interaction association network maps. The strength of connections is based on co-expression, gene fusion, co-occurrence, neighbourhood, databases, experiments, and text-mining collated in the STRING database. The strength of interactions is indicated by the thickness of the lines. STRING analysis revealed 4 protein clusters involved in metabolic (red), synaptic (blue), structural (purple) and signalling (green). Collectively, these proteins and their associations suggest that CPZ induced metabolic dysregulations and mitochondrial dysfunction.

Although the addition of PT injections during CPZ-feeding did not enhance the extent of OLG degeneration, demyelination and gliosis, it did produce several proteoform changes (Figures 5 and 7; Supplementary Figure S10a–c, Table S2). This is the first study to assess changes in the mouse brain proteome profile when the BBB is compromised by PT. Following 5 weeks of CPZ-feeding, PT injection more than doubled the number of significant CPZ-associated changes (18–46%), with PT alone contributing 15% of all proteoform changes, including fructose-bisphosphate aldolase C, NADH dehydrogenase iron-sulfur protein 2, ATP synthase subunit-α, rab GDP dissociation inhibitor-β, septin-2 and 5, guanine nucleotide-binding protein G(o) subunit-α, adipocyte plasma membrane-associated protein and leukocyte elastase inhibitor A. Likewise, in the 12-week study, the small number of proteins identified following CPZ-feeding (6%) increased following the addition of PT (52%), with PT alone contributing 21% including aconitate hydratase, succinate dehydrogenase flavoprotein subunit, NADH dehydrogenase iron-sulphur protein 2, isovaleryl-CoA dehydrogenase, tyrosine-tRNA ligase, syntaxin-binding protein 1, glial fibrillary acidic protein, neurofilament light polypeptide, septin-2 and calreticulin. Statistical comparisons of the CPZ(+PT) groups with the PT only group (Supplementary Table S2) revealed that at least 50% of the PT-mediated proteoform changes were either increased (or decreased) when combined with CPZ-feeding; in the 5-week study, 25% increased (27% decreased) whereas, in the 12-week study 52% increased (6% decreased).

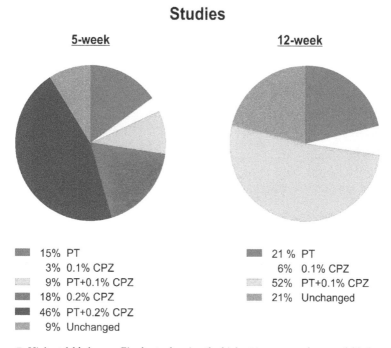

Figure 7. Highest fold change. Pie charts showing the highest increase or decrease fold change in abundance of the 33 proteoforms relative to Ctrls in the 5- and 12-week studies. PT alone or when PT was combined with CPZ showed greater change than CPZ alone. Quantification is based on ≥1.5-fold changes.

4. Discussion

The present study was designed to test whether CPZ-induced oligodendrocytosis, when combined with PT-induced BBB disruption, could induce an 'inside-out' activation of the immune system causing infiltration and detection of CD4/8 immune cells into the brain parenchyma. The effectiveness of CPZ-feeding was confirmed by the reduced weight gain in these groups, the almost complete demyelination of the corpus callosum using the standard feeding paradigm of 0.2% CPZ for 5 weeks, and changes to the brain proteome [58,59,108,109]. The results also demonstrated that 0.2% or 0.1% CPZ(±PT) produced comparable oligodendrocytosis but dose- and time-dependent demyelination and gliosis within the corpus callosum, indicating that 0.1% CPZ is as effective as higher doses when fed for a longer period of time. The presence of comparable oligodendrocytosis, but less marked demyelination and gliosis following 5 weeks of 0.1% CPZ-feeding, suggested a slower transition between oligodendrocytosis and demyelination and/or better clearance of myelin debris. In this transition state, limited gliosis may impede subsequent remyelination [110] and induce a slow, progressive demyelination reminiscent of MS. CPZ-induced a dose dependent (0.2% > 0.1%) atrophy of the spleen and extending the CPZ-feeding resulted in time dependent atrophy in 0.1% CPZ(±PT)-fed groups as well. Likewise, CPZ produced a dose-dependent (0.2% > 0.1%) suppression of CD4/8 in the spleen. Prolonged low-dose feeding of CPZ for 12 weeks did not cause a further decrease in CD4 compared to 5 weeks; however, with prolonged feeding, the reduction of CD8 deteriorated to levels seen at 5 weeks using the standard CPZ-feeding. Our results are thus consistent with a previous study where apoptosis of CD4 and CD8 positive cells and atrophy of thymus were observed following 0.2% CPZ-feeding [50]. In contrast, 5 weeks of 0.1% CPZ-feeding produced less CD4 and CD8 suppression and no splenic atrophy, suggesting this lower dose may be a better model to examine the aetiology of MS.

Following CPZ-feeding(±PT) in either experiment, there were no detectable CD4 or CD8 signals in brain tissue. There are several possible reasons for this: firstly, insensitivity of the immunohistological or biochemical assays. This is unlikely, as WB sensitivity was confirmed by spiking whole brain homogenates with biological (spleen) and commercially available CD4/8 recombinant proteins, and both high protein loads (60–120 µg) and high antibody titres were used [111]. Additionally, these results are consistent with previous studies using immunofluorescence and flow cytometry [48,112] as well as BBB disruption using ethidium bromide or lysolecithin [48]. Secondly, the BBB was breached during weeks 1–2 of CPZ-feeding, and this is a transient event [113]. However, whether infiltration occurred at this early time point, but resolved before the 5- or 12-week endpoints remains unknown. Was the lack of a detectable adaptive immune response the result of failure to generate an antigen? This seems unlikely, as marked gliosis, demyelination and oligodendrocytosis occurred. Alternatively, was there a direct effect of CPZ on peripheral immune function? This would be consistent with the observed suppression of CD4/8 signal in the spleens of CPZ-fed animals. It was recently reported that combining two weeks of 0.2% CPZ-feeding with an 'immune booster' (CFA) produced a secondary $CD3^+$ (pan T-cell marker) response, a response that was not observed with 0.2% CPZ-feeding alone or when CPZ-feeding was continued [31]. Consequently, it can be expected that prolonged CPZ-feeding is unlikely to facilitate infiltration of immune cells into the CNS as continued feeding would result in sustained immune suppression (and reduction of mitochondrial ATP production) which is known to lead to declining health and death of mice. Likewise, extending observations beyond the cessation of CPZ-feeding is unlikely to reveal immune cell infiltration since cessation of CPZ-feeding results in spontaneous remyelination, reduction in myelin debris (a potential antigen), and cessation of glial activation (i.e., the antigen presenting cells) (reviewed in [2,39,57,58,114]).

Notably, the prolonged CPZ-feeding preferentially suppressed CD8 signal intensity in the spleen, and these are the cells that predominate in human MS CNS pathology [22]. These observations are further supported by the large number of changes in spleen (n = 22) and peripheral blood mononuclear cell-derived (n = 5) proteoforms in the CPZ-fed animals [49]. In the spleen, the vast majority (87%) of these changes included membrane associated structural and metabolic proteins suggesting perturbation of adaptive immune function [49]. The spleen and thymus are responsible for the maturation, selection and proliferation of T-cells [115,116], key functions that are impeded by the ion dishomeostasis induced by CPZ [41–44]. Such effects on the peripheral immune system may explain why the severity of disease and extent of peripheral immune involvement in EAE and Theiler's murine encephalomyelitis were suppressed by CPZ [117–119]. Although, there is no evidence as to whether T-cells become functionally inactivated ('T-cell anergy'; [120,121]) due to the suppression of adaptive immune organs (e.g., thymus) following CPZ-feeding, there is some evidence indicating that CPZ may otherwise affect T-cell functions. Copper plays both direct and indirect roles (via interleukin-2) in the maturation and production of functional T-cells [122]. Consequently, it can be argued that CPZ, like other copper chelators (e.g., 2,3,2-tetraamine), can reduce T-cell function [123,124]. Moreover, it has been shown that when lymph node cells harvested from CPZ-fed mice are exposed to neuroantigens (e.g., concanavalin

A and MBP) in vitro, there is reduced cellular proliferation. Likewise, reduced T-helper cell 1 cytokines (e.g., interferon-γ and tumor necrosis factor-α) and CD4 T-cell response to interleukin-17 are observed in EAE mice fed with CPZ [119]. Together, these findings indicate that CPZ can modify the functional capacity of T-cells, and this may potentially affect migration into the CNS.

The suppression of CD4/8 in the spleen indicated that CPZ has direct effects on the peripheral immune system that may limit the maturation and migration of adaptive immune cells to the CNS. Moreover, calcium/calmodulin-dependent protein kinase type II subunit-α, a protein known to play a role in CD8 T-cell proliferation and the transition to a cytotoxic phenotype [125,126] decreased in abundance following CPZ-feeding. Likewise, leukocyte elastase inhibitor A, a protein that supresses proteases including those released by T-cells [127], also decreased in abundance. These findings extend earlier proteomic analyses [49] that found CPZ reduced the abundance of protein disulphide isomerase (subunits A2, A3 and A6) in spleen, a protein involved with folding and assembly of functional major histocompatibility complex class I molecules [128]. In addition, the CPZ-induced dysregulation of mitochondria identified in the present study presumably extends to the immune system as several studies have shown that mitochondrial dysfunction leads to the suppression of T-cell function and a compromised immune system [129,130]. Collectively, these findings indicate that CPZ has suppressive effects on the capacity of the peripheral immune system to launch a response, making it more difficult to address the 'inside-out hypothesis'. Therefore, strategies to overcome peripheral immune organ suppression should be the next stage of investigation in evaluating the 'inside-out' hypothesis of MS using the CPZ model.

While there is growing evidence that MS may initiate as a slow, low-grade primary oligodendrocytosis, a substantial gap still exists in our understanding of the molecular mechanisms underlying the fundamental susceptibility to and initiation of oligodendrocytosis [1–4]. Therefore, a quantitative 'top-down' proteomic approach was carried out to resolve protein species from whole brains of the CPZ-fed(±PT) mice. To date, proteome analyses of MS have mainly assessed tissue from late-stage post mortem brain samples and cerebrospinal fluid [89,96] or EAE animals [88,93,100,105]; these analyses provide useful insight into the 'final readout' of the disease or its autoimmune aspects [131–133] but not necessarily insight into the processes involved in disease aetiology and progression, including the role of oligodendrocytosis [1,49].

Here, CPZ-mediated oligodendrocytosis and glial activation were confirmed using histology, and the changes in protein abundance were quantified using a high sensitivity 'top-down' analysis rather than the more common 'bottom-up' (i.e., 'shotgun') analyses; this is critical, as the 'top-down' approach has the highest inherent capacity to resolve intact proteoforms (i.e., isoforms, splice variants, and post-translationally modified species) as well as provide better sequence coverage, and assessment of lower molecular weight species, with a high degree of consistency across technical and biological replicates [54,55,134]. This approach thus detected changes in 33 proteoforms (16 metabolic, 7 synaptic, 5 structural and 5 signalling), of which 8 have not been previously associated with CPZ, EAE or MS. Furthermore, it also provided high quality confirmation of changes in the abundance of 25 proteoforms of proteins previously related to MS or animal models of the disorder. In contrast to earlier work [49,91,97,99,135], this study used three technical replicates for each of the 5 biological replicates per experimental group, ensuring that only the most reproducible changes in proteoform abundance were assessed. Consequently, in contrast to the detection of ~700–1200 spots in previous studies [49,88,100,105], ~1650 protein spots per condition were resolved in this study.

The most striking observation to emerge from the proteomic analysis was the large number of changes in metabolic proteoforms involved in glycolysis, Krebs cycle and oxidative phosphorylation pathways and the regulation of mitochondrial function. Bioinformatics platforms (DAVID, PANTHER, UniProt and STRING) further revealed the likely association between metabolic dysregulation and CPZ-feeding. Understanding the molecular pathways and potential PPI is important to understanding how dysregulation of biological processes may lead to a disease [132,136]. It is thus notable that over 80% of proteins do not function alone but in complexes [136]. The data here suggested that 79% of the identified proteoforms may be clustered in two complexes, with 58% being metabolic in origin, which is consistent with previous observations [136,137]. The strong PPI highlight the likely cross-talk and shared biochemical reactions among the metabolic proteoforms [136]. Interestingly, protein-to-protein connection analysis revealed that malate dehydrogenase may have 11 connections with other identified proteoforms, which was the highest seen, and that the ATP synthase subunit-α had the second highest with 7 connections (Figure 6d). The central placement and increased inter-connectedness of malate dehydrogenase or ATP synthase subunit-α with other proteoforms implies that these proteoforms may be pivotal for CPZ-induced metabolic dysregulation. This explanation is further supported by the increased abundance of ATP synthase subunit-α and malate dehydrogenase (in contrast to the decreased abundance of most of the other identified metabolic proteoforms), perhaps as a compensatory response to the CPZ-induced metabolic perturbations. The increased abundance of ATP synthase subunit-α after CPZ-feeding leads to the accumulation of protons in the inter-membrane space of mitochondria and increased generation of reactive oxygen species [91]. Likewise, an increase in the abundance of aconitate hydratase (6 PPI connections, Figure 6d), an iron-sulphur protein containing 4Fe-4S clusters, is likely to regulate iron-induced oxidative stress in the CPZ model. Disruption of iron metabolism is linked to increased oxidative stress and lipid peroxidation [138]; CPZ-feeding is associated with the dysregulation of iron [43] and thus increased oxidative stress results in accentuation of mitochondrial perturbations [139]. Moreover, we observed a synergy between CPZ-feeding(±PT) with regard to synaptic protein dynamin 1 (increased at 5 weeks but decreased at 12 weeks) suggesting that CPZ-feeding interferes with the fission and fusion dynamics of mitochondria. The initial increase in dynamin 1 is consistent with mitochondrial response to metabolic stress, wherein the mitochondria divide, generating new functional mitochondria, and target damaged mitochondria for autophagy [140]. Perturbations to the fission and fusion dynamics of mitochondria are consistent with the formation of mega-mitochondria reported in the CNS, in particular in OLG [141,142]. The current data are consistent with previous investigations [49,50,91,99] corroborating the hypothesised link between mitochondrial dysregulation with CPZ-feeding and the emergence of structural and functional abnormalities that may predispose OLG to degeneration and death [1].

This susceptibility of OLG may be explained, in part, by the intense energy requirements associated with the production and maintenance of the expansive myelin sheath [143,144]. The high metabolic rate of OLG means that the increased production of reactive oxygen (and nitrogen) species, coupled with their low levels of anti-oxidants (e.g., glutathione [145], metallothionein [146] and manganese superoxide dismutase [147]) predisposes them to oxidative injury [45,148]. In addition, increased abundance (in 5- and 12-week studies) of the endoplasmic reticulum (ER) stress-related chaperone protein calreticulin may be associated with unfolded protein responses that enhance oligodendrocytosis [144,149]. This finding extends previously observed changes in ER proteins such as ribosome-binding protein 1, endoplasmin [49] and heat shock protein [99] reinforcing the role of ER stress in the CPZ-fed mice. The presence of heat shock protein and activating transcription factor 4 in MS [89,150,151] and EAE [105,150] indicate the association of protein misfolding and ER stress with OLG degeneration and demyelination.

The apparent oligomerization of some proteoforms in response to CPZ-feeding(±PT) may be indicative of toxic protein aggregation [152,153], effects that have also been observed in an EAE study [154] and MS patients [155]. The oligomerization of proteoforms has been documented in other studies, as with calreticulin and dynamin, assessed using SDS-PAGE or crystallographic analysis [156–158]. Likewise, the divergence of molecular weight was also observed in other proteomic studies of CPZ-fed mice (protein phosphatase 1G, 59.38 vs. 104.7 KD) and MS cerebrospinal fluid (albumin, 67 vs. 180 KD and alpha 1 antitrypsin, 47 vs. 100 KD) [49,159]. Indeed, changes in theoretical vs. experimentally observed isoelectric point have also been reported in previous studies on CPZ (ornithine carbamoyltransferase, 8.81 vs. 6.9 and ribosome-binding protein 1, 9.35 vs. 5.0) [49], EAE (septin-8, 5.7 vs. 6.4 and cytochrome c oxidase, 9.2 vs. 5.8) [100] and MS (Ig kappa chain NIG 93 precursor, 8.1 vs. 6.5 and albumin, 5.5 vs. 9) [159,160]. Naturally, none of this key data, upon which to better design future studies to identify critical proteoforms (rather than just amino acid sequences), would be routinely available using any other current approach.

Within the demyelinated regions, the hypertrophy of astrocytes and microglia, compounded by increased microglia numbers, may intensify the local energy imbalance and further compromise the function of OLG. Moreover, hypertrophied microglia and astrocytes may diminish the supportive roles played by glia at synapses leading to a pre-disposition to excitotoxicity [161–163]—a disturbance further implicated by changes in other synaptic proteoforms identified in this study.

The abundance of synaptic-regulatory proteoforms was either decreased (rab GDP dissociation inhibitor-α/β and vesicle-fusing ATPase), elevated (charged multivesicular body protein) or displayed opposing changes (syntaxin-binding protein 1) across the two time points—changes in synaptic function that may contribute to alteration of mood or behaviour in humans or animal models [164–166]. Likewise, the changes of abundance of proteins such as calcineurin, calbindin 2 and parvalbumin-α, involved in neurotransmitter release, has also been reported in EAE [100]. Although CPZ-feeding induce a wide range of behavioural deficits including motor, anxiety and cognition [reviewed by 2]; the present proteome analysis did not seek correlation with behavioural phenotypes, other studies have argued that CPZ-induced alteration of proteoforms involved in neurotransmitter release can result in cognitive decline [166] or increased climbing, rearing and ambulatory behaviours [99,108,109]. This study also revealed a marked change in neuronal and glial structural proteoforms, including Gfap, neurofilament light polypeptide, actin-related protein 2/3 complex subunit 5, and septins-2 and -5, at both time points, indicating axonal and glial remodelling. Consistent with the capacity for CPZ to induce the hypertrophy of astrocytes and increase microglia in the innate immune system (Figure 2a; Supplementary Figures S1–3), a marked increase in proteoforms involved in structural (Gfap), signalling, and inflammatory pathways were observed.

For some proteoforms, breaching the BBB negated the effects of 5 weeks of CPZ-feeding (e.g., septin-5, creatine kinase U-type and 14-3-3 protein epsilon). The effects of PT alone, or in combination with CPZ, indicate that PT did have an effect on the brain proteome, likely, in part at least, by altering the capacity of the BBB to regulate access to the CNS [60,113,167]. This is the first study to document the effects of giving PT alone on the CD4/8$^+$ cell migration and the whole brain proteome. In EAE, when PT is given together with adjuvant (CFA) and antigen stimulation (e.g., MOG), increased disruption of tight junctions at sites of perivascular inflammation and demyelination occurs [63]. In addition, increased rates of relapsing-remitting episodes [60], increased infiltration of serum albumin in the spinal cord [131] and suppression of the anti-inflammatory interleukin-10 [168] are observed. However, the extent to which these changes are attributable to PT alone, or the combined treatments used to induce EAE, remains un-documented. However, other studies have shown that PT alone evokes changes in BBB function leading to increased protein infiltration into the brain (~15 days; [169]), or disruption of G-protein function (~40 days; [170]) following PT administration. In the present study, repeated injections of PT during the second and third weeks of CPZ-feeding resulted in proteoform changes after 5 or 12 weeks indicating that PT injections alone have long-term effects at least on the brain proteome. Moreover, the proteomic analysis highlighted proteoform differences between the CPZ vs. EAE models; specifically, CPZ-feeding resulted in increased guanine nucleotide-binding protein G(o) subunit-α, glutamate dehydrogenase 1 and malate dehydrogenase, which decrease in EAE [88,94,105], perhaps reflecting the different underlying aetiologies (potentially including changes in specific proteoforms). Whether these opposing changes result specifically from the use of peripherally administered exogenous myelin antigens (e.g., MBP, PLP and MOG) in EAE or endogenously generated antigens (i.e., myelin debris) in the brain of CPZ-fed mice remains unclear but would seem likely.

Despite our rigorous efforts to minimize the experimental variables, we acknowledge certain inherent limitations in the analytical approach. This study relied on only two time points (i.e., 5 and 12 weeks) of CPZ-feeding(±PT), which did not allow us to determine if or when the identified proteoforms returned to baseline nor to correlate proteome changes with the initiation of oligodendrocytosis, demyelination and gliosis. Moreover, the sub-femtomole in-gel detection sensitivity may well have missed significant changes in very low abundance species, although these remain a substantial issue with all available analytical approaches if high quality final identifications are a serious expectation [78]. Furthermore, reliance on existing databases that address only amino acid sequences, as well as an apparent developing reliance in the field for online bioinformatics platforms that also largely address only what is known about canonical proteins [171–173], tends to further emphasize the fundamental importance of developing even more sensitive analytical approaches to routinely quantifying and fully characterizing proteoforms in order to provide the most direct understanding of the molecular mechanism underlying human diseases like MS.

5. Conclusions

This study confirmed that CPZ-feeding(±PT) in mice induced dose- and time-dependent oligodendrocytosis, demyelination and gliosis, but was not associated with any detectable invasion of peripheral adaptive (CD4/8) immune cells into the CNS. In the periphery, CPZ-feeding induced a dose-dependent suppression of splenic CD4/8 and organ mass, suggesting that this peripheral action of CPZ was a major impediment to studying the role of the peripheral immune system following demyelination and disruption of the BBB. Notably, oligodendrocytosis, demyelination and gliosis with the low dose of CPZ for 5 weeks resulted in minimal splenic atrophy and less severe adaptive immune system suppression, indicating that this might be a better model to test the 'inside-out' theory of MS. Moreover, using a highly sensitive 'top-down' proteomic approach, changes in 33 brain proteoforms were identified in the CPZ-fed mice, the majority of which were found to be associated with mitochondrial function. This strongly suggests that mitochondrial perturbations may elicit oligodendrocytosis and demyelination.

Supplementary Materials: The following are available online at http://www.mdpi.com/2073-4409/8/11/1314/s1. The Stereology and LC/MS/MS files associated with this article can be found in the following link http:$\delimiter"026E30F$$\delimiter"026E30F$galen.uws.edu.au$\delimiter"026E30F$mmrg$\delimiter"026E30F$sen.

Author Contributions: M.K.S., E.G., S.J.M., P.J.S., D.A.M. and J.R.C. conceived the study and provided all resources. Lab work and preliminary analysis were carried out by M.K.S. and M.S.M.A. All authors reviewed the analyses and draft manuscripts, and approved the final version.

Funding: This study was supported by funding from the Rotary Club of Narellan.

Acknowledgments: The authors acknowledge the Western Sydney University School of Medicine Animal Care and Mass Spectrometry Facilities for their help with this project. MKS was the recipient of a WSU-International Postgraduate Research Scholarship. MSMA was the recipient of a PhD sponsorship from the Higher Committee for Education Development in Iraq.

Conflicts of Interest: All authors declare no conflict of interest.

References

1. Partridge, M.A.; Myers, S.J.; Gopinath, S.; Coorssen, J.R. Proteomics of a conundrum: Thoughts on addressing the aetiology versus progression of multiple sclerosis. *Proteom. Clin. Appl.* **2015**, *9*, 838–843. [CrossRef] [PubMed]
2. Sen, M.K.; Mahns, D.A.; Coorssen, J.R.; Shortland, P.J. Behavioural phenotypes in the cuprizone model of central nervous system demyelination. *Neurosci. Biobehav. Rev.* **2019**, *107*, 23–46. [CrossRef] [PubMed]
3. Stys, P.K. Pathoetiology of multiple sclerosis: are we barking up the wrong tree? *F1000Prime Rep.* **2013**, *5*, 20. [CrossRef]
4. Stys, P.K.; Zamponi, G.W.; van Minnen, J.; Geurts, J.J. Will the real multiple sclerosis please stand up? *Nat. Rev. Neurosci.* **2012**, *13*, 507–514. [CrossRef] [PubMed]

5. Constantinescu, C.S.; Farooqi, N.; O'Brien, K.; Gran, B. Experimental autoimmune encephalomyelitis (EAE) as a model for multiple sclerosis (MS). *Br. J. Pharm.* **2011**, *164*, 1079–1106. [CrossRef] [PubMed]
6. Gold, R.; Linington, C.; Lassmann, H. Understanding pathogenesis and therapy of multiple sclerosis via animal models: 70 years of merits and culprits in experimental autoimmune encephalomyelitis research. *Brain* **2006**, *129*, 1953–1971. [CrossRef] [PubMed]
7. Glatigny, S.; Bettelli, E. Experimental Autoimmune Encephalomyelitis (EAE) as Animal Models of Multiple Sclerosis (MS). *Cold Spring Harb. Perspect. Med.* **2018**, *8*. [CrossRef]
8. Bittner, S.; Afzali, A.M.; Wiendl, H.; Meuth, S.G. Myelin oligodendrocyte glycoprotein (MOG35-55) induced experimental autoimmune encephalomyelitis (EAE) in C57BL/6 mice. *J. Vis. Exp.* **2014**. [CrossRef]
9. Patel, J.; Balabanov, R. Molecular mechanisms of oligodendrocyte injury in multiple sclerosis and experimental autoimmune encephalomyelitis. *Int. J. Mol. Sci.* **2012**, *13*, 10647–10659. [CrossRef]
10. Lu, J.; Kurejova, M.; Wirotanseng, L.N.; Linker, R.A.; Kuner, R.; Tappe-Theodor, A. Pain in experimental autoimmune encephalitis: a comparative study between different mouse models. *J. Neuroinflammation.* **2012**, *9*, 233. [CrossRef]
11. Evonuk, K.S.; Baker, B.J.; Doyle, R.E.; Moseley, C.E.; Sestero, C.M.; Johnston, B.P.; De Sarno, P.; Tang, A.; Gembitsky, I.; Hewett, S.J.; et al. Inhibition of System Xc(-) Transporter Attenuates Autoimmune Inflammatory Demyelination. *J. Immunol.* **2015**, *195*, 450–463. [CrossRef]
12. Hart, B.A.; Gran, B.; Weissert, R. EAE: imperfect but useful models of multiple sclerosis. *Trends Mol. Med.* **2011**, *17*, 119–125. [CrossRef] [PubMed]
13. Trapp, B.D.; Nave, K.A. Multiple sclerosis: an immune or neurodegenerative disorder? *Annu. Rev. Neurosci.* **2008**, *31*, 247–269. [CrossRef] [PubMed]
14. Lassmann, H.; Bradl, M. Multiple sclerosis: experimental models and reality. *Acta Neuropathol.* **2017**, *133*, 223–244. [CrossRef] [PubMed]
15. Lovett-Racke, A.E. Contribution of EAE to understanding and treating multiple sclerosis. *J. Neuroimmunol.* **2017**, *304*, 40–42. [CrossRef]
16. Krishnamoorthy, G.; Wekerle, H. EAE: an immunologist's magic eye. *Eur. J. Immunol.* **2009**, *39*, 2031–2035. [CrossRef]
17. Sriram, S.; Steiner, I. Experimental allergic encephalomyelitis: a misleading model of multiple sclerosis. *Ann. Neurol.* **2005**, *58*, 939–945. [CrossRef]
18. Behan, P.O.; Chaudhuri, A. EAE is not a useful model for demyelinating disease. *Mult. Scler. Relat. Disord.* **2014**, *3*, 565–574. [CrossRef]
19. Flugel, A.; Berkowicz, T.; Ritter, T.; Labeur, M.; Jenne, D.E.; Li, Z.; Ellwart, J.W.; Willem, M.; Lassmann, H.; Wekerle, H. Migratory activity and functional changes of green fluorescent effector cells before and during experimental autoimmune encephalomyelitis. *Immunity* **2001**, *14*, 547–560. [CrossRef]
20. Lassmann, H.; van Horssen, J. The molecular basis of neurodegeneration in multiple sclerosis. *FEBS Lett.* **2011**, *585*, 3715–3723. [CrossRef]
21. Friese, M.A.; Fugger, L. Autoreactive CD8+ T cells in multiple sclerosis: a new target for therapy? *Brain* **2005**, *128*, 1747–1763. [CrossRef] [PubMed]
22. Hauser, S.L.; Bhan, A.K.; Gilles, F.; Kemp, M.; Kerr, C.; Weiner, H.L. Immunohistochemical analysis of the cellular infiltrate in multiple sclerosis lesions. *Ann. Neurol.* **1986**, *19*, 578–587. [CrossRef]
23. Babbe, H.; Roers, A.; Waisman, A.; Lassmann, H.; Goebels, N.; Hohlfeld, R.; Friese, M.; Schroder, R.; Deckert, M.; Schmidt, S.; et al. Clonal expansions of CD8(+) T cells dominate the T cell infiltrate in active multiple sclerosis lesions as shown by micromanipulation and single cell polymerase chain reaction. *J. Exp. Med.* **2000**, *192*, 393–404. [CrossRef] [PubMed]
24. Ransohoff, R.M. Animal models of multiple sclerosis: the good, the bad and the bottom line. *Nat. Neurosci.* **2012**, *15*, 1074–1077. [CrossRef] [PubMed]
25. Gilmore, C.P.; Donaldson, I.; Bo, L.; Owens, T.; Lowe, J.; Evangelou, N. Regional variations in the extent and pattern of grey matter demyelination in multiple sclerosis: a comparison between the cerebral cortex, cerebellar cortex, deep grey matter nuclei and the spinal cord. *J. Neurol. Neurosurg. Psychiatry.* **2009**, *80*, 182–187. [CrossRef]
26. Day, M.J. Histopathology of EAE. In *Experimental Models of Multiple Sclerosis*; Lavi, E., Constantinescu, C.S., Eds.; Springer US: Boston, MA, USA, 2005; pp. 25–43.

27. Tanuma, N.; Shin, T.; Matsumoto, Y. Characterization of acute versus chronic relapsing autoimmune encephalomyelitis in DA rats. *J. Neuroimmunol.* **2000**, *108*, 171–180. [CrossRef]
28. Broadwater, L.; Pandit, A.; Clements, R.; Azzam, S.; Vadnal, J.; Sulak, M.; Yong, V.W.; Freeman, E.J.; Gregory, R.B.; McDonough, J. Analysis of the mitochondrial proteome in multiple sclerosis cortex. *BBA* **2011**, *1812*, 630–641. [CrossRef]
29. Rosenling, T.; Attali, A.; Luider, T.M.; Bischoff, R. The experimental autoimmune encephalomyelitis model for proteomic biomarker studies: from rat to human. *Clin. Chim. Acta.* **2011**, *412*, 812–822. [CrossRef] [PubMed]
30. Vargas, D.L.; Tyor, W.R. Update on disease-modifying therapies for multiple sclerosis. *J. Investig. Med.* **2017**, *65*, 883–891. [CrossRef] [PubMed]
31. Caprariello, A.V.; Rogers, J.A.; Morgan, M.L.; Hoghooghi, V.; Plemel, J.R.; Koebel, A.; Tsutsui, S.; Dunn, J.F.; Kotra, L.P.; Ousman, S.S.; et al. Biochemically altered myelin triggers autoimmune demyelination. *PNAS* **2018**, *115*, 5528–5533. [CrossRef]
32. Barnett, M.H.; Prineas, J.W. Relapsing and remitting multiple sclerosis: pathology of the newly forming lesion. *Ann. Neurol.* **2004**, *55*, 458–468. [CrossRef] [PubMed]
33. Henderson, A.P.; Barnett, M.H.; Parratt, J.D.; Prineas, J.W. Multiple sclerosis: distribution of inflammatory cells in newly forming lesions. *Ann. Neurol.* **2009**, *66*, 739–753. [CrossRef] [PubMed]
34. Lucchinetti, C.; Bruck, W.; Parisi, J.; Scheithauer, B.; Rodriguez, M.; Lassmann, H. Heterogeneity of multiple sclerosis lesions: implications for the pathogenesis of demyelination. *Ann. Neurol.* **2000**, *47*, 707–717. [CrossRef]
35. Lucchinetti, C.F.; Bruck, W.; Rodriguez, M.; Lassmann, H. Distinct patterns of multiple sclerosis pathology indicates heterogeneity on pathogenesis. *Brain Pathol.* **1996**, *6*, 259–274. [CrossRef] [PubMed]
36. Rodriguez, M.; Scheithauer, B. Ultrastructure of multiple sclerosis. *Ultrastruct. Pathol.* **1994**, *18*, 3–13. [CrossRef] [PubMed]
37. Traka, M.; Podojil, J.R.; McCarthy, D.P.; Miller, S.D.; Popko, B. Oligodendrocyte death results in immune-mediated CNS demyelination. *Nat. Neurosci.* **2016**, *19*, 65–74. [CrossRef]
38. Carlton, W.W. Studies on the induction of hydrocephalus and spongy degeneration by cuprizone feeding and attempts to antidote the toxicity. *Life Sci.* **1967**, *6*, 11–19. [CrossRef]
39. Matsushima, G.K.; Morell, P. The neurotoxicant, cuprizone, as a model to study demyelination and remyelination in the central nervous system. *Brain Pathol.* **2001**, *11*, 107–116. [CrossRef]
40. Komoly, S.; Jeyasingham, M.D.; Pratt, O.E.; Lantos, P.L. Decrease in oligodendrocyte carbonic anhydrase activity preceding myelin degeneration in cuprizone induced demyelination. *J. Neurol. Sci.* **1987**, *79*, 141–148. [CrossRef]
41. Zatta, P.; Raso, M.; Zambenedetti, P.; Wittkowski, W.; Messori, L.; Piccioli, F.; Mauri, P.L.; Beltramini, M. Copper and zinc dismetabolism in the mouse brain upon chronic cuprizone treatment. *Cell. Mol. Life Sci.* **2005**, *62*, 1502–1513. [CrossRef]
42. Moldovan, N.; Al-Ebraheem, A.; Lobo, L.; Park, R.; Farquharson, M.J.; Bock, N.A. Altered transition metal homeostasis in the cuprizone model of demyelination. *Neurotoxicol* **2015**, *48*, 1–8. [CrossRef] [PubMed]
43. Varga, E.; Pandur, E.; Abraham, H.; Horvath, A.; Acs, P.; Komoly, S.; Miseta, A.; Sipos, K. Cuprizone Administration Alters the Iron Metabolism in the Mouse Model of Multiple Sclerosis. *Cell. Mol. Neurobiol.* **2018**, *38*, 1081–1097. [CrossRef] [PubMed]
44. Venturini, G. Enzymic activities and sodium, potassium and copper concentrations in mouse brain and liver after cuprizone treatment in vivo. *J. Neurochem.* **1973**, *21*, 1147–1151. [CrossRef] [PubMed]
45. McTigue, D.M.; Tripathi, R.B. The life, death, and replacement of oligodendrocytes in the adult CNS. *J. Neurochem.* **2008**, *107*, 1–19. [CrossRef]
46. McLaurin, J.A.; Yong, V.W. Oligodendrocytes and myelin. *Neurol. Clin.* **1995**, *13*, 23–49. [CrossRef]
47. Liblau, R.; Fontaine, B.; Baron-Van Evercooren, A.; Wekerle, H.; Lassmann, H. Demyelinating diseases: from pathogenesis to repair strategies. *Trends Neurosci.* **2001**, *24*, 134–135. [CrossRef]
48. Tejedor, L.S.; Wostradowski, T.; Gingele, S.; Skripuletz, T.; Gudi, V.; Stangel, M. The Effect of Stereotactic Injections on Demyelination and Remyelination: a Study in the Cuprizone Model. *J. Mol. Neurosci.* **2017**, *61*, 479–488. [CrossRef]
49. Partridge, M.A.; Gopinath, S.; Myers, S.J.; Coorssen, J.R. An initial top-down proteomic analysis of the standard cuprizone mouse model of multiple sclerosis. *J. Chem. Biol.* **2016**, *9*, 9–18. [CrossRef]

50. Solti, I.; Kvell, K.; Talaber, G.; Veto, S.; Acs, P.; Gallyas, F.; Illes, Z.; Fekete, K.; Zalan, P.; Szanto, A.; et al. Thymic Atrophy and Apoptosis of CD4+CD8+ Thymocytes in the Cuprizone Model of Multiple Sclerosis. *PLoS ONE* **2015**, *10*, e0129217. [CrossRef]
51. Butt, R.H.; Coorssen, J.R. Postfractionation for enhanced proteomic analyses: routine electrophoretic methods increase the resolution of standard 2D-PAGE. *J. Proteome Res.* **2005**, *4*, 982–991. [CrossRef]
52. Wright, E.P.; Partridge, M.A.; Padula, M.P.; Gauci, V.J.; Malladi, C.S.; Coorssen, J.R. Top-down proteomics: enhancing 2D gel electrophoresis from tissue processing to high-sensitivity protein detection. *Proteomics* **2014**, *14*, 872–889. [CrossRef] [PubMed]
53. D'Silva, A.M.; Hyett, J.A.; Coorssen, J.R. Proteomic analysis of first trimester maternal serum to identify candidate biomarkers potentially predictive of spontaneous preterm birth. *J. Proteom.* **2018**, *178*, 31–42. [CrossRef]
54. Oliveira, B.M.; Coorssen, J.R.; Martins-de-Souza, D. 2DE: the phoenix of proteomics. *J. Proteom.* **2014**, *104*, 140–150. [CrossRef] [PubMed]
55. Coorssen, J.R.; Yergey, A.L. Proteomics Is Analytical Chemistry: Fitness-for-Purpose in the Application of Top-Down and Bottom-Up Analyses. *Proteomes* **2015**, *3*, 440–453. [CrossRef] [PubMed]
56. Kurgan, N.; Noaman, N.; Pergande, M.R.; Cologna, S.M.; Coorssen, J.R.; Klentrou, P. Changes to the Human Serum Proteome in Response to High Intensity Interval Exercise: A Sequential Top-Down Proteomic Analysis. *Front. Physiol.* **2019**, *10*. [CrossRef]
57. Kipp, M.; Clarner, T.; Dang, J.; Copray, S.; Beyer, C. The cuprizone animal model: new insights into an old story. *Acta Neuropathol.* **2009**, *118*, 723–736. [CrossRef]
58. Gudi, V.; Gingele, S.; Skripuletz, T.; Stangel, M. Glial response during cuprizone-induced de- and remyelination in the CNS: lessons learned. *Front. Cell. Neurosci.* **2014**, *8*, 73. [CrossRef]
59. Hiremath, M.M.; Saito, Y.; Knapp, G.W.; Ting, J.P.; Suzuki, K.; Matsushima, G.K. Microglial/macrophage accumulation during cuprizone-induced demyelination in C57BL/6 mice. *J. Neuroimmunol.* **1998**, *92*, 38–49. [CrossRef]
60. Mohajeri, M.; Sadeghizadeh, M.; Javan, M. Pertussis toxin promotes relapsing-remitting experimental autoimmune encephalomyelitis in Lewis rats. *J. Neuroimmunol.* **2015**, *289*, 105–110. [CrossRef]
61. Hofstetter, H.H.; Shive, C.L.; Forsthuber, T.G. Pertussis toxin modulates the immune response to neuroantigens injected in incomplete Freund's adjuvant: induction of Th1 cells and experimental autoimmune encephalomyelitis in the presence of high frequencies of Th2 cells. *J. Immunol.* **2002**, *169*, 117–125. [CrossRef]
62. Gao, Z.; Nissen, J.C.; Ji, K.; Tsirka, S.E. The experimental autoimmune encephalomyelitis disease course is modulated by nicotine and other cigarette smoke components. *PLoS ONE* **2014**, *9*, e107979. [CrossRef] [PubMed]
63. Bennett, J.; Basivireddy, J.; Kollar, A.; Biron, K.E.; Reickmann, P.; Jefferies, W.A.; McQuaid, S. Blood-brain barrier disruption and enhanced vascular permeability in the multiple sclerosis model EAE. *J. Neuroimmunol.* **2010**, *229*, 180–191. [CrossRef] [PubMed]
64. Almuslehi, M.S.M.; Sen, M.K.; Mahns, D.A.; Shortland, P.J.; Coorssen, J.R. Blood Brain Barrier Disruption Facilitates CD8 T Cells infiltration into the CNS of Orchiectomized Cuprizone Treated Mice. In Proceedings of the Australasian Neuroscience Society, Brisbane Convention & Exhibition Centre, Brisbane, Australia, 3–6 December 2018; p. 163.
65. Gyengesi, E.; Liang, H.; Millington, C.; Sonego, S.; Sirijovski, D.; Gunawardena, D.; Dhananjayan, K.; Venigalla, M.; Niedermayer, G.; Munch, G. Investigation Into the Effects of Tenilsetam on Markers of Neuroinflammation in GFAP-IL6 Mice. *Pharm. Res.* **2018**, *35*, 22. [CrossRef] [PubMed]
66. Gyengesi, E.; Calabrese, E.; Sherrier, M.C.; Johnson, G.A.; Paxinos, G.; Watson, C. Semi-automated 3D segmentation of major tracts in the rat brain: comparing DTI with standard histological methods. *Brain Struct. Funct.* **2014**, *219*, 539–550. [CrossRef]
67. Pistorio, A.L.; Hendry, S.H.; Wang, X. A modified technique for high-resolution staining of myelin. *J. Neurosci. Methods.* **2006**, *153*, 135–146. [CrossRef]
68. Paxinos, G.; Franklin, K. *Paxinos and Franklin's the Mouse Brain in Stereotaxic Coordinates*, Fourth ed.; Academic Press: Cambridge, MA, USA, 2012.
69. Taylor, L.C.; Gilmore, W.; Matsushima, G.K. SJL mice exposed to cuprizone intoxication reveal strain and gender pattern differences in demyelination. *Brain Pathol.* **2009**, *19*, 467–479. [CrossRef]

70. Gumusoglu, S.B.; Fine, R.S.; Murray, S.J.; Bittle, J.L.; Stevens, H.E. The role of IL-6 in neurodevelopment after prenatal stress. *Brain Behav. Immun.* **2017**, *65*, 274–283. [CrossRef]
71. Butt, R.H.; Coorssen, J.R. Pre-extraction sample handling by automated frozen disruption significantly improves subsequent proteomic analyses. *J. Proteome Res.* **2006**, *5*, 437–448. [CrossRef]
72. Butt, R.H.; Pfeifer, T.A.; Delaney, A.; Grigliatti, T.A.; Tetzlaff, W.G.; Coorssen, J.R. Enabling coupled quantitative genomics and proteomics analyses from rat spinal cord samples. *Mol. Cell. Proteom.* **2007**, *6*, 1574–1588. [CrossRef]
73. Wright, E.P.; Prasad, K.A.; Padula, M.P.; Coorssen, J.R. Deep imaging: how much of the proteome does current top-down technology already resolve? *PLoS ONE* **2014**, *9*, e86058. [CrossRef]
74. Gauci, V.J.; Padula, M.P.; Coorssen, J.R. Coomassie blue staining for high sensitivity gel-based proteomics. *J. Proteom.* **2013**, *90*, 96–106. [CrossRef] [PubMed]
75. Noaman, N.; Coorssen, J.R. Coomassie does it (better): A Robin Hood approach to total protein quantification. *Anal. Biochem.* **2018**, *556*, 53–56. [CrossRef] [PubMed]
76. Butt, R.H.; Coorssen, J.R. Coomassie blue as a near-infrared fluorescent stain: a systematic comparison with Sypro Ruby for in-gel protein detection. *Mol. Cell. Proteom.* **2013**, *12*, 3834–3850. [CrossRef] [PubMed]
77. Harris, L.R.; Churchward, M.A.; Butt, R.H.; Coorssen, J.R. Assessing detection methods for gel-based proteomic analyses. *J. Proteome Res.* **2007**, *6*, 1418–1425. [CrossRef]
78. Noaman, N.; Abbineni, P.S.; Withers, M.; Coorssen, J.R. Coomassie staining provides routine (sub)femtomole in-gel detection of intact proteoforms: Expanding opportunities for genuine Top-down Proteomics. *Electrophoresis* **2017**, *38*, 3086–3099. [CrossRef]
79. D'Silva, A.M.; Hyett, J.A.; Coorssen, J.R. A Routine 'Top-Down' Approach to Analysis of the Human Serum Proteome. *Proteomes* **2017**, *5*, 13. [CrossRef]
80. Stimpson, S.E.; Coorssen, J.R.; Myers, S.J. Mitochondrial protein alterations in a familial peripheral neuropathy caused by the V144D amino acid mutation in the sphingolipid protein, SPTLC1. *J. Chem. Biol.* **2015**, *8*, 25–35. [CrossRef]
81. Stroud, L.J.; Slapeta, J.; Padula, M.P.; Druery, D.; Tsiotsioras, G.; Coorssen, J.R.; Stack, C.M. Comparative proteomic analysis of two pathogenic Tritrichomonas foetus genotypes: there is more to the proteome than meets the eye. *Int. J. Parasitol.* **2017**, *47*, 203–213. [CrossRef]
82. UniProt Consortium, T. UniProt: the universal protein knowledgebase. *Nucleic Acids Res.* **2018**, *46*, 2699. [CrossRef]
83. Sharma, S.; Ray, S.; Moiyadi, A.; Sridhar, E.; Srivastava, S. Quantitative proteomic analysis of meningiomas for the identification of surrogate protein markers. *Sci. Rep.* **2014**, *4*, 7140. [CrossRef]
84. Szklarczyk, D.; Franceschini, A.; Wyder, S.; Forslund, K.; Heller, D.; Huerta-Cepas, J.; Simonovic, M.; Roth, A.; Santos, A.; Tsafou, K.P.; et al. STRING v10: protein-protein interaction networks, integrated over the tree of life. *Nucleic Acids Res.* **2015**, *43*, D447–D452. [CrossRef] [PubMed]
85. Hossain, M.U.; Khan, M.A.; Hashem, A.; Islam, M.M.; Morshed, M.N.; Keya, C.A.; Salimullah, M. Finding Potential Therapeutic Targets against Shigella flexneri through Proteome Exploration. *Front. Microbiol.* **2016**, *7*, 1817. [CrossRef] [PubMed]
86. De Las Rivas, J.; Fontanillo, C. Protein-protein interactions essentials: key concepts to building and analyzing interactome networks. *PLoS Comput. Biol.* **2010**, *6*, e1000807. [CrossRef] [PubMed]
87. De Riccardis, L.; Ferramosca, A.; Danieli, A.; Trianni, G.; Zara, V.; De Robertis, F.; Maffia, M. Metabolic response to glatiramer acetate therapy in multiple sclerosis patients. *BBA* **2016**, *6*, 131–137. [CrossRef] [PubMed]
88. Farias, A.S.; Martins-de-Souza, D.; Guimaraes, L.; Pradella, F.; Moraes, A.S.; Facchini, G.; Novello, J.C.; Santos, L.M. Proteome analysis of spinal cord during the clinical course of monophasic experimental autoimmune encephalomyelitis. *Proteomics* **2012**, *12*, 2656–2662. [CrossRef]
89. Menon, K.N.; Steer, D.L.; Short, M.; Petratos, S.; Smith, I.; Bernard, C.C. A novel unbiased proteomic approach to detect the reactivity of cerebrospinal fluid in neurological diseases. *Mol. Cell. Proteom.* **2011**, *10*, M110.000042. [CrossRef] [PubMed]
90. Almeras, L.; Lefranc, D.; Drobecq, H.; de Seze, J.; Dubucquoi, S.; Vermersch, P.; Prin, L. New antigenic candidates in multiple sclerosis: identification by serological proteome analysis. *Proteomics* **2004**, *4*, 2184–2194. [CrossRef]

91. Gat-Viks, I.; Geiger, T.; Barbi, M.; Raini, G.; Elroy-Stein, O. Proteomics-level analysis of myelin formation and regeneration in a mouse model for Vanishing White Matter disease. *J. Neurochem.* **2015**, *134*, 513–526. [CrossRef]
92. Noben, J.P.; Dumont, D.; Kwasnikowska, N.; Verhaert, P.; Somers, V.; Hupperts, R.; Stinissen, P.; Robben, J. Lumbar cerebrospinal fluid proteome in multiple sclerosis: characterization by ultrafiltration, liquid chromatography, and mass spectrometry. *J. Proteome Res.* **2006**, *5*, 1647–1657. [CrossRef]
93. Jastorff, A.M.; Haegler, K.; Maccarrone, G.; Holsboer, F.; Weber, F.; Ziemssen, T.; Turck, C.W. Regulation of proteins mediating neurodegeneration in experimental autoimmune encephalomyelitis and multiple sclerosis. *Proteom. Clin. Appl.* **2009**, *3*, 1273–1287. [CrossRef]
94. Linker, R.A.; Brechlin, P.; Jesse, S.; Steinacker, P.; Lee, D.H.; Asif, A.R.; Jahn, O.; Tumani, H.; Gold, R.; Otto, M. Proteome profiling in murine models of multiple sclerosis: identification of stage specific markers and culprits for tissue damage. *PLoS ONE* **2009**, *4*, e7624. [CrossRef] [PubMed]
95. Werner, P.; Pitt, D.; Raine, C.S. Multiple sclerosis: altered glutamate homeostasis in lesions correlates with oligodendrocyte and axonal damage. *Ann. Neurol.* **2001**, *50*, 169–180. [CrossRef] [PubMed]
96. Newcombe, J.; Eriksson, B.; Ottervald, J.; Yang, Y.; Franzen, B. Extraction and proteomic analysis of proteins from normal and multiple sclerosis postmortem brain. *J. Chromatogr. B Anal. Technol. Biomed. Life Sci.* **2005**, *815*, 191–202. [CrossRef] [PubMed]
97. Martin, N.A.; Molnar, V.; Szilagyi, G.T.; Elkjaer, M.L.; Nawrocki, A.; Okarmus, J.; Wlodarczyk, A.; Thygesen, E.K.; Palkovits, M.; Gallyas, F., Jr.; et al. Experimental Demyelination and Axonal Loss Are Reduced in MicroRNA-146a Deficient Mice. *Front. Immunol.* **2018**, *9*, 490. [CrossRef]
98. Liu, T.; Donahue, K.C.; Hu, J.; Kurnellas, M.P.; Grant, J.E.; Li, H.; Elkabes, S. Identification of differentially expressed proteins in experimental autoimmune encephalomyelitis (EAE) by proteomic analysis of the spinal cord. *J. Proteome Res.* **2007**, *6*, 2565–2575. [CrossRef]
99. Werner, S.R.; Saha, J.K.; Broderick, C.L.; Zhen, E.Y.; Higgs, R.E.; Duffin, K.L.; Smith, R.C. Proteomic analysis of demyelinated and remyelinating brain tissue following dietary cuprizone administration. *J. Mol. Neurosci.* **2010**, *42*, 210–225. [CrossRef]
100. Fazeli, A.S.; Nasrabadi, D.; Pouya, A.; Mirshavaladi, S.; Sanati, M.H.; Baharvand, H.; Salekdeh, G.H. Proteome analysis of post-transplantation recovery mechanisms of an EAE model of multiple sclerosis treated with embryonic stem cell-derived neural precursors. *J. Proteom.* **2013**, *94*, 437–450. [CrossRef]
101. Axelsson, M.; Malmestrom, C.; Nilsson, S.; Haghighi, S.; Rosengren, L.; Lycke, J. Glial fibrillary acidic protein: a potential biomarker for progression in multiple sclerosis. *J. Neurol.* **2011**, *258*, 882–888. [CrossRef]
102. Jain, M.R.; Bian, S.; Liu, T.; Hu, J.; Elkabes, S.; Li, H. Altered proteolytic events in experimental autoimmune encephalomyelitis discovered by iTRAQ shotgun proteomics analysis of spinal cord. *Proteome Sci.* **2009**, *7*, 25. [CrossRef]
103. Gresle, M.M.; Butzkueven, H.; Shaw, G. Neurofilament proteins as body fluid biomarkers of neurodegeneration in multiple sclerosis. *Mult. Scler. Int.* **2011**, *2011*, 315406. [CrossRef]
104. De Masi, R.; Vergara, D.; Pasca, S.; Acierno, R.; Greco, M.; Spagnolo, L.; Blasi, E.; Sanapo, F.; Trianni, G.; Maffia, M. PBMCs protein expression profile in relapsing IFN-treated multiple sclerosis: A pilot study on relation to clinical findings and brain atrophy. *J. Neuroimmunol.* **2009**, *210*, 80–86. [CrossRef] [PubMed]
105. Fazeli, A.S.; Nasrabadi, D.; Sanati, M.H.; Pouya, A.; Ibrahim, S.M.; Baharvand, H.; Salekdeh, G.H. Proteome analysis of brain in murine experimental autoimmune encephalomyelitis. *Proteomics* **2010**, *10*, 2822–2832. [CrossRef] [PubMed]
106. Ni Fhlathartaigh, M.; McMahon, J.; Reynolds, R.; Connolly, D.; Higgins, E.; Counihan, T.; Fitzgerald, U. Calreticulin and other components of endoplasmic reticulum stress in rat and human inflammatory demyelination. *Acta Neuropathol. Commun.* **2013**, *1*, 37. [CrossRef] [PubMed]
107. Colucci, M.; Roccatagliata, L.; Capello, E.; Narciso, E.; Latronico, N.; Tabaton, M.; Mancardi, G.L. The 14-3-3 protein in multiple sclerosis: a marker of disease severity. *Mult. Scler.* **2004**, *10*, 477–481. [CrossRef]
108. Chang, H.; Liu, J.; Zhang, Y.; Wang, F.; Wu, Y.; Zhang, L.; Ai, H.; Chen, G.; Yin, L. Increased central dopaminergic activity might be involved in the behavioral abnormality of cuprizone exposure mice. *Behav. Brain Res.* **2017**, *331*, 143–150. [CrossRef]
109. Franco-Pons, N.; Torrente, M.; Colomina, M.T.; Vilella, E. Behavioral deficits in the cuprizone-induced murine model of demyelination/remyelination. *Toxicol. Lett.* **2007**, *169*, 205–213. [CrossRef]

110. Lampron, A.; Larochelle, A.; Laflamme, N.; Prefontaine, P.; Plante, M.M.; Sanchez, M.G.; Yong, V.W.; Stys, P.K.; Tremblay, M.E.; Rivest, S. Inefficient clearance of myelin debris by microglia impairs remyelinating processes. *J. Exp. Med.* **2015**, *212*, 481–495. [CrossRef]
111. Coorssen, J.R.; Blank, P.S.; Albertorio, F.; Bezrukov, L.; Kolosova, I.; Backlund, P.S., Jr.; Zimmerberg, J. Quantitative femto- to attomole immunodetection of regulated secretory vesicle proteins critical to exocytosis. *Anal. Biochem.* **2002**, *307*, 54–62. [CrossRef]
112. Remington, L.T.; Babcock, A.A.; Zehntner, S.P.; Owens, T. Microglial recruitment, activation, and proliferation in response to primary demyelination. *Am. J. Pathol.* **2007**, *170*, 1713–1724. [CrossRef]
113. Kugler, S.; Bocker, K.; Heusipp, G.; Greune, L.; Kim, K.S.; Schmidt, M.A. Pertussis toxin transiently affects barrier integrity, organelle organization and transmigration of monocytes in a human brain microvascular endothelial cell barrier model. *Cell. Microbiol.* **2007**, *9*, 619–632. [CrossRef]
114. Praet, J.; Guglielmetti, C.; Berneman, Z.; Van der Linden, A.; Ponsaerts, P. Cellular and molecular neuropathology of the cuprizone mouse model: clinical relevance for multiple sclerosis. *Neurosci. Biobehav. Rev.* **2014**, *47*, 485–505. [CrossRef] [PubMed]
115. Cesta, M.F. Normal structure, function, and histology of the spleen. *Toxicol. Pathol.* **2006**, *34*, 455–465. [CrossRef] [PubMed]
116. Pearse, G. Normal structure, function and histology of the thymus. *Toxicol. Pathol.* **2006**, *34*, 504–514. [CrossRef] [PubMed]
117. Emerson, M.R.; Biswas, S.; LeVine, S.M. Cuprizone and piperonyl butoxide, proposed inhibitors of T-cell function, attenuate experimental allergic encephalomyelitis in SJL mice. *J. Neuroimmunol.* **2001**, *119*, 205–213. [CrossRef]
118. Herder, V.; Hansmann, F.; Stangel, M.; Schaudien, D.; Rohn, K.; Baumgartner, W.; Beineke, A. Cuprizone inhibits demyelinating leukomyelitis by reducing immune responses without virus exacerbation in an infectious model of multiple sclerosis. *J. Neuroimmunol.* **2012**, *244*, 84–93. [CrossRef]
119. Mana, P.; Fordham, S.A.; Staykova, M.A.; Correcha, M.; Silva, D.; Willenborg, D.O.; Linares, D. Demyelination caused by the copper chelator cuprizone halts T cell mediated autoimmune neuroinflammation. *J. Neuroimmunol.* **2009**, *210*, 13–21. [CrossRef]
120. Schwartz, R.H. T cell anergy. *Annu. Rev. Immunol.* **2003**, *21*, 305–334. [CrossRef]
121. Macián, F.; Im, S.-H.; García-Cózar, F.J.; Rao, A. T-cell anergy. *Curr. Opin. Immunol.* **2004**, *16*, 209–216. [CrossRef]
122. Nelson, B.H. IL-2, regulatory T cells, and tolerance. *J. Immunol.* **2004**, *172*, 3983–3988. [CrossRef]
123. Bala, S.; Failla, M.L. Copper deficiency reversibly impairs DNA synthesis in activated T lymphocytes by limiting interleukin 2 activity. *PNAS* **1992**, *89*, 6794–6797. [CrossRef]
124. Hopkins, R.G.; Failla, M.L. Transcriptional regulation of interleukin-2 gene expression is impaired by copper deficiency in Jurkat human T lymphocytes. *J. Nutr.* **1999**, *129*, 596–601. [CrossRef] [PubMed]
125. Lin, M.Y.; Zal, T.; Ch'en, I.L.; Gascoigne, N.R.; Hedrick, S.M. A pivotal role for the multifunctional calcium/calmodulin-dependent protein kinase II in T cells: from activation to unresponsiveness. *J. Immunol.* **2005**, *174*, 5583–5592. [CrossRef] [PubMed]
126. Bui, J.D.; Calbo, S.; Hayden-Martinez, K.; Kane, L.P.; Gardner, P.; Hedrick, S.M. A role for CaMKII in T cell memory. *Cell* **2000**, *100*, 457–467. [CrossRef]
127. Weyer, A.D.; Stucky, C.L. Repurposing a leukocyte elastase inhibitor for neuropathic pain. *Nat. Med.* **2015**, *21*, 429–430. [CrossRef] [PubMed]
128. Kang, K.; Park, B.; Oh, C.; Cho, K.; Ahn, K. A role for protein disulfide isomerase in the early folding and assembly of MHC class I molecules. *Antioxid. Redox Signal.* **2009**, *11*, 2553–2561. [CrossRef]
129. Desdin-Mico, G.; Soto-Heredero, G.; Mittelbrunn, M. Mitochondrial activity in T cells. *Mitochondrion* **2018**, *41*, 51–57. [CrossRef]
130. Sukumar, M.; Liu, J.; Mehta, G.U.; Patel, S.J.; Roychoudhuri, R.; Crompton, J.G.; Klebanoff, C.A.; Ji, Y.; Li, P.; Yu, Z.; et al. Mitochondrial Membrane Potential Identifies Cells with Enhanced Stemness for Cellular Therapy. *Cell. Metab.* **2016**, *23*, 63–76. [CrossRef]
131. Farias, A.S.; Pradella, F.; Schmitt, A.; Santos, L.M.; Martins-de-Souza, D. Ten years of proteomics in multiple sclerosis. *Proteomics.* **2014**, *14*, 467–480. [CrossRef]

132. Dagley, L.F.; Croft, N.P.; Isserlin, R.; Olsen, J.B.; Fong, V.; Emili, A.; Purcell, A.W. Discovery of novel disease-specific and membrane-associated candidate markers in a mouse model of multiple sclerosis. *Mol. Cell. Proteom.* **2014**, *13*, 679–700. [CrossRef]
133. Elkabes, S.; Li, H. Proteomic strategies in multiple sclerosis and its animal models. *Proteom. Clin. Appl.* **2007**, *1*, 1393–1405. [CrossRef]
134. Zhan, X.; Yang, H.; Peng, F.; Li, J.; Mu, Y.; Long, Y.; Cheng, T.; Huang, Y.; Li, Z.; Lu, M.; et al. How many proteins can be identified in a 2DE gel spot within an analysis of a complex human cancer tissue proteome? *Electrophoresis* **2018**, *39*, 965–980. [CrossRef] [PubMed]
135. Oveland, E.; Nystad, A.; Berven, F.; Myhr, K.M.; Torkildsen, O.; Wergeland, S. 1,25-Dihydroxyvitamin-D3 induces brain proteomic changes in cuprizone mice during remyelination involving calcium proteins. *Neurochem. Int.* **2018**, *112*, 267–277. [CrossRef] [PubMed]
136. Berggard, T.; Linse, S.; James, P. Methods for the detection and analysis of protein-protein interactions. *Proteomics* **2007**, *7*, 2833–2842. [CrossRef] [PubMed]
137. Turvey, M.E.; Koudelka, T.; Comerford, I.; Greer, J.M.; Carroll, W.; Bernard, C.C.; Hoffmann, P.; McColl, S.R. Quantitative proteome profiling of CNS-infiltrating autoreactive CD4+ cells reveals selective changes during experimental autoimmune encephalomyelitis. *J. Proteome Res.* **2014**, *13*, 3655–3670. [CrossRef] [PubMed]
138. Puntarulo, S. Iron, oxidative stress and human health. *Mol. Asp. Med.* **2005**, *26*, 299–312. [CrossRef] [PubMed]
139. Faizi, M.; Salimi, A.; Seydi, E.; Naserzadeh, P.; Kouhnavard, M.; Rahimi, A.; Pourahmad, J. Toxicity of cuprizone a Cu(2+) chelating agent on isolated mouse brain mitochondria: a justification for demyelination and subsequent behavioral dysfunction. *Toxicol. Mech. Methods.* **2016**, *26*, 276–283. [CrossRef] [PubMed]
140. Youle, R.J.; van der Bliek, A.M. Mitochondrial fission, fusion, and stress. *Science* **2012**, *337*, 1062–1065. [CrossRef]
141. Acs, P.; Komoly, S. Selective ultrastructural vulnerability in the cuprizone-induced experimental demyelination. *Ideggyogy Sz.* **2012**, *65*, 266–270.
142. Biancotti, J.C.; Kumar, S.; de Vellis, J. Activation of inflammatory response by a combination of growth factors in cuprizone-induced demyelinated brain leads to myelin repair. *Neurochem. Res.* **2008**, *33*, 2615–2628. [CrossRef]
143. Harris, J.J.; Attwell, D. The energetics of CNS white matter. *J. Neurosci.* **2012**, *32*, 356–371. [CrossRef]
144. Bradl, M.; Lassmann, H. Oligodendrocytes: biology and pathology. *Acta Neuropathol.* **2010**, *119*, 37–53. [CrossRef] [PubMed]
145. Carvalho, A.N.; Lim, J.L.; Nijland, P.G.; Witte, M.E.; Van Horssen, J. Glutathione in multiple sclerosis: more than just an antioxidant? *Mult. Scler.* **2014**, *20*, 1425–1431. [CrossRef]
146. Kang, Y.J. Metallothionein redox cycle and function. *Exp. Biol. Med.* **2006**, *231*, 1459–1467. [CrossRef] [PubMed]
147. Pinteaux, E.; Perraut, M.; Tholey, G. Distribution of mitochondrial manganese superoxide dismutase among rat glial cells in culture. *Glia* **1998**, *22*, 408–414. [CrossRef]
148. Lassmann, H.; van Horssen, J. Oxidative stress and its impact on neurons and glia in multiple sclerosis lesions. *BBA.* **2016**, *1862*, 506–510. [CrossRef] [PubMed]
149. Stone, S.; Lin, W. The unfolded protein response in multiple sclerosis. *Front. Neurosci.* **2015**, *9*, 264. [CrossRef] [PubMed]
150. Cwiklinska, H.; Mycko, M.P.; Luvsannorov, O.; Walkowiak, B.; Brosnan, C.F.; Raine, C.S.; Selmaj, K.W. Heat shock protein 70 associations with myelin basic protein and proteolipid protein in multiple sclerosis brains. *Int. Immunol.* **2003**, *15*, 241–249. [CrossRef] [PubMed]
151. Mycko, M.P.; Papoian, R.; Boschert, U.; Raine, C.S.; Selmaj, K.W. Microarray gene expression profiling of chronic active and inactive lesions in multiple sclerosis. *Clin. Neurol. Neurosurg.* **2004**, *106*, 223–229. [CrossRef]
152. Michaels, T.C.; Lazell, H.W.; Arosio, P.; Knowles, T.P. Dynamics of protein aggregation and oligomer formation governed by secondary nucleation. *J. Chem. Phys.* **2015**, *143*, 054901. [CrossRef]
153. Choi, M.L.; Gandhi, S. Crucial role of protein oligomerization in the pathogenesis of Alzheimer's and Parkinson's diseases. *FEBS J.* **2018**, *285*, 3631–3644. [CrossRef]

154. Dasgupta, A.; Zheng, J.; Perrone-Bizzozero, N.I.; Bizzozero, O.A. Increased carbonylation, protein aggregation and apoptosis in the spinal cord of mice with experimental autoimmune encephalomyelitis. *ASN Neuro.* **2013**, *5*, e00111. [CrossRef] [PubMed]
155. David, M.A.; Tayebi, M. Detection of Protein Aggregates in Brain and Cerebrospinal Fluid Derived from Multiple Sclerosis Patients. *Front. Neurol.* **2014**, *5*. [CrossRef] [PubMed]
156. Frohlich, C.; Grabiger, S.; Schwefel, D.; Faelber, K.; Rosenbaum, E.; Mears, J.; Rocks, O.; Daumke, O. Structural insights into oligomerization and mitochondrial remodelling of dynamin 1-like protein. *EMBO J.* **2013**, *32*, 1280–1292. [CrossRef]
157. Clinton, R.W.; Francy, C.A.; Ramachandran, R.; Qi, X.; Mears, J.A. Dynamin-related Protein 1 Oligomerization in Solution Impairs Functional Interactions with Membrane-anchored Mitochondrial Fission Factor. *J. Biol. Chem.* **2016**, *291*, 478–492. [CrossRef] [PubMed]
158. Jorgensen, C.S.; Ryder, L.R.; Steino, A.; Hojrup, P.; Hansen, J.; Beyer, N.H.; Heegaard, N.H.; Houen, G. Dimerization and oligomerization of the chaperone calreticulin. *Eur. J. Biochem.* **2003**, *270*, 4140–4148. [CrossRef] [PubMed]
159. Hammack, B.N.; Fung, K.Y.; Hunsucker, S.W.; Duncan, M.W.; Burgoon, M.P.; Owens, G.P.; Gilden, D.H. Proteomic analysis of multiple sclerosis cerebrospinal fluid. *Mult. Scler.* **2004**, *10*, 245–260. [CrossRef] [PubMed]
160. Lehmensiek, V.; Sussmuth, S.D.; Tauscher, G.; Brettschneider, J.; Felk, S.; Gillardon, F.; Tumani, H. Cerebrospinal fluid proteome profile in multiple sclerosis. *Mult. Scler.* **2007**, *13*, 840–849. [CrossRef]
161. Ziegler-Waldkirch, S.; Meyer-Luehmann, M. The Role of Glial Cells and Synapse Loss in Mouse Models of Alzheimer's Disease. *Front. Cell. Neurosci.* **2018**, *12*, 473. [CrossRef]
162. Bisht, K.; Sharma, K.P.; Lecours, C.; Sanchez, M.G.; El Hajj, H.; Milior, G.; Olmos-Alonso, A.; Gomez-Nicola, D.; Luheshi, G.; Vallieres, L.; et al. Dark microglia: A new phenotype predominantly associated with pathological states. *Glia* **2016**, *64*, 826–839. [CrossRef]
163. Matute, C.; Alberdi, E.; Ibarretxe, G.; Sanchez-Gomez, M.V. Excitotoxicity in glial cells. *Eur. J. Pharm.* **2002**, *447*, 239–246. [CrossRef]
164. Hanin, I. Central neurotransmitter function and its behavioral correlates in man. *Env. Health Perspect.* **1978**, *26*, 135–141. [CrossRef] [PubMed]
165. Dutta, R.; Chang, A.; Doud, M.K.; Kidd, G.J.; Ribaudo, M.V.; Young, E.A.; Fox, R.J.; Staugaitis, S.M.; Trapp, B.D. Demyelination causes synaptic alterations in hippocampi from multiple sclerosis patients. *Ann. Neurol.* **2011**, *69*, 445–454. [CrossRef] [PubMed]
166. Dutta, R.; Chomyk, A.M.; Chang, A.; Ribaudo, M.V.; Deckard, S.A.; Doud, M.K.; Edberg, D.D.; Bai, B.; Li, M.; Baranzini, S.E.; et al. Hippocampal demyelination and memory dysfunction are associated with increased levels of the neuronal microRNA miR-124 and reduced AMPA receptors. *Ann. Neurol.* **2013**, *73*, 637–645. [CrossRef] [PubMed]
167. Schellenberg, A.E.; Buist, R.; Del Bigio, M.R.; Toft-Hansen, H.; Khorooshi, R.; Owens, T.; Peeling, J. Blood-brain barrier disruption in CCL2 transgenic mice during pertussis toxin-induced brain inflammation. *Fluids Barriers CNS* **2012**, *9*, 10. [CrossRef] [PubMed]
168. Arimoto, H.; Tanuma, N.; Jee, Y.; Miyazawa, T.; Shima, K.; Matsumoto, Y. Analysis of experimental autoimmune encephalomyelitis induced in F344 rats by pertussis toxin administration. *J. Neuroimmunol.* **2000**, *104*, 15–21. [CrossRef]
169. Amiel, S.A. The effects of Bordetella pertussis vaccine on cerebral vascular permeability. *Br. J. Exp. Pathol.* **1976**, *57*, 653–662.
170. Shah, S.; Breivogel, C.; Selly, D.; Munirathinam, G.; Childers, S.; Yoburn, B.C. Time-dependent effects of in vivo pertussis toxin on morphine analgesia and G-proteins in mice. *Pharm. Biochem. Behav.* **1997**, *56*, 465–469. [CrossRef]
171. Armirotti, A.; Damonte, G. Achievements and perspectives of top-down proteomics. *Proteomics* **2010**, *10*, 3566–3576. [CrossRef]

172. Kachuk, C.; Doucette, A.A. The benefits (and misfortunes) of SDS in top-down proteomics. *J. Proteom.* **2018**, *175*, 75–86. [CrossRef]
173. Perkel, J.M. Tearing the top off 'Top-Down' Proteomics. *Biotechniques* **2012**, *53*, 75–78. [CrossRef]

© 2019 by the authors. Licensee MDPI, Basel, Switzerland. This article is an open access article distributed under the terms and conditions of the Creative Commons Attribution (CC BY) license (http://creativecommons.org/licenses/by/4.0/).

Article

The Neutrophil-to-Lymphocyte Ratio is Related to Disease Activity in Relapsing Remitting Multiple Sclerosis

Emanuele D'Amico [1,*], Aurora Zanghì [1], Alessandra Romano [2], Mariangela Sciandra [3], Giuseppe Alberto Maria Palumbo [4] and Francesco Patti [1]

1. Department "G.F. Ingrassia", MS Center, University of Catania, Via Santa Sofia 78, 95123 Catania, Italy; aurora.zanghi@studium.unict.it (A.Z.); patti@unict.it (F.P.)
2. Department of Surgery and Medical Specialties, Division of Hematology-A.O.U. Policlinico-OVE, Catania, Via Santa Sofia 78, 95123 Catania, Italy; sandrina.romano@gmail.com
3. Department of Economics, Business and Statistics, University of Palermo, 90128 Palermo, Italy; mariangela.sciandra@unipa.it
4. Department "G.F. Ingrassia", Division of Hematology-A.O.U. Policlinico-OVE, Catania, University of Catania, Via Santa Sofia 78, 95123 Catania, Italy; giuseppealberto.palumbo@gmail
* Correspondence: emanuele.damico@unict.it; Tel.: +39-095-378-2754

Received: 13 August 2019; Accepted: 17 September 2019; Published: 20 September 2019

Abstract: Background: The role of the neutrophil-to-lymphocyte ratio (NLR) of peripheral blood has been investigated in relation to several autoimmune diseases. Limited studies have addressed the significance of the NLR in terms of being a marker of disease activity in multiple sclerosis (MS). Methods: This is a retrospective study in relapsing–remitting MS patients (RRMS) admitted to the tertiary MS center of Catania, Italy during the period of 1 January to 31 December 2018. The aim of the present study was to investigate the significance of the NLR in reflecting the disease activity in a cohort of early diagnosed RRMS patients. Results: Among a total sample of 132 patients diagnosed with RRMS, 84 were enrolled in the present study. In the association analysis, a relation between the NLR value and disease activity at onset was found (V-Cramer 0.271, $p = 0.013$). In the logistic regression model, the variable NLR ($p = 0.03$ ExpB 3.5, CI 95% 1.089–11.4) was related to disease activity at onset. Conclusion: An elevated NLR is associated with disease activity at onset in RRMS patients. More large-scale studies with a longer follow-up are needed.

Keywords: neutrophils; lymphocytes; NLR; multiple sclerosis; disease activity

1. Introduction

Multiple sclerosis (MS) is a chronic autoimmune disease of the central nervous system (CNS), in which inflammation, demyelination, and axonal loss coexist, manifesting in a plethora of clinical signs and symptoms [1]. Regarding its pathogenesis, it has been suggested that autoreactive myelin-specific T cells may trigger and modulate the access and trafficking of inflammatory leukocytes to the CNS. Increasing evidence also suggests a fundamental role of B cells in the pathogenesis and development of MS [2].

Neutrophils are bone-marrow-derived cells that represent the most abundant peripheral blood leucocyte and are able to generate extracellular traps, which can be proinflammatory and provide a potential source of autoantigens, triggering autoimmunity [3]. The contribution of neutrophils to CNS autoimmune diseases was first suggested by studies on experimental autoimmune encephalomyelitis (EAE), the animal model of MS, in which neutrophils delayed the onset and decreased the severity of EAE [3–6].

The role of the neutrophil-to-lymphocyte ratio (NLR) of the peripheral blood in several autoimmune diseases has recently been suggested as a potential cheap and effective surrogate marker for the systemic inflammatory state and thus disease activity [5]. Few studies have addressed the significance of the NLR in terms of being a marker of disease activity in MS [7–9].

The aim of the present study was to investigate the significance of the NLR in reflecting the level of disease activity in a cohort of early diagnosed relapsing–remitting MS (RRMS) patients.

2. Materials and Methods

The iMED© software (6.5.6, Merck Serono SA, Geneva, Switzerland) was used as the data entry portal, and we followed a rigorous quality assurance procedure to ensure data quality.

The inclusion criteria were: 1) in the age range of 18 to 55 years old; 2) a diagnosis of RRMS according to the 2017 McDonald criteria (retrospectively evaluated at the end of the diagnostic-iter) [10].

The exclusion criteria were: 1) short-term steroid use in the last 30 days; 2) recent infections (≤1 month); 3) stressful concomitant events in the last 6 months (e.g., traumatic bone fractures); 4) a history of tumors; 5) pregnancy; 6) confirmed autoimmune comorbidities (rheumatoid arthritis, psoriasis, Sjögren syndrome, etc.) (Figure 1).

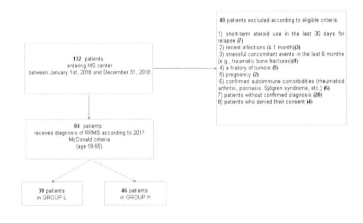

Figure 1. Patients' selection flow chart. RRMS; relapsing–remitting multiple sclerosis.

2.1. Clinical Data

Data were collected from each patient at the time of enrollment, defined as the first access to a diagnostic MS center prior to setting a confirmed MS diagnosis or starting any disease-modifying therapy (DMT).

The data collected were the following: demographic data, including patient age and gender, clinical data including the date of MS onset, type of MS, number of relapses in the year prior to diagnosis, and level of disability assessed by the Expanded Disability Status Scale (EDSS) in the year prior to diagnosis [11]. As a radiological measure of disease activity, we considered the number of brain and spinal lesions detected on T1 gadolinium (Gd) magnetic resonance imaging (MRI) sequences (Ingenia 1.5 Tesla-Philips®). Lesions and imaging were compared with the support of available software (Carestream® Healthcare Information Systems software).

Patients consecutively included in the present study were divided into two groups—low (L) and high disease activity (H), according to the level of disease activity in the year prior to diagnosis, and considering the definition of highly active MS in patients naïve to DMT [12–15].

High disease activity at onset was classified as: ≥2 relapses in the year prior to study entry and ≥1 Gd-enhancing lesion at the time of study [12–15].

2.2. Blood Tests

Blood cell counts of neutrophils and lymphocytes were obtained via routine blood sampling, as part of the clinical practice in each enrolled patient. The last blood test values from immediately prior to commencement of the first DMT were used, and blood tests taken within 30 days of the start of short-term steroid treatment for relapse were excluded.

Blood sampling was performed between 8.00 and 8.30 a.m. and collected in dipotassium-ethylenediaminetetraacetic acid tubes. Samples were analyzed within 40 min to 1 h of collection. A complete blood count (CBC) was obtained using a blood count machine (ABX micros 60, Horiba medical®).

2.3. Outcomes

The primary aim of the present study was to investigate, in a real-world setting, whether the NLR is related to high disease activity at the time of disease onset.

2.4. Ethical Standards

The study protocol was approved by the Local Ethics Committee (Comitato Etico Catania 1, n.17/2019/PO). All patients provided written informed consent. The present study was conducted in accordance with the ethical principles of the Declaration of Helsinki and appropriate national regulations.

2.5. Statistical Analysis

All patient characteristics summary statistics are reported in terms of frequencies (%) for categorical variables, mean ± standard deviation (SD), or median with interquartile range (IQR) for continuous variables. The NLR was calculated as the quotient between the neutrophil and lymphocyte cell counts in blood. The distribution of the NLR was, as expected for a quotient, highly skewed. Therefore, in order to correct for skewness, a log transformation of the NLR was considered. The resulting transformed data show an approximatively Gaussian distribution. Normality assumption has been assessed by using both visual methods (Q-Q plot) and the Kolmogorov–Smirnov (K–S) normality test. Associations between qualitative variables were analyzed using Pearson's X^2 tests. In order to express results as a percentage of the maximum possible variation, the V-Cramer index was used instead of the X^2 test. This index can vary between 0 (in the case of independence) and 1 (in the case of maximum association). Several associations were evaluated: between the log-NLR (ln-NLR) and gender, age, EDSS value, and disease activity, respectively. Each variable was treated as categorical according to literature cut-off and the quantile of the observed distributions. In particular, the ln-NLR was dichotomized as 0 for ln-NLR values of ≤0.6 and 1 when ln-NLR was >0.6; gender was defined as 0 for Male and 1 for Female; 40 was the cut-off used for age (0 when age was ≤40 and 1 when age was >40); the EDSS categories were 0 (when the EDSS value was ≤3.5) and 1 (when the EDSS value was >3.5); disease activity as 0 (group L/low disease activity) and 1 (group H/high disease activity). A logistic regression model was used to study the relationship between disease activity (0 = group L, 1 = group H) and the following variables: gender, age (expressed as a continuous variable), ln-NLR expressed as a continuous variable, and the EDSS value. Logistic regression assumptions were verified; the outcome was a binary variable and no high correlations among predictors were observed. All analysis was performed using the SPSS version 21 statistical software (IBM SPSS Statistics 21, IBM, Armonk, NY, USA).

3. Results

Among a total sample of 132 RRMS patients, 84 fulfilled the required inclusion criteria and were enrolled in the present study (Figure 1). Of these, 38 were in group L and 46 in group H. Patient demographic and clinical characteristics are shown in Table 1. Patients in group H showed a higher EDSS score at disease onset ($p = 0.020$). Blood tests showed that the ln-NLR ratio was highest in patients in group H (1 ± 0.4 vs. 0.7 ± 0.4, $p = 0.032$) (Figure 2).

Table 1. Demographical and clinical characteristics of the two groups at disease onset.

	Group L (n = 38)	Group H (n = 46)	P-Value
Men	19 (50%)	19 (41.3%)	0.42
Women	19 (50%)	27 (58.7%)	
AGE	43 ± 13.4	36.9 ± 12.5	0.23
L%	29 ± 8.5	27.8 ± 6.6	0.95
N%	62.3 ± 7.7	63.3 ± 7.7	0.57
lnNLR	0.7 ± 0.4	1 ± 0.4	0.032
EDSS value	1.5 (1.0–3.0)	2.5 (1.0–5.0)	0.020

Results are expressed as mean ± SD, median (IQR) and No. (%). IQR = interquartile range; SD = standard deviation. Abbreviations: EDSS, Expanded Disability Status Scale; L, lymphocytes; N, neutrophils; ln-NLR, logarithmic-neutrophils/lymphocyte ratio.

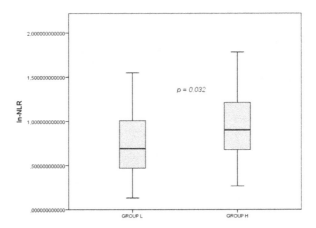

Figure 2. Box plot of the ln-NLR in the two groups. Group L, Low disease activity; Group H, High disease activity; ln.NLR, log transformation of the neutrophil to lymphocyte ratio.

The association analysis revealed a V-Cramer of 0.271 with a p-value of 0.013 between disease activity and the ln-NLR value (Figure 3). No associations were found between the ln-NLR value and other demographical and clinical variables (Table 2). In the regression logistic regression model, the relation between the highest ln-NLR and disease activity was maintained ($p = 0.03$ expB 3.5, CI 95% 1.089–11.4) (Table 3).

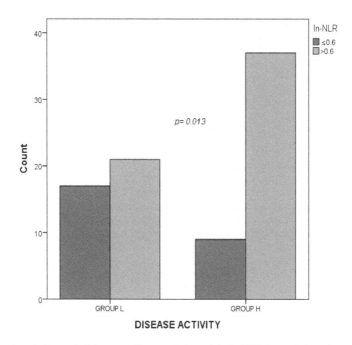

Figure 3. Association analysis between disease activity and the ln-NLR. Group L, Low disease activity; Group H, High disease activity; ln.NLR log transformation of the neutrophil to lymphocyte ratio.

Table 2. Association analysis between the ln-NLR and demographical/clinical parameters.

Patient Characteristics vs. ln-NLR	V-Cramer Index	P-Value
Gender	0.016	0.289
Age	0.052	0.637
Disease activity	0.271	*0.013*
EDSS value at onset	0.062	0.567

Table 3. Logistic regression model for disease activity at onset.

	Exp B	P-Value	Confidence	Interval (95%)
			Lower	Upper
Gender	0.887	0.778	0.352	2.18
Age	1.00	0.555	1.00	1.00
ln-NLR	3.5	0.03	1.08	11.4
EDSS value at onset	1.00	0.926	1.00	1.00

4. Discussion

We found that a higher NLR value increased the risk of disease activity in our cohort. The NLR is employed as a marker of systemic inflammation in other autoimmune diseases and cancers, since it is a low-cost analysis [5,6]. A study using the EAE animal model of MS showed that neutrophil depletion led to a reduction in the severity of the disease [16].

Neutrophils are short-lived, bone-marrow-derived cells, and are the most abundant peripheral blood leucocytes. The number and lifespan of neutrophils are under tight control. Neutrophils are phagocytic and microbiocidal, releasing tissue-remodeling enzymes and reactive oxidative species, as well as proinflammatory cytokines and chemokines, which can provide a potential source of autoantigens, triggering autoimmunity [3,5]. In other fields, such as rheumatoid arthritis, several studies have suggested that joint destruction and disease activity are directly correlated with the recruitment of neutrophils in the synovium [17].

Few studies have investigated the association of the NLR with disease activity in MS patients in a real-world setting. The role of neutrophil counts in MS is a recent and interesting challenge, focusing on the involvement of the innate immune system [18].

Findings are available for MS and optic neuritis (ON) patients as compared with healthy controls (HC) in a large Danish cohort. The NLR was found to be higher in MS and ON patients as compared to HC, indicating the occurrence of chronic inflammation. The NLR may be an inexpensive and easily accessible supplemental marker of disease activity in RRMS patients, revealing an association between a high NLR value and MS occurrence as compared with controls [7,9,18]. Furthermore, a higher NLR during relapse as compared to that during remission was shown in a study by Demirci et al., in which the NLR predicted activity with 67% sensitivity and 97% specificity [7], indicating the value of NLR use in measuring disease activity [7]. Furthermore, NLR has also been investigated as a possible marker relating to the depression, anxiety, and stress score in MS patients [19].

Naegele et al. demonstrated that the increase in neutrophil count in RRMS patients is most likely due to a decrease in apoptosis, and that neutrophils have an altered cell surface expression of certain proteins, which may enhance recruitment to sites of inflammation [4].

However, there are a lack of available data regarding the prediction of long-term disease progression, according to a high NLR ratio. Two previous studies found that the NLR correlates with the EDSS value [7,8], whilst another found no association [18]. Interestingly, Hemond et al. investigated the association of the NLR ratio with clinical, neuroimaging, and psycho-neuro-immunological associations in a large cohort of MS patients, also including patient-reported outcomes. Here, NLR strongly predicted increased MS-related disability independent of all demographic, clinical, treatment-related, and psychosocial variables ($p < 0.001$) [20]. Moreover, Giovannoni et al. failed to find a correlation between the mean levels of several proinflammatory immunological markers and clinical disease progression [21]. Therefore, to further clarify whether the NLR is associated with MS disability, studies with long follow-up times are needed, since this would enable measurement of disease progression. The use of a marker obtained from a simple routine blood exam could make the NLR ratio a useful supplemental biomarker of disease activity in MS patients in clinical practice.

The present study has some limitations, many of which are related to the retrospective design and small sample size. The results build on simple data from routine laboratory tests, and, therefore, may be biased toward the availability of certain laboratory data. Furthermore, there exists no data regarding concomitant diseases that could alter the neutrophil count, or other data such as smoking status. Prior to any possible use of the NLR in clinical practice, more data, preferably from multi-center, long-term studies, are needed. Studies comparing the NLR with MRI and cerebrospinal fluid markers would also be desirable. Additionally, the impact of disease-modifying drugs on the NLR is required to establish the clinical application of this parameter. The NLR has gained academic relevance, contributing to the knowledge of MS immunology.

Author Contributions: Conceptualization, E.D., A.Z., A.R., G.A.M.P. and F.P.; Formal analysis, A.Z.; Investigation, G.A.M.P.; Methodology, E.D., A.Z., M.S. and A.R.; Resources, F.P.; Supervision, E.D., M.S., G.A.M.P. and F.P.; Visualization, F.P.; Writing—original draft, E.D.; Writing—draft and editing, M.S.

Funding: This research received no external funding.

Acknowledgments: Authors are grateful to Laura Parrinello for her contribute as co-investigator.

Conflicts of Interest: Emanuele D'Amico has nothing to disclose related to the manuscript. Aurora Zanghì has nothing to disclose related to the manuscript. Alessandra Romano has nothing to disclose related to the manuscript. Mariangela Sciandra has nothing to disclose related to the manuscript. Giuseppe Alberto Maria Palumbo has nothing to disclose related to the manuscript. Francesco Patti has nothing to disclose related to the manuscript.

References

1. Reich, D.S.; Lucchinetti, C.F.; Calabresi, P.A. Multiple Sclerosis. *N. Engl. J. Med.* **2018**, *378*, 169–180. [CrossRef] [PubMed]
2. D'Amico, E.; Zanghì, A.; Gastaldi, M.; Patti, F.; Zappia, M.; Franciotta, D. Placing CD20-targeted B cell depletion in multiple sclerosis therapeutic scenario: Present and future perspectives. *Autoimmun. Rev.* **2019**, *18*, 665–672. [CrossRef] [PubMed]
3. Woodberry, T.; Bouffler, S.E.; Wilson, A.S.; Buckland, R.L.; Brustle, A. The Emerging Role of Neutrophil Granulocytes in Multiple Sclerosis. *J. Clin. Med.* **2018**, *7*, 511. [CrossRef] [PubMed]
4. Naegele, M.; Tillack, K.; Reinhardt, S.; Schippling, S.; Martin, R.; Sospedra, M. Neutrophils in multiple sclerosis are characterized by a primed phenotype. *J. Neuroimmunol.* **2012**, *242*, 60–71. [CrossRef] [PubMed]
5. Wang, X.; Qiu, L.; Li, Z.; Wang, X.Y.; Yi, H. Understanding the Multifaceted Role of Neutrophils in Cancer and Autoimmune Diseases. *Front. Immunol.* **2018**, *9*, 2456. [CrossRef] [PubMed]
6. Casserly, C.S.; Nantes, J.C.; Hawkins, R.F.W.; Vallieres, L. Neutrophil perversion in demyelinating autoimmune diseases: Mechanisms to medicine. *Autoimmun. Rev.* **2017**, *16*, 294–307. [CrossRef] [PubMed]
7. Demirci, S.; Demirci, S.; Kutluhan, S.; Koyuncuoglu, H.R.; Yurekli, V.A. The clinical significance of the neutrophil-to-lymphocyte ratio in multiple sclerosis. *Int. J. Neurosci.* **2016**, *126*, 700–706. [CrossRef] [PubMed]
8. Guzel, I.; Mungan, S.; Oztekin, Z.N.; Ak, F. Is there an association between the Expanded Disability Status Scale and inflammatory markers in multiple sclerosis? *J. Chin. Med. Assoc. JCMA* **2016**, *79*, 54–57. [CrossRef]
9. Hasselbalch, I.C.; Sondergaard, H.B.; Koch-Henriksen, N.; Olsson, A.; Ullum, H.; Sellebjerg, F.; Otural, A.B. The neutrophil-to-lymphocyte ratio is associated with multiple sclerosis. *Mult. Scler. J. Exp. Transl. Clin.* **2018**, *4*. [CrossRef]
10. Thompson, A.J.; Banwell, B.L.; Barkhof, F.; Carroll, W.M.; Coetzee, T.; Comi, G.; Correale, J.; Fazekas, F.; Filippi, M.; Freedman, M.S.; et al. Diagnosis of multiple sclerosis: 2017 revisions of the McDonald criteria. *Lancet Neurol.* **2018**, *17*, 162–173. [CrossRef]
11. Kurtzke, J.F. Rating neurologic impairment in multiple sclerosis: An expanded disability status scale (EDSS). *Neurology* **1983**, *33*, 1444–1452. [CrossRef]
12. Derfuss, T.; Bergvall, N.K.; Sfikas, N.; Tomic, D.L. Efficacy of fingolimod in patients with highly active relapsing–remitting multiple sclerosis. *Curr. Med. Res. Opin.* **2015**, *31*, 1687–1691. [CrossRef] [PubMed]
13. Fernández, Ó. Is there a change of paradigm towards more effective treatment early in the course of apparent high-risk MS? *Mult. Scler. Relat. Disord.* **2017**, *17*, 75–83. [CrossRef] [PubMed]
14. Hutchinson, M.; Kappos, L.; Calabresi, P.A.; Confavreux, C.; Giovannoni, G.; Galetta, S.L.; Havrdova, E.; Lublin, F.D.; Miller, D.H.; O'Connor, P.W.; et al. The efficacy of natalizumab in patients with relapsing multiple sclerosis: Subgroup analyses of AFFIRM and SENTINEL. *J. Neurol.* **2009**, *256*, 405–415. [CrossRef] [PubMed]
15. Krieger, S.; Singer, B.; Freedman, M.; Lycke, J.; Berkovich, R.; Margolin, D.; Thangavelu, K.; Havrdova, E. Treatment-Naive Patients with Highly Active RRMS Demonstrated Durable Efficacy with Alemtuzumab over 5 Years (S51.003). *Neurology* **2016**, *86*, S51.003.
16. Zeng, Z.; Wang, C.; Wang, B.; Wang, N.; Yang, Y.; Guo, S.; Du, Y. Prediction of neutrophil-to-lymphocyte ratio in the diagnosis and progression of autoimmune encephalitis. *Neurosci. Lett.* **2019**, *694*, 129–135. [CrossRef]
17. Erre, G.L.; Paliogiannis, P.; Castagna, F.; Mangoni, A.A.; Carru, C.; Passiu, G.; Zinellu, A. Meta-analysis of neutrophil-to-lymphocyte and platelet-to-lymphocyte ratio in rheumatoid arthritis. *Eur. J. Clin. Investig.* **2019**, *49*, e13037. [CrossRef]
18. Bisgaard, A.K.; Pihl-Jensen, G.; Frederiksen, J.L. The neutrophil-to-lymphocyte ratio as disease actvity marker in multiple sclerosis and optic neuritis. *Mult. Scler. Relat. Disord.* **2017**, *18*, 213–217. [CrossRef]
19. Al-Hussain, F.; Alfallaj, M.M.; Alahmari, A.N.; Almazyad, A.N.; Alsaeed, T.K.; Abdurrahman, A.A.; Murtaza, G.; Bashir, S. Relationship between Neutrophil-to-Lymphocyte Ratio and Stress in Multiple Sclerosis Patients. *J. Clin. Diagn. Res. JCDR* **2017**, *11*, CC01–CC044.

20. Hemond, C.C.; Glanz, B.I.; Bakshi, R.; Chitnis, T.; Healy, B.C. The neutrophil-to-lymphocyte and monocyte-to-lymphocyte ratios are independently associated with neurological disability and brain atrophy in multiple sclerosis. *BMC Neurol.* **2019**, *19*, 23. [CrossRef]
21. Giovannoni, G.; Miller, D.H.; Losseff, N.A.; Sailer, M.; Lewellyn-Smith, N.; Thompson, A.J.; Thompson, E.J. Serum inflammatory markers and clinical/MRI markers of disease progression in multiple sclerosis. *J. Neurol.* **2001**, *248*, 487–495. [CrossRef] [PubMed]

© 2019 by the authors. Licensee MDPI, Basel, Switzerland. This article is an open access article distributed under the terms and conditions of the Creative Commons Attribution (CC BY) license (http://creativecommons.org/licenses/by/4.0/).

Review

Memory CD4$^+$ T Cells in Immunity and Autoimmune Diseases

Itay Raphael [1,*], **Rachel R. Joern** [2] **and Thomas G. Forsthuber** [2,*]

[1] Department of Neurological Surgery, University of Pittsburgh, UPMC Children's Hospital, Pittsburgh, PA 15224, USA
[2] Department of Biology, University of Texas at San Antonio, San Antonio, TX 78249, USA; Rachel.Joern@utsa.edu
* Correspondence: i.raphael@pitt.edu (I.R.); thomas.forsthuber@utsa.edu (T.G.F.)

Received: 28 January 2020; Accepted: 20 February 2020; Published: 25 February 2020

Abstract: CD4$^+$ T helper (Th) cells play central roles in immunity in health and disease. While much is known about the effector function of Th cells in combating pathogens and promoting autoimmune diseases, the roles and biology of memory CD4$^+$ Th cells are complex and less well understood. In human autoimmune diseases such as multiple sclerosis (MS), there is a critical need to better understand the function and biology of memory T cells. In this review article we summarize current concepts in the field of CD4$^+$ T cell memory, including natural history, developmental pathways, subsets, and functions. Furthermore, we discuss advancements in the field of the newly-described CD4$^+$ tissue-resident memory T cells and of CD4$^+$ memory T cells in autoimmune diseases, two major areas of important unresolved questions in need of answering to advance new vaccine design and development of novel treatments for CD4$^+$ T cell-mediated autoimmune diseases.

Keywords: CD4$^+$ T cells; memory T cells; autoimmune disease; effector memory T cell; central memory T cell; tissue-resident T cell

1. Introduction

CD4$^+$ T helper (Th) cells play a central role in the immune system and carry out multiple functions including activation, coordination, modulation, and regulation of innate and adaptive immune responses. These various functions of Th cells are necessary to attain effective immune responses against a variety of different pathogens, while maintaining self-tolerance and avoiding undesired attacks against self-tissues [1–3]. The regulation of immune responses by Th cells is accomplished through the secretion of specific cytokines, which, together with a "master" regulatory transcription factor, define the respective Th cell subset and its specialized functions and attributes [1–4]. While much is known about the effector function of Th cells, the roles and biology of memory CD4$^+$ T cells are more complex and less well understood. Moreover, the function of memory CD4$^+$ T cells in mounting an immune response can only partially be defined by the precursor Th subset from which the primary immune response originated. An additional layer of complexity is added by the different memory CD4$^+$ T cell subsets generated during the primary immune response [3,5]. Memory T cells are generally subdivided into three main populations: central memory T cells (T$_{CM}$), effector memory T cells (T$_{EM}$), and tissue-resident memory T cells (T$_{RM}$). At present, these memory T cell subsets are primarily characterized by their phenotype, migratory properties, and tissue homing patterns, which in many instances imply unique functional attributes [5–7].

Antigen (Ag)-specific naïve CD4$^+$ T cells are first activated in lymphatic tissues by professional Ag-presenting cells (APCs) presenting their specific antigen on major histocompatibility complex (MHC) class II molecules and providing costimulatory signals, for example in the context of an infection with microbial pathogens. Activated CD4$^+$ T cells proliferate and differentiate into specific

Th subsets, which will mount distinct immune responses directed against specific pathogens [8]. After the infectious pathogen has been cleared, the majority of effector Th cells will undergo apoptosis, while the remaining cells contribute to the CD4$^+$ memory T cell pool [9]. The importance of memory T cell generation centers on the ability to provide a faster and augmented immune response upon secondary exposure to previously encountered microbial pathogens. The ability of memory T cells to respond faster and more efficiently is based on several essential characteristics, which render them superior in their ability to alter the outcome of infections. First, memory T cells have a lower activation threshold and are less co-stimulation dependent [10]. Therefore, upon re-challenge, memory T cells will generate robust effector responses more effectively and faster as compared with a primary T cell response [11–13]. Similar to the effector response generated by recently-activated naïve T cells, the effector response of memory T cells is dependent on the nature and context of their encounter with their cognate antigen, for example provided by cues such as cytokines in the microenvironment [14]. Second, the frequencies of Ag-experienced memory T cells are much higher than those of naïve T cells during a primary immune response [14–16]. The higher frequencies of Ag-specific memory T cells increase their likelihood to encounter their cognate antigen faster upon re-infection, and to more rapidly generate a larger effector T cell pool [17,18]. Third, unlike naïve T cells, which circulate between secondary lymphatic tissues (i.e., lymph nodes) and blood, memory T cells circulate between lymphatic, blood, and peripheral tissues (e.g., lungs, gut, or skin) [19,20]. This allows memory T cells to directly and rapidly respond to the presence of pathogens in peripheral tissues. More recently a subset of non-circulating and tissue-homing memory T cells was identified and termed T_{RM} [21–23]. Both CD8$^+$ and CD4$^+$ T_{RM} have been described [reviewed in 24]. T_{RM} cells migrate to specific peripheral tissues locations, including the skin, liver, and lungs, and take up permanent residency. Importantly, the strategic location of T_{RM} at many barrier sites such as the skin and mucosal tissues, where pathogens preferentially seek access, enhances the likelihood that they rapidly encounter pathogens upon infection. CD8$^+$ T_{RM} have been reviewed elsewhere [24–26] and this review will focus on CD4$^+$ T_{RM}.

Taken together, the unique characteristics of memory T cells enhance their ability for in situ immune surveillance, increase their likelihood for faster encounter of pathogens at the site of infection, and facilitate the generation of more rapid and generally superior effector responses.

2. Memory T Cell Development

2.1. Development Pathways

The developmental pathways of memory T cells are still not fully understood, and there are some controversies as to the different models proposed (reviewed in [27]), and compelling evidence exists for each of the models. The models are shown in Figure 1 and are summarized below.

The linear model of memory T cell development (Figure 1a) suggests that during the contraction phase of a primary T cell immune response, the surviving effector T cells differentiate into T_{EM}, which will then give rise to T_{CM} [27,28]. In contrast, the asymmetrical model of memory T cell development (Figure 1b), also called the bifurcative model, suggests that two daughter cells of the same T cell clone can undergo different fates: the daughter cell proximal to the immunological synapse can give rise to both terminally-differentiated effector cells and T_{EM}, while the distal daughter cell gives rise to T_{CM} [29]. a third model (Figure 1c, self-renewal model) proposes that naïve T cells first give rise to either self-renewing T_{CM} or effector T cells, and that those can further differentiate to T_{EM}, which give rise to terminally-differentiated effector cells in non-lymphoid tissues [27]. Pepper and Jenkins suggested an additional model for the generation of memory CD4$^+$ T cells. In this "simultaneous" model (Figure 1d), the effector T cell subset (e.g., Th1 or Th2) determines the fate of the generated memory T cell subset. For instance, Th1 cell or Th17 cell subsets will give rise to T_{EM}, whereas T follicular helper (T_{FH}) cell subset will generate T_{CM} upon help from B cells [30].

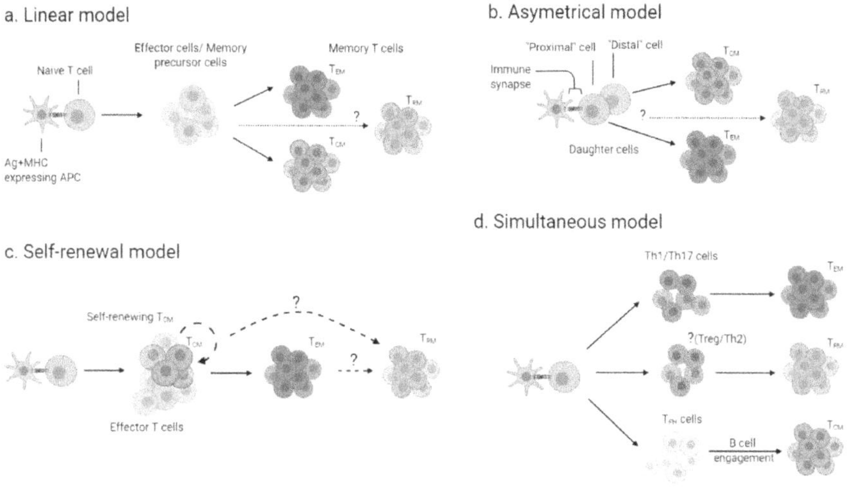

Figure 1. Developmental models of memory T cells: (**a**) In the linear model, effector T cells or memory precursor cells (yellow) are generated following activation of naïve T cell by antigen-presenting cells (APCs) presenting peptide on major histocompatibility complex (MHC) molecules. The intermediate effectors or memory precursors give rise to mature effector memory T cells (T_{EM}) (red) and central memory T cells (T_{CM}) (blue). It remains to be answered if the intermediate effectors/precursors also give rise to tissue-resident memory T cells (T_{RM}) (green). (**b**) In the asymmetrical model, the proximal daughter cells to the immune-synapse (naïve T cell- T cell receptor (TCR) + peptide and MHC-APC) develop into T_{EM}, while the distal daughter cells develop into T_{CM}. It is currently unknown which cells give rise to T_{RM} (green). (**c**) In the self-renewal model self-renewing effector T cells or T_{CM} are generated from naïve T cells. These self-renewing cells can then give rise to T_{EM} cells. It is unresolved if T_{RM} are generated from self-renewing T_{CM}/effector cells or from T_{EM}. (**d**) In the simultaneous model naïve T cells first differentiate into different T cell subsets. T cell subsets give rise to different memory subsets as follows: Th1 and Th17 cells (dark gray) generate T_{EM}, while T_{FH} cells (light gray) generate T_{CM}. The T helper cell subset(s) that generate T_{RM} has not yet been identified.

2.2. Role of T Cell Receptor Signaling Strength and Precursor Frequencies for Memory T Cell Development

Central to memory T cell development are the roles of T cell receptor (TCR) signaling strength and T cell precursor frequencies. The TCR signaling strength is determined by the affinity of the TCR for Ag and MHC molecules, the density of antigen presented on APCs, and the duration of the TCR interaction with peptide-loaded MHC [31,32]. The degree of these parameters affects TCR-dependent biochemical pathways which result in changes to T cell transcriptional profiles, to either promote or inhibit memory T cell generation [33,34]. This appears to be regulated via differences in the ratios of several transcription factors, including Bcl-6, Blimp-1, Eomes, and T-bet [33,35,36], their upstream regulators nuclear factor of activated T-cells (NFAT) and nuclear factor kappa-light-chain-enhancer of activated B cells (NF-κB), or by directly affecting the expression of cytokines and cytokine receptors, such as IL-2 and the IL-2 receptor alpha chain (CD25), which synergize with TCR signaling to affect memory-cell development [35,37,38]. For memory CD8$^+$ T cells, it is generally accepted that TCR signaling strength inversely correlates with memory-cell formation [33]. However, for the generation of memory CD4$^+$ T cells the data are less clear, as described below [38–41].

CD4$^+$ T cells have been shown in several systems to compete for Ag, and a higher TCR affinity for Ag may affect the access of T cells to Ag [42–44]. Thus, TCRs with higher affinity for Ag are likely to promote greater expansion of Ag-specific T cells, and thereby increase the reservoir of T cells that could potentially enter the memory T cell pool. Furthermore, stable and sustained interaction with antigens is a critical determining factor for promoting the differentiation of memory CD4+ T cells during acute viral infection [45]. In contrast, low affinity TCRs for MHC plus Ag results in T cell memory with shorter lifespan and with impaired secondary responses upon re-challenge [46,47]. Not only may TCR affinity for MHC and Ag influence the generation of memory T cells, but it may also affect their recruitment into different memory T cell subsets and their lifespan. McKinstry et al. reported that CD4+ T cells require signals from MHC class-II molecules and CD70 during the effector phase of an immune response, and are dependent on IL-2 signaling in order to generate long-lived memory cells [48]. It has been proposed that T_{CM} may require relatively stronger TCR stimuli for their generation, while relatively weaker antigenic stimuli may generate T_{EM}. Consistent with this notion, naïve CD4+ T cells require prolonged exposure to antigen during the expansion phase to generate T cell memory [39].

However, memory T cell formation is not only dependent on TCR affinity/antigen accessibility, but also on the frequencies of naïve T cell precursors [49]. The naïve T cell precursor frequency is defined by the fraction of naive T cells capable of responding to a specific antigenic epitope and to enter the effector/memory Ag-specific T cell pool. For example, excessively high precursor frequencies of Ag-specific T cells seem to have a negative impact on memory T cell formation. Accordingly, high precursor frequencies of adoptively-transferred TCR-transgenic CD4+ T cells reduced the proliferation and differentiation of these cells upon infection, and thereby resulted in impaired memory T cell formation [50]. This observation is further supported by the inverse relationship between T cell precursor frequencies and the survival of both naïve and memory CD4+ T cells [51]. Similarly, Blair and Lefrançois showed that transfer of high precursor frequencies of TCR-transgenic naïve CD4+ T cells resulted in a lack of T cell memory, which was linked to impaired effector T cell induction, reduced proliferation, and cytokine production [52]. Importantly, this phenomenon was independent of IL-7R expression by the responding memory T cells. Additionally, they showed that competition for antigen during CD4+ T cell priming is a major confounding factor for the development of the memory T cell pool [52]. However, in these studies, during the initial priming of naïve CD4+ T cells, the availability of Ag rather than the frequency of precursor cells per se appeared to be pivotal for the formation of CD4+ T cell memory. Nevertheless, Ag persistence negatively affects the function of CD4+ memory T lymphocytes and impairs their ability to produce effector cytokines, perhaps by promoting the generation of T_{CM} rather than T_{EM} [53].

2.3. Role of Transcription Factors and Cytokine Signaling for Memory T Cell Development

T-bet, the master-regulator of Th1 cells, acts in an expression-level-dependent manner to regulate formation of memory and effector CD4+ T cells, such that high expression of T-bet promotes terminal effector cells, while intermediate to low expression promotes memory development [36]. Furthermore, T-bet appears to regulate the generation of memory T cell subsets from effector cells: T-bet^{low} effector cells express the chemokine receptor (CCR) 7 and give rise to T_{CM} cells, whereas T-bet^{high} effector cells rapidly produce interferon (IFN)-γ, lack CCR7 expression and give rise to T_{EM} cells [54]. Similar to CD4+ memory T cells, high levels of T-bet expression by CD8+ effector T cells generate short-lived memory cells and favor terminally-differentiated effector cells, while low expression levels generate long-lived memory cells [55]. Along these lines, the cytokine IFN-γ, which induces T-bet expression in a signal transducer and activator of transcription (STAT) 1-dependent manner, has been shown to enhance the development of memory CD4+ T cells generated from both naïve and effector cells [47,56,57]. Additionally, IFN-γ producing effector cells give rise to long-lived T_{CM} and T_{EM} [58]. Thus, it appears that effector cytokines present during the primary response, and in particular IFN-γ, may tip the balance in favor of the generation of specific memory T cell subsets. Moreover, since T_{EM} cells are generated from progenitor cells which express a master regulator (e.g., T-bet for Th1), it appears that T_{EM} cells maintain lineage integrity, whereas T_{CM} cells that are generated from progenitor cells that lack (or express at low levels) a master regulator show more lineage plasticity and can generate different effector responses upon re-challenge [30,54]. Similarly, IL-12 induces the expression of T-bet in a STAT4-dependent manner in CD4$^+$ T cells [59], and IL-12 promotes development of hepatitis B virus (HBV)-specific CD8$^+$ T$_{EM}$ cells [60]. Along these lines, although no data exists for the role of IL-12 in development of CD4$^+$ memory T cells, mice lacking both STAT4 and T-bet have marginally reduced virus-specific CD4$^+$ memory T cell frequencies as compared with mice lacking either one of these transcription factors (STAT4 or T-bet) [61]. These data suggest that early activation of STAT4 and T-bet may not be required for development of CD4$^+$ T cell memory; however, their expression may affect the quality of memory responses during recall and the type (subset) of memory generated. Although CD4$^+$ T$_{RM}$ were reported to express low levels of T-bet [62], suggesting that this pathway is dispensable for CD4 T$_{RM}$ maintenance, future work should investigate the role of T-bet-promoting cytokines for regulating CD4$^+$ T$_{RM}$ development, since this has not yet been fully resolved.

IL-2, via regulation of Bcl-6 expression, also plays fundamental roles in memory CD4 T cell development [63–65]. Current dogma suggests that naïve CD4$^+$ T cells require high levels of IL-2 to differentiate into memory precursor cells; however, these precursor cells are then dependent on low levels of IL-2 to develop into long-lived memory cells. IL-2 signals sustain the expression of STAT5 and Bcl-6, and promote expression of IL-7 receptor, thereby enabling the survival and maturation of CD4$^+$ memory T cells [48,66]. Interestingly, however, IL-2-deficient memory CD4$^+$ T cells generate a more vigorous and effective recall response against influenza virus than wild-type memory cells that produce IL-2 [67], suggesting that IL-2 and Bcl-6 may have regulatory or inhibitory roles once Ag-specific T cell memory has established, perhaps to prevent immunopathology during chronic antigen exposure. Of note, IL-2, IL-7, and IL-15, also known as common-gamma chain receptor cytokines, play key roles in memory (as well as in naïve and effector) CD4$^+$ and CD8$^+$ T cell development, homeostasis, and maintenance [68–70]. The functions of these cytokines in CD4$^+$ T cell memory remain somewhat controversial, and this has been reviewed elsewhere [10,68].

Shown in Figure 2 is the integration of TCR signaling- and inflammation-dependent factors influencing memory T cell development.

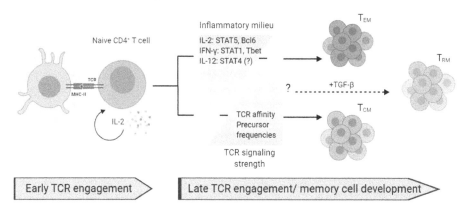

Figure 2. Factors influencing memory T cell development: Naïve T cells are activated by peptide and MHC via TCR. High levels of interleukin (IL)-2 are critical during early TCR engagement for memory-cell development. The memory T cell development fate is dependent on the inflammatory milieu and TCR signaling strength. Increased levels of inflammatory signals favor T$_{EM}$ generation and decreased levels of these signals favor T$_{CM}$. Conversely, increased levels of TCR affinity and precursor frequencies favor T$_{CM}$ development while decreased levels favor T$_{EM}$. TGF-β is important for generating T$_{RM}$. How other inflammatory signals and TCR signaling strength affect T$_{RM}$ generation remains unresolved.

Naturally, models of T cell memory development became more complex with the discovery of T$_{RM}$, and there are still many open questions, in particular for the generation of CD4$^+$ T$_{RM}$ [15,71].

Nevertheless, it is widely accepted that T_{RM} require transforming growth factor (TGF)-β for their generation, which can be produced by a variety of cells in tissues, including fibroblasts, epithelial cells, keratinocytes, and enterocytes [72,73]. While CD8$^+$ T_{RM} have been at the center of research in this area, as much as 20% of CD4$^+$ T cells in certain tissues, such as the intestinal epithelium, also express CD103, the prototypical marker for tissue-resident memory cells [74]. TGF-β induces the expression of CD103 by T_{RM} precursors and promotes their entry and retention in epithelial sites [72,73]. Moreover, a sizable fraction of CD4$^+$ T_{RM} express Foxp3$^+$, a key transcription factor of regulatory T cells, which is induced by TGF-β signaling [75]. Therefore, these data suggest that TGF-β is a pivotal factor in CD4$^+$ T_{RM} generation and that CD4$^+$ T_{RM} may assume regulatory functions [76]. Future investigations will elucidate the mechanisms guiding the generation of CD4$^+$ T_{RM}, as well as characterize their functions.

Taken together, several models have been proposed for the generation of memory T lymphocytes. It is conceivable that more than one of these models contribute to memory T cell development in vivo, and may be influenced by factors such as TCR signal strength and inflammatory conditions in the microenvironment (cytokines), and in addition, costimulatory signals (e.g., CD28 family) [77] and other environmental cues, such as chemokines and even the microbiota from the tissue in which a T cell is activated [78–80].

3. Memory T Cell Subsets and Function

3.1. Memory T Cell Phenotype

Memory T cells are generally subdivided into three main populations: T_{CM}, T_{EM}, and T_{RM}. Human memory T cells share many phenotypic characteristics with other species such as mouse. The phenotype of the different memory T cell subsets is discussed below and summarized in Figure 3.

Human memory T cells can be distinguished from naïve T cells by expression changes of the CD45 isoforms and increased capacity for effector cytokine production upon antigen recall [81–83]. Human memory T cells were initially distinguished from naïve T cells by being CD45RO$^+$ and CD45RA$^-$, whereas naïve T cells show the CD45RO$^-$ CD45RA$^+$ phenotype [84–86]. Sallusto et al. subsequently showed that CD45RA$^-$ memory T cells could be further divided into two subpopulations based on the expression of the chemokine receptor CCR7 [28]. They proposed a model in which CD45RA$^-$ CCR7$^+$ T cells recirculated to lymphatic tissues and termed these cells T_{CM}, whereas CD45RA$^-$ CCR7$^-$ T cells, which remained in the immune periphery, were termed T_{EM} [28,87].

For mice, Bradley et al. showed that mouse naïve and memory T cells can be phenotypically distinguished based on the expression of CD62L (L-selectin) and CD44, and increased expression of effector cytokines upon re-challenge. Naïve cells are CD62L$^+$ and CD44^{low}, whereas memory cells (T_{EM}) are CD62L$^-$ and CD44^{high} [88,89]. Additional differences in cell surface marker expression in mouse (and human) T cells have emerged to allow classification of naïve vs. memory T cells, such as CD69 and chemokine receptors [90–92]. Reinhardt et al. showed that the CD4$^+$ T_{CM} and T_{EM} paradigm also translates to mice, and that these memory-cell subsets can be distinguished based on the expression of CD62L and CCR7 [87,93,94]. Thus, both mouse and human T_{CM} are CD62L$^+$ CCR7$^+$, while T_{EM} are CD62L$^-$ CCR7$^-$ [95]. Therefore, in addition to classifying memory vs. naïve T cells based on CD45 isoforms, both human and mouse naïve T cells can be characterized by a CD44^{low} CCR7$^+$ CD62L^{high} phenotype, whereas T_{CM} are CD44^{high} CCR7$^+$ CD62L^{high}, and T_{EM} cells are CD44^{high} CCR7$^-$ CD62L^{low} [96,97]. Unlike human T cells, mouse naïve and memory T cells cannot be distinguished based on the expression of CD45 isoforms [98].

Recently, Lefrançois and colleagues described an additional memory T cell subset, termed T_{RM} [99]. This subset resides in peripheral tissues during or after infectious encounters and does not recirculate between blood, lymphatics, or other peripheral tissues, as do T_{CM} or T_{EM} [100–102]. T_{RM} constitutively express CD69, CD103 (integrin alpha E, ITGAE; most prominent in CD8$^+$ T_{RM}), and S1PR1 (sphingosine-1-phosphate receptor 1), but they do not express CCR7 or CD62L [23,24,102]. The expression of CD69, CD103, and S1PR1 by T_{RM}, together with the absence of CCR7 expression, promotes their tissue homing and impedes their tissue egress [24,100,101]. Interestingly, the expression of CD69 and CD103 by T_{RM} is independent of TCR signaling and Ag persistence, and may be dependent on constitutive signaling of "alarm cytokines" such as IL-33 and type I interferons [103,104]. Human T_{RM} are also CD45RO$^+$ CD45RA$^-$ [105]. Interestingly, Kumar et al. recently identified a core transcriptional profile for human CD4$^+$ and CD8$^+$ T_{RM} at various sites that shows increased expression of specific adhesion molecules (such as CD103) and production of both pro-inflammatory and regulatory cytokines and chemokines (such as IL-2 and IL-10) [106].

Recently, the dogma of T_{RM} as a non-circulating subset was challenged by identifying a population of circulating CD4$^+$ T_{RM} in human blood, which was designated as "ex-T_{RM}" [107]. These CD4$^+$ ex-T_{RM} express the skin homing and retention glycan cutaneous lymphocyte-associated antigen (CLA), have similar phenotypic and transcriptional attributes as skin resident CD4$^+$ CD103$^+$ CLA$^+$ T_{RM}, and share a clonal origin with CD4$^+$ CD103$^+$ CLA$^+$ T_{RM} in the skin, based on TCR sequencing [108]. Thus, these new data suggest that CD4$^+$ T_{RM} may reside in tissues for prolonged periods of time, but possibly not for all of their lifespan.

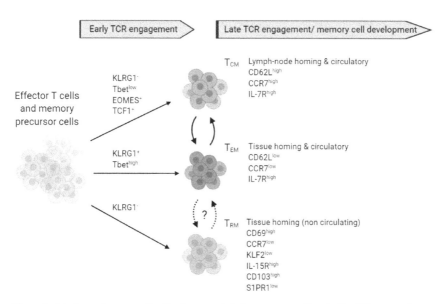

Figure 3. Markers of memory T cell subsets and their precursors: Terminally differentiated T_{CM} (blue) and T_{EM} (red) can be distinguished based on the expression of CD62L, CCR7, IL7R, and other markers not shown. T_{RM} can be further characterized based on the expression of KLF2, IL-15R, CD103, and S1PR1. T_{CM} can give rise to T_{EM} and vice versa. *Recent data suggested that T_{RM} can re-enter the circulation.

3.2. Memory T Cell Subset Function

All memory CD4$^+$ T cell subsets play a pivotal role in defending against pathogens [109–111]. However, their individual contributions vary, which is to some degree a function of their respective migratory properties and tissue homing [112]. Along these lines, the expression of CD62L and CCR7 by T_{CM} endows them with the ability to migrate and home to secondary lymphatic tissues, and thereby facilitates immune surveillance of antigens collected via lymphatic drainage or dendritic cells (DCs) from peripheral tissues [113,114]. The frequencies of Ag-specific T_{CM} are as much as thousand-fold increased as compared with Ag-specific naïve T cells, and thereby they are able to rapidly generate a robust effector T cell pool upon secondary encounter of cognate antigens [16]. Moreover, while T_{CM} show a lower capacity for the production of some cytokines, such as IFN-γ and IL-4, as compared with T$_{EM}$, they produce more IL-2 and have an overall greater capacity for proliferation [93,115]. Thus, T$_{CM}$ have a higher likelihood to become activated by APCs in lymphatic tissues upon re-infection with a previously encountered pathogen to provide a stronger and more rapid response and proliferate rapidly to generate a large pool of pathogen-specific effector T cells. Furthermore, CD4$^+$ T$_{CM}$ provide superior B cell help, which results in faster B cell expansion, more rapid class switching, and increased antibody production [89,116]. Indeed, CD4$^+$ T$_{CM}$ with T$_{FH}$ cell phenotype persist in germinal centers of draining lymph nodes following vaccination to regulate memory B cell development and maintenance and support rapid generation of long-lived plasma cells upon re-exposure to antigen [117,118].

In contrast to T$_{CM}$, T$_{EM}$ preferentially recirculate between blood and peripheral tissues. As indicated by their designation, T$_{EM}$ rapidly exhibit effector functions, such as the production of cytokines upon activation. Interestingly, IL-1β promotes effector-cytokine production, such as IL-17 and IFN-γ, by CD4 T$_{EM}$ cells by stabilizing the cytokine transcripts upon Ag-encounter [119]. T$_{EM}$ have a longer lifespan as compared with effector T cells and provide a readily-available pool for effector T cells in the immune periphery [113]. Thus, T$_{EM}$ can quickly supply Ag-specific effector T cells in peripheral tissues when the need arises, as compared with the longer time required for the differentiation of naïve T cells into effector T cells.

3.3. CD4$^+$ T$_{RM}$ Subset Function

CD4$^+$ T$_{RM}$, which reside in peripheral tissues, function as the first line of defense at these sites together with T$_{EM}$. However, T$_{RM}$ exhibit some unique functions different from those of T$_{EM}$, and in some cases, show more vigorous responses during secondary Ag-encounter [120–123]. For instance, lung-resident memory CD4$^+$ T$_{RM}$ cells provide optimal protection against secondary respiratory viral challenge with influenza virus, whereas protection provided by influenza-specific circulating memory CD4$^+$ T cells is weaker despite their ability to expand and migrate to the lungs upon infection with the same pathogens [121]. Likewise, optimal protection against *Chlamydia* infection is dependent on the generation of mucosa (genital tract) homing CD4$^+$ T$_{RM}$, while protection provided by circulating memory T cells is less effective [124] suggesting that mucosal CD4$^+$ T$_{RM}$ are critical for optimal protection against pathogens entering via the mucosal entry sites. Of note, tumor-homing CD4$^+$ T$_{RM}$ are more potent producers of TNF and IFN-γ compared with other tumor infiltrating T cells [125]. Additionally, CD4$^+$ T$_{RM}$ directed against certain pathogens emerge and persist in peripheral tissues following infection, such as influenza-virus-specific CD4$^+$ T$_{RM}$ in the lungs and *Leishmania*-specific memory CD4$^+$ T$_{RM}$ in the skin [126,127]. These CD4$^+$ T$_{RM}$ cells rapidly produce effector cytokines such as IFN-γ and IL-17 upon re-challenge [126,127]. Using a new strategy for mucosal vaccination, Stary et al. showed that IFN-γ producing CD4$^+$ T$_{RM}$ are pivotal for protection against *Chlamydia trachomatis* [124,128]. Similar findings have been reported for genital tract herpes simplex virus (HSV) vaccination [120], and gastric subserous vaccination with *Helicobacter pylori* vaccine [129]. Therefore, identification of mechanisms which promote the generation and retention of CD4$^+$ T$_{RM}$ should be further explored for development of more effective vaccines against a range of human pathogens [130]. Of note, female lower genital tract CD4$^+$ T$_{RM}$ were identified to serve as primary targets of HIV infection and persistence, thus providing an HIV cellular sanctuary [131]. Thus, HIV treatment strategies and vaccines may consider targeting T$_{RM}$ [131].

The mechanisms by which CD4$^+$ T$_{RM}$ provide enhanced protection is an area of intense research, and some evidence suggests that they may differ from those used by circulating effector/memory CD4$^+$ T cells. Along this line, CD4$^+$ T$_{RM}$ provide rapid protection by promoting the recruitment of immune cells into the affected tissues [121,122,132–134]. In addition, CD4$^+$ T$_{RM}$ are important for the maintenance, distribution, and homing of CD8$^+$ T$_{RM}$ in situ [135,136]. Since it was shown that CD4$^+$ T cells can foster the development of lung CD8$^+$ T$_{RM}$ cells during infection with influenza virus [137], it is conceivable that CD4$^+$ T$_{RM}$ may also contribute to the generation of CD8$^+$ T$_{RM}$. Interestingly, CD4$^+$ T$_{RM}$ outnumber CD8$^+$ T$_{RM}$ in many tissues [23,123], suggesting a critical role for CD4$^+$ T$_{RM}$ in tissue-specific immunity and barrier function. For instance, approximately 70% to 85% of total T$_{RM}$ in the human skin are CD4$^+$ cells.

Mechanistically, CD4$^+$ T$_{RM}$ cells in the skin proliferate more extensively and produce significantly higher levels of IFN-γ, TNF, and IL-22 (and to a lesser extent IL-17 and IL-4) as compared with circulating memory CD4$^+$ T cells [123]. In fact, immunosurveillance of non-lymphoid tissues is orchestrated by CD4$^+$ T$_{RM}$ cells rather than by CD8$^+$ T$_{RM}$ [138]. Notably, CD4$^+$ T$_{RM}$ share overlapping transcriptional, phenotypic, and location-specific functional properties with CD8$^+$ T$_{RM}$ and orchestrate local recall responses [138]. In contrast to CD8$^+$ T$_{RM}$, the human skin is populated with CD4$^+$ T$_{RM}$ which are either CD103$^+$ and reside primarily in the epidermis, or CD103$^-$ which mainly reside in the dermis [123]. Interestingly, CD103$^+$ CD4$^+$ T$_{RM}$ in skin show lower proliferative capacity but increased effector function as compared with CD103$^-$ CD4$^+$ T$_{RM}$, independent of their location in the dermis or epidermis [123]. These data suggest that CD103$^+$ and CD103$^-$ CD4$^+$ T$_{RM}$ cells encompass unique functional attributes in which CD103$^+$ T$_{RM}$ cells provide robust effector responses (cytokine production), while CD103$^-$ T$_{RM}$ cells proliferate extensively to supply the Ag-specific CD4$^+$ T$_{RM}$ cell pool. Future studies should investigate the cross-regulation between these two populations and whether CD103$^-$ CD4$^+$ T$_{RM}$ can give rise to CD103$^+$ T$_{RM}$ cells or vice versa. Finally, the generation and retention of skin CD4$^+$ T$_{RM}$ was shown to be dependent on skin-resident CD8$^+$ T cells or CD11b$^+$ skin-resident macrophages [139], adding to the complexity in this system.

In addition to providing enhanced tissue protection, CD4$^+$ T$_{RM}$ have also been implicated in undesired immunopathology of inflammatory diseases and they may contribute to the persistence of inflammatory cells and chronic inflammation in the affected tissues [140–142]. Nevertheless, approximately 10% of CD4$^+$ T$_{RM}$ express the transcription factor Foxp3 and are thought to have regulatory functions [143]. In this context, Foxp3$^+$ CD4$^+$ T cells enter and reside in the skin during the neonatal period and mediate tolerance to commensal, non-pathogenic microbes [144]. Therefore, it will be critical to elucidate the mechanisms of CD4$^+$ T$_{RM}$ developmental pathways, generation and maintenance, and their intersection with anti-microbial, regulatory, or pathologic functions to elicit optimal protection, while avoiding tissue damage.

In summary, memory CD4$^+$ T cell responses and their unique functional attributes provide critical contributions to protection against microbial pathogens [111]. These include increased cytokine production, regulation of innate immune cell functions, mobilization of immune cells to sites of infection, providing B cell help, and enhancing cytotoxic T cell responses [111]. Robust and rapid local responses are provided at entry sites of infection by CD4$^+$ T$_{RM}$, which tailor immune-responses to specific tissues and the local microenvironment by providing local cues, mediating the rapid recruitment of other immune cells, and by regulating and maintaining other tissue-resident cells, including CD8$^+$ T$_{RM}$.

4. Memory CD4$^+$ T Cells in Autoimmunity

Memory CD4$^+$ T cells are of great interest in the context of autoimmune diseases because of their long-lived nature, efficient responses to antigens, and unique potential to mediate recurring autoimmune responses. However, until now, T cell memory has been more extensively investigated in the context of infectious diseases and its role in autoimmune diseases is not fully elucidated. Here, we will summarize and discuss some of the most pertinent findings on memory CD4$^+$ T cells in autoimmune diseases, with special focus on multiple sclerosis (MS) and its animal model experimental autoimmune encephalomyelitis (EAE).

4.1. Persistence of Autoreactive Memory T Cells in Autoimmune Diseases and the Role of Immunoscenescence

Autoreactive memory CD4$^+$ T cells have been studied in patients in several autoimmune disease conditions [145,146]. For instance, patients with MS and psoriasis show increased numbers of memory CD4$^+$ T cells as compared with healthy individuals, suggesting that memory CD4$^+$ T cell are critical mediators of autoimmune disease [147]. Subsequent work in animal models of autoimmune diseases further highlighted the roles which memory CD4$^+$ T cell play in promoting autoimmunity. Along these lines, adoptive transfer of autoreactive memory CD4$^+$ T cells is sufficient to induce disease, for example in animal models of MS, diabetes, and uveoretinitis [148–150]. However, while autoreactive CD4$^+$ memory T cells are sufficient to induce autoimmune pathology, the context that these cells are transferred into is important. For example, in an elegant study in EAE, neonatal mice were injected with myelin basic protein (MBP)-specific CD4$^+$ T cells to allow the generation of MBP-specific memory CD4$^+$ T cells prior to adulthood [151]. Importantly, these animals remained healthy despite the presence of memory CD4$^+$ T cells in lymphoid tissues when they reached adulthood. However, mice that had received MBP-specific CD4$^+$ T cells developed earlier and had more severe EAE disease when they were immunized with MBP in complete Freud's adjuvant (CFA) as compared with mice that received ovalbumin (OVA)-specific CD4$^+$ T cells. These data suggest that memory CD4$^+$ T cells that develop at early ages (prior to adulthood) may have regulatory functions, or that autoreactive memory T cells can form and persist in healthy individuals but may require additional events in order to become activated and induce disease. Nevertheless, these data demonstrate that autoreactive CD4$^+$ memory T cells generate a more vigorous response upon exposure to autoantigen as compared with Ag-inexperienced naïve autoreactive T cells. These data also suggest that the re-activation of autoimmune memory T cells requires a lower Ag threshold than that of naïve T cells, similar to pathogen-reactive memory T cells.

Along these lines, a critical unresolved question for future studies is how T cell senescence and aging affects memory T cell functions. The effect of aging on CD4$^+$ naïve and memory T cells in the context of infectious diseases has been studied and extensively reviewed elsewhere [152–155]. Immunoscenescence (i.e., ageing of the immune system) is characterized by a decline of adoptive and innate immune cell functions, including of CD4$^+$ T cells [152,156]. However, aging is also associated with autoimmune phenomena, and certain autoimmune disease conditions are more frequently observed in elderly individuals [157]. Moreover, aging and T cell immunoscenescence has been associated with MS and EAE [158]), as well as with other autoimmune diseases [159,160]). Furthermore, treatment efficacy and development of progressive multifocal leukoencephalopathy in MS patients has been linked and associated with immunoscenescence [161–163]. As the role of the aging immune system in infectious diseases and autoimmunity appears somewhat contradictory [160], an interesting question is how does the inflammatory environment during aging/immunoscenescence contribute to and modulate autoimmune memory? Along these lines, a recent interesting study showed that aging promotes the development and accumulation of extreme pro-inflammatory cytotoxic CD4$^+$ T cells, as well as anti-inflammatory regulatory T cells (Tregs) (~30% of the total T cell pool) [164]. How these findings affect autoreactive memory T cells will be an interesting question for further research. Furthermore, many pressing questions remain to be further explored, for instance, which antigens these cells recognize and/or their clonotypes both in healthy individuals as well as in autoimmune disease, what are the mechanisms that maintain or revoke the activation/quiescence of these cells, and what is their relationship to memory T cells? Taken together, immunoscenescence may have important implications for the approach to treating autoimmune diseases in the elderly.

4.2. Role of Autoantigen for Memory T Cells

A question central to autoimmune memory is which autoantigens the memory CD4$^+$ T cells recognize in human autoimmune diseases. The answer to this question could provide insights into mechanisms that promote escape from immune tolerance. For instance, rheumatoid arthritis (RA) patients exhibit memory CD4$^+$ T cells specific for various citrullinated antigens, including citrullinated aggrecan and citrullinated vimentin, which correspond to autoantibodies directed against citrullinated antigens and proteins in these patients [165–167]. Furthermore, CD4$^+$ memory cells with a Th17 phenotype and specific for a citrullinated vimentin epitope expanded more significantly in RA patients with active disease and significantly decreased upon anti-TNF treatment [168]. Additionally, memory CD4$^+$ T cells recognizing glycosylated type II collagen peptides have been identified in RA patients [169]. Together, these data suggest that post-translationally modified autoantigens, and particularly citrullinated autoantigens, may be critical to drive RA progression and increase the number of autoreactive memory T cell clones. Memory CD4+ T cells from MS patients have been reported to recognize several neuroantigens, particularly myelin antigens, including myelin oligodendrocyte glycoprotein (MOG) and MBP [170,171], and T cell responses against these autoantigens are pathogenic in its animal model EAE [151]. It remains to be determined whether modifications of myelin antigens also play a role in MS patients, similar to RA, and whether this results in generation of autoreactive memory CD4+ T cells. This may provide important clues for the role of autoreactive memory T cells in disease etiology and progression. Taken together, strong evidence supports a key role for memory CD4+ T cells in the pathogenesis of autoimmune diseases. Important remaining questions will center on the mechanisms that activate, sustain, and regulate these autoreactive T cells.

4.3. Autoreactive Memory T Cells with Th17 Cell Phenotype

Th17 cells are key drivers of many chronic autoimmune disease conditions, including MS, RA, and psoriasis [2,172,173]. Thus, memory T cells with a Th17 cell phenotype have been reported in many of these conditions. Th17 cells can give rise to self-renewing CD4+ T_{EM} that maintain their Th17 cell phenotype [174]. Moreover, CD4+ T_{CM} with Th1 and Th17 phenotype were reported as selectively increased in blood of MS patients and to correlate with disease severity. Interestingly, the transcriptional profile of blood Th1 T_{CM} and Th17 T_{CM} strongly resembled conventional effector Th17 cells (and not Th1 cells) with more pathogenic features. The cerebrospinal fluid (CSF) of these patients contained mainly CXCR3-expressing Th1/Th17 T_{CM} cells. However, CSF Th17 T_{CM} cells of MS patients reacted strongly to myelin-derived self-antigens (including MOG and MBP), while Th1 cells responded consistently only to virus antigens (such as Epstein–Barr virus). Additionally, the CSF Th1 T_{CM} and Th17 T_{CM} from MS patients had the capacity to produce high levels of pathogenic cytokines upon activation, including IFN-γ, IL-17, GM-CSF, and IL-22 [170]. Moreover, IL-23, which is essential for terminal differentiation of pathogenic Th17 cells [175], has been proposed to regulate memory Th17 cell generation and function. In support of this view, IL-23 drives the proliferation and expansion of memory Th17 cells from MS patients and promotes expression of IL-17 and IFN-γ in these cells [176]. Similarly, IL-23 signaling is critical for development and proliferation of memory Th17 cells in EAE [177], and memory CD4+ T cells in EAE mice proliferate more and produce more IFN-γ but less IL-17 as compared with effector T cells [148]. Furthermore, a population of gut resident CD161+ Th17 cells was identified in patients with Crohn's disease. These gut-homing memory Th17 cells expressed high levels of pro-inflammatory cytokines, including IL-17, IL-22, and IFN-γ upon re-activation with αCD3 and αCD28 in presence of IL-1β and IL-23 [178]. Moreover, in an animal model of colitis, IL-23 promotes the proliferation of memory CD4+ T cells and increases the expression of IFN-γ and IL-17 to promote inflammation [179]. Thought-provokingly, IL-21, a cytokine that is expressed by Th17 cells and enhances their development and expansion in an autocrine fashion [180,181], was reported to inhibit de novo generation of pathogenic Th17 (and Th1) effector T cells from IL-21-expressing T_{CM} cells. These data indicate a potentially protective role for IL-21+ T_{CM} in the context of autoimmunity [182].

4.4. Autoreactive T_{CM} and T_{EM} Subsets and Disease-Modifying Therapies

A critical remaining question centers on the functional attributes of the different memory CD4+ T cell subsets (i.e., T_{CM}, T_{EM}, and T_{RM}) in promoting autoimmune disease. Although both autoantigen- specific CD4+ T_{CM} and T_{EM} cells have been reported, the generation of CD4+ T_{EM} appears more common [183]. This phenomenon is attributed to the presence of chronic autoantigen exposure, which appears to favor the development of CD4+ T_{EM} while hindering CD4+ T_{CM} cell formation, similar to chronic infection settings [174,178,183]. Along these lines, MS and systemic lupus erythematosus (SLE) patients show higher frequencies of autoreactive CD4+ T_{EM} and lower CD4+ T_{CM} in peripheral blood compared with healthy controls [183]. Similar results were reported in patients with colitis, type 1 diabetes, and other autoimmune diseases [174,178,183]. Interestingly, unlike pathogen-specific memory T cells, which are long-lived and highly proliferative, memory CD4+ T cells from autoimmune disease patients are more likely to undergo apoptosis and are less likely to proliferate, most notably for CD4$^+$ T$_{CM}$ [183]. These data suggest that chronic autoimmune disease conditions promote memory CD4$^+$ T cell death and inhibit their proliferation and survival. Determining how differentiation and survival of different subsets of memory CD4$^+$ T cells in autoimmune disease conditions are affected by factors such as cytokine milieu and the presence of autoantigens may lead to potential new avenues for treatment of disease progression and relapses.

Further implicating memory T cells in MS are the results of treating MS patients with fingolimod (also known as FTY720), an S1P receptor (S1PR) antagonist which is thought to act by downregulating the expression of S1PR1 on lymphocytes and is now approved for the treatment of relapsing MS [184,185]. Responsiveness to S1P (via S1PRs) and S1P-dependent tissue trafficking from lymphoid tissues to inflamed tissues are complex and reviewed elsewhere [186,187]. Briefly, S1P acts to promote lymph node egress by overcoming retention signals mediated by factors such as CCR7 [188]. Additionally, T$_{CM}$ largely depend on the S1P/S1PR-axis to traffic/exit from lymphoid tissues to the blood circulation, while T$_{EM}$ which are CCR7^{low}, have already egressed to the circulation and do no-longer rely on S1P/S1PR1 [187,188]. Thus, fingolimod prevents the circulation of T$_{CM}$ but not T$_{EM}$ to the CNS and promotes (CCR7-mediated) T$_{CM}$ retention in secondary lymphoid tissues [189]. T$_{RM}$ cells do not express S1PR1 as they do not express its transcription factor KLF2 (Figure 3) [24]. Along these lines, fingolimod treatment of MS patients showed a marked reduction in blood-circulating CD4$^+$ T$_{CM}$ but not in T$_{EM}$ [189,190]. Subsequent studies showed that fingolimod affected primarily IL-17-producing CD4$^+$ T$_{CM}$ [191]. Interestingly, fingolimod-treated relapsed MS patients showed greater percentages of CD4$^+$ T$_{CM}$ (and naïve cells) but not T$_{EM}$, suggesting that CD4$^+$ T$_{CM}$ may be involved in promoting relapses following fingolimod treatment in MS patients [192]. Furthermore, fingolimod treatment was associated with elevated frequencies of CD56$^+$ memory T cells, and increased granzyme (GZM) B, perforin, and Fas ligand expression in memory T cells in MS patients, and interestingly, this T cell phenotype was also associated with clinical relapses [192]. Additionally, Herich et al. demonstrated that CD4$^+$ T$_{EM}$ expressing high levels of CCR5 and GZMK are involved in CNS immune surveillance in healthy individuals, but that this subset was dominant in peripheral blood mononuclear cells of MS patients, and that natalizumab (anti-α4-integrin) treatment significantly decreased these cells [193]. Furthermore, the CCR5^{high} GZMK$^+$ CD4$^+$ T$_{EM}$ subset shares many transcriptional features with T$_{RM}$ and Th17 cells, suggesting that it could play a central role in CNS pathology [193].

Therapies aimed at modulating the function of autoreactive memory T cells should take advantage of the current understanding of mechanism that regulate effector T cells (Table 1). For example, are immune checkpoints similarly effective in memory T cells as compared to effector T cells? Furthermore, what is the impact of regulatory T cells on memory T cell pool and function? At least some of the mechanisms known to regulate effector T cells act differently on memory T cells, which could potentially be explored therapeutically. a better understanding of these critical mechanisms may have major implications for therapeutic intervention for autoimmune diseases, for instance by targeting the common gamma-chain-receptor cytokines (e.g., IL-7, IL-15), which greatly affect naïve and memory CD4$^+$ T cell development and homeostasis in healthy individuals and infectious diseases [194,195].Thus, future work should focus on further unraveling the different mechanisms by which memory T cell subsets contribute to autoimmune inflammatory diseases, and elucidate mechanisms by which regulatory mechanisms and therapeutic drugs may affect memory T cell subset effector functions and migratory capacities.

Table 1. Regulation of memory and effector CD4$^+$ T cells. Abbreviations used: programmed cell death protein 1 (PD-1), B-and T-lymphocyte attenuator (BTLA), lymphocyte-activation gene 3 (LAG-3), T-cell immunoglobulin, mucin-domain containing-3 (TIM-3), and dendritic cell (DC).

Mechanism of Regulation	Effector T Cells	Memory T Cells	Reviewed in Reference(s)
Immune checkpoints/ T cell exhaustion	PD-1 [196] BTLA [197] LAG-3 [198] TIM-3 promotes effector responses [199]	CTLA-4 [200] LAG-3 [198] TIM-3 suppresses memory differentiation [199]	[196–200]
Antigen persistence	Low antigen dose results in suboptimal activation High antigen and prolonged exposure results in exhaustion	Lower activation threshold Lower co-stimulation dependence may facilitate exhaustion	[53,201,202]
Regulatory T cells	Secretion of inhibitory cytokines —IL-10, TGF-β Metabolic regulation (indirectly) via modulation of DC functions	Secretion of inhibitory cytokines —IL-10, TGF-β Metabolic regulation (indirectly) via modulation of (DC) functions	[203–205]
Cytokines in the maintenance/development of effector and memory cells	Depends on the subset [2,206]: Th1: IL-2, IFN-γ, IL-12 Th2: IL-4, IL-2 Th17: TGF-β, IL-6, IL-1, IL-23 Tregs: TGF-β	IL-7 [207–209] IL-15 [210,211]	[2,206–211]

5. Concluding Remarks

A better understanding of the immunobiology of CD4$^+$ memory T cells in chronic autoimmune diseases is critical to develop better treatments. While it is understood that there are differences in memory T cell populations and subpopulations in autoimmune disease conditions, there are important gaps in the current understanding of how these cells develop and how the host microenvironment, including antigen exposure and cytokine milieu, affect the function and maintenance of these cells. Additionally, it remains incompletely understood if and how these cells differ from memory T cells directed against infectious pathogens in terms of activation thresholds, cytokine secretion, and long-term survival, and how regulatory mechanism apply to these cells as compared with naïve/effector T cells.

CD8$^+$ and CD4$^+$ T$_{RM}$ cells were identified in human brain and in lesions of MS patients [212,213], and recent research primarily focused on the role of CD8$^+$ T$_{RM}$ cells and their contribution to autoimmune pathology in MS and EAE [212–216]. However, autoimmune CD4$^+$ T$_{RM}$ are not as well studied, and many critical questions remain as to their potential contribution to autoimmune disease pathology. Moreover, a better understanding of the role of autoreactive, pathogenic CD4$^+$ T cells in relapses and progression of autoimmune diseases could have major therapeutic implications. Addressing these questions will be paramount to develop better treatments for CD4$^+$ T cell-driven autoimmune diseases.

Funding: This research was supported by grants G12MD007591 and NS084201 from the National Institute of Health (TGF) and grants RG5501 and RG1602 from the National Multiple Sclerosis Society (TGF).

Acknowledgments: We would like to thank Saisha Nalawade and Carol Chase for reading the manuscript. Figures created with BioRender.com.

Conflicts of Interest: The authors declare no conflict of interest.

References

1. Jiang, S.; Dong, C. a complex issue on CD4(+) T-cell subsets. *Immunol. Rev.* **2013**, *252*, 5–11. [CrossRef]
2. Raphael, I.; Nalawade, S.; Eagar, T.N.; Forsthuber, T.G. T cell subsets and their signature cytokines in autoimmune and inflammatory diseases. *Cytokine* **2015**, *74*, 5–17. [CrossRef]
3. Kumar, B.V.; Connors, T.J.; Farber, D.L. Human T Cell Development, Localization, and Function throughout Life. *Immunity* **2018**, *48*, 202–213. [CrossRef] [PubMed]

4. Raphael, I.; Forsthuber, T.G. Stability of T-cell lineages in autoimmune diseases. *Expert Rev. Clin. Immunol.* **2012**, *8*, 299–301. [CrossRef] [PubMed]
5. Mueller, S.N.; Gebhardt, T.; Carbone, F.R.; Heath, W.R. Memory T cell subsets, migration patterns, and tissue residence. *Annu. Rev. Immunol.* **2013**, *31*, 137–161. [CrossRef] [PubMed]
6. Baaten, B.J.; Cooper, A.M.; Swain, S.L.; Bradley, L.M. Location, location, location: The impact of migratory heterogeneity on T cell function. *Front. Immunol.* **2013**, *4*, 311. [CrossRef]
7. Mackay, C.R. T-cell memory: The connection between function, phenotype and migration pathways. *Immunol Today* **1991**, *12*, 189–192. [CrossRef]
8. Jenkins, M.K.; Khoruts, A.; Ingulli, E.; Mueller, D.L.; McSorley, S.J.; Reinhardt, R.L.; Itano, A.; Pape, K.A. In vivo activation of antigen-specific CD4 T cells. *Annu Rev Immunol* **2001**, *19*, 23–45. [CrossRef]
9. Sprent, J.; Surh, C.D. T cell memory. *Annu Rev Immunol* **2002**, *20*, 551–579. [CrossRef]
10. MacLeod, M.K.; Kappler, J.W.; Marrack, P. Memory CD4 T cells: Generation, reactivation and re-assignment. *Immunology* **2010**, *130*, 10–15. [CrossRef]
11. Croft, M. Activation of naive, memory and effector T cells. *Curr. Opin. Immunol.* **1994**, *6*, 431–437. [CrossRef]
12. Cho, B.K.; Wang, C.; Sugawa, S.; Eisen, H.N.; Chen, J. Functional differences between memory and naive CD8 T cells. *Proc. Natl. Acad. Sci. USA* **1999**, *96*, 2976–2981. [CrossRef] [PubMed]
13. Mishima, T.; Toda, S.; Ando, Y.; Matsunaga, T.; Inobe, M. Rapid proliferation of activated lymph node CD4(+) T cells is achieved by greatly curtailing the duration of gap phases in cell cycle progression. *Cell Mol. Biol. Lett.* **2014**, *19*, 638–648. [CrossRef] [PubMed]
14. Berard, M.; Tough, D.F. Qualitative differences between naive and memory T cells. *Immunology* **2002**, *106*, 127–138. [CrossRef]
15. Farber, D.L.; Yudanin, N.A.; Restifo, N.P. Human memory T cells: Generation, compartmentalization and homeostasis. *Nat. Rev. Immunol.* **2014**, *14*, 24–35. [CrossRef]
16. Blattman, J.N.; Antia, R.; Sourdive, D.J.; Wang, X.; Kaech, S.M.; Murali-Krishna, K.; Altman, J.D.; Ahmed, R. Estimating the precursor frequency of naive antigen-specific CD8 T cells. *J. Exp. Med.* **2002**, *195*, 657–664. [CrossRef]
17. Mackay, C.R. Dual personality of memory T cells. *Nature* **1999**, *401*, 659–660. [CrossRef]
18. MacLeod, M.K.; Clambey, E.T.; Kappler, J.W.; Marrack, P. CD4 memory T cells: What are they and what can they do? *Semin. Immunol.* **2009**, *21*, 53–61. [CrossRef]
19. Mora, J.R.; von Andrian, U.H. T-cell homing specificity and plasticity: New concepts and future challenges. *Trends Immunol.* **2006**, *27*, 235–243. [CrossRef]
20. Woodland, D.L.; Kohlmeier, J.E. Migration, maintenance and recall of memory T cells in peripheral tissues. *Nat. Rev. Immunol.* **2009**, *9*, 153–161. [CrossRef]
21. Gebhardt, T.; Wakim, L.M.; Eidsmo, L.; Reading, P.C.; Heath, W.R.; Carbone, F.R. Memory T cells in nonlymphoid tissue that provide enhanced local immunity during infection with herpes simplex virus. *Nat. Immunol* **2009**, *10*, 524–530. [CrossRef] [PubMed]
22. Turner, D.L.; Gordon, C.L.; Farber, D.L. Tissue-resident T cells, in situ immunity and transplantation. *Immunol. Rev.* **2014**, *258*, 150–166. [CrossRef] [PubMed]
23. Sathaliyawala, T.; Kubota, M.; Yudanin, N.; Turner, D.; Camp, P.; Thome, J.J.; Bickham, K.L.; Lerner, H.; Goldstein, M.; Sykes, M.; et al. Distribution and compartmentalization of human circulating and tissue-resident memory T cell subsets. *Immunity* **2013**, *38*, 187–197. [CrossRef] [PubMed]
24. Schenkel, J.M.; Masopust, D. Tissue-resident memory T cells. *Immunity* **2014**, *41*, 886–897. [CrossRef]
25. Topham, D.J.; Reilly, E.C. Tissue-Resident Memory CD8(+) T Cells: From Phenotype to Function. *Front Immunol.* **2018**, *9*, 515. [CrossRef]
26. Wu, X.; Wu, P.; Shen, Y.; Jiang, X.; Xu, F. CD8(+) Resident Memory T Cells and Viral Infection. *Front Immunol.* **2018**, *9*, 2093. [CrossRef]
27. Ahmed, R.; Bevan, M.J.; Reiner, S.L.; Fearon, D.T. The precursors of memory: Models and controversies. *Nat. Rev. Immunol.* **2009**, *9*, 662–668. [CrossRef]
28. Sallusto, F.; Lenig, D.; Forster, R.; Lipp, M.; Lanzavecchia, A. Two subsets of memory T lymphocytes with distinct homing potentials and effector functions. *Nature* **1999**, *401*, 708–712. [CrossRef]
29. Chang, J.T.; Palanivel, V.R.; Kinjyo, I.; Schambach, F.; Intlekofer, A.M.; Banerjee, A.; Longworth, S.A.; Vinup, K.E.; Mrass, P.; Oliaro, J.; et al. Asymmetric T lymphocyte division in the initiation of adaptive immune responses. *Science* **2007**, *315*, 1687–1691. [CrossRef]

30. Pepper, M.; Jenkins, M.K. Origins of CD4(+) effector and central memory T cells. *Nat. Immunol.* **2011**, *12*, 467–471. [CrossRef]
31. Corse, E.; Gottschalk, R.A.; Allison, J.P. Strength of TCR-peptide/MHC interactions and in vivo T cell responses. *J. Immunol.* **2011**, *186*, 5039–5045. [CrossRef] [PubMed]
32. Baumgartner, C.K.; Yagita, H.; Malherbe, L.P. a TCR affinity threshold regulates memory CD4 T cell differentiation following vaccination. *J. Immunol.* **2012**, *189*, 2309–2317. [CrossRef] [PubMed]
33. Daniels, M.A.; Teixeiro, E. TCR Signaling in T Cell Memory. *Front. Immunol.* **2015**, *6*, 617. [CrossRef] [PubMed]
34. Kuhns, M.S.; Davis, M.M. TCR Signaling Emerges from the Sum of Many Parts. *Front. Immunol.* **2012**, *3*, 159. [CrossRef]
35. Pepper, M.; Pagan, A.J.; Igyarto, B.Z.; Taylor, J.J.; Jenkins, M.K. Opposing signals from the Bcl6 transcription factor and the interleukin-2 receptor generate T helper 1 central and effector memory cells. *Immunity* **2011**, *35*, 583–595. [CrossRef]
36. Marshall, H.D.; Chandele, A.; Jung, Y.W.; Meng, H.; Poholek, A.C.; Parish, I.A.; Rutishauser, R.; Cui, W.; Kleinstein, S.H.; Craft, J.; et al. Differential expression of Ly6C and T-bet distinguish effector and memory Th1 CD4(+) cell properties during viral infection. *Immunity* **2011**, *35*, 633–646. [CrossRef]
37. Richer, M.J.; Nolz, J.C.; Harty, J.T. Pathogen-specific inflammatory milieux tune the antigen sensitivity of CD8(+) T cells by enhancing T cell receptor signaling. *Immunity* **2013**, *38*, 140–152. [CrossRef]
38. Snook, J.P.; Kim, C.; Williams, M.A. TCR signal strength controls the differentiation of CD4(+) effector and memory T cells. *Sci. Immunol.* **2018**, *3*. [CrossRef]
39. Obst, R.; van Santen, H.M.; Mathis, D.; Benoist, C. Antigen persistence is required throughout the expansion phase of a CD4(+) T cell response. *J. Exp. Med.* **2005**, *201*, 1555–1565. [CrossRef]
40. Jelley-Gibbs, D.M.; Brown, D.M.; Dibble, J.P.; Haynes, L.; Eaton, S.M.; Swain, S.L. Unexpected prolonged presentation of influenza antigens promotes CD4 T cell memory generation. *J. Exp. Med.* **2005**, *202*, 697–706. [CrossRef]
41. Gasper, D.J.; Tejera, M.M.; Suresh, M. CD4 T-cell memory generation and maintenance. *Crit. Rev. Immunol.* **2014**, *34*, 121–146. [CrossRef] [PubMed]
42. Smith, A.L.; Wikstrom, M.E.; Fazekas de St Groth, B. Visualizing T cell competition for peptide/MHC complexes: a specific mechanism to minimize the effect of precursor frequency. *Immunity* **2000**, *13*, 783–794. [CrossRef]
43. Butz, E.A.; Bevan, M.J. Massive expansion of antigen-specific CD8+ T cells during an acute virus infection. *Immunity* **1998**, *8*, 167–175. [CrossRef]
44. Busch, D.H.; Pamer, E.G. T cell affinity maturation by selective expansion during infection. *J. Exp. Med.* **1999**, *189*, 701–710. [CrossRef]
45. Kim, C.; Wilson, T.; Fischer, K.F.; Williams, M.A. Sustained interactions between T cell receptors and antigens promote the differentiation of CD4(+) memory T cells. *Immunity* **2013**, *39*, 508–520. [CrossRef]
46. Williams, M.A.; Ravkov, E.V.; Bevan, M.J. Rapid culling of the CD4+ T cell repertoire in the transition from effector to memory. *Immunity* **2008**, *28*, 533–545. [CrossRef]
47. Whitmire, J.K.; Benning, N.; Eam, B.; Whitton, J.L. Increasing the CD4+ T cell precursor frequency leads to competition for IFN-gamma thereby degrading memory cell quantity and quality. *J. Immunol.* **2008**, *180*, 6777–6785. [CrossRef]
48. McKinstry, K.K.; Strutt, T.M.; Bautista, B.; Zhang, W.; Kuang, Y.; Cooper, A.M.; Swain, S.L. Effector CD4 T-cell transition to memory requires late cognate interactions that induce autocrine IL-2. *Nat Commun.* **2014**, *5*, 5377. [CrossRef]
49. Jenkins, M.K.; Moon, J.J. The role of naive T cell precursor frequency and recruitment in dictating immune response magnitude. *J. Immunol.* **2012**, *188*, 4135–4140. [CrossRef]
50. Foulds, K.E.; Shen, H. Clonal competition inhibits the proliferation and differentiation of adoptively transferred TCR transgenic CD4 T cells in response to infection. *J. Immunol.* **2006**, *176*, 3037–3043. [CrossRef]
51. Hataye, J.; Moon, J.J.; Khoruts, A.; Reilly, C.; Jenkins, M.K. Naive and memory CD4+ T cell survival controlled by clonal abundance. *Science* **2006**, *312*, 114–116. [CrossRef] [PubMed]
52. Blair, D.A.; Lefrancois, L. Increased competition for antigen during priming negatively impacts the generation of memory CD4 T cells. *Proc. Natl. Acad. Sci. USA* **2007**, *104*, 15045–15050. [CrossRef] [PubMed]

53. Han, S.; Asoyan, A.; Rabenstein, H.; Nakano, N.; Obst, R. Role of antigen persistence and dose for CD4+ T-cell exhaustion and recovery. *Proc. Natl. Acad. Sci. USA* **2010**, *107*, 20453–20458. [CrossRef] [PubMed]
54. Pepper, M.; Linehan, J.L.; Pagan, A.J.; Zell, T.; Dileepan, T.; Cleary, P.P.; Jenkins, M.K. Different routes of bacterial infection induce long-lived TH1 memory cells and short-lived TH17 cells. *Nat. Immunol.* **2010**, *11*, 83–89. [CrossRef]
55. Joshi, N.S.; Cui, W.; Chandele, A.; Lee, H.K.; Urso, D.R.; Hagman, J.; Gapin, L.; Kaech, S.M. Inflammation directs memory precursor and short-lived effector CD8(+) T cell fates via the graded expression of T-bet transcription factor. *Immunity* **2007**, *27*, 281–295. [CrossRef]
56. Whitmire, J.K.; Eam, B.; Benning, N.; Whitton, J.L. Direct interferon-gamma signaling dramatically enhances CD4+ and CD8+ T cell memory. *J. Immunol.* **2007**, *179*, 1190–1197. [CrossRef]
57. Afkarian, M.; Sedy, J.R.; Yang, J.; Jacobson, N.G.; Cereb, N.; Yang, S.Y.; Murphy, T.L.; Murphy, K.M. T-bet is a STAT1-induced regulator of IL-12R expression in naive CD4+ T cells. *Nat. Immunol.* **2002**, *3*, 549–557. [CrossRef]
58. Harrington, L.E.; Janowski, K.M.; Oliver, J.R.; Zajac, A.J.; Weaver, C.T. Memory CD4 T cells emerge from effector T-cell progenitors. *Nature* **2008**, *452*, 356–360. [CrossRef]
59. Thieu, V.T.; Yu, Q.; Chang, H.C.; Yeh, N.; Nguyen, E.T.; Sehra, S.; Kaplan, M.H. Signal transducer and activator of transcription 4 is required for the transcription factor T-bet to promote T helper 1 cell-fate determination. *Immunity* **2008**, *29*, 679–690. [CrossRef]
60. Xiong, S.Q.; Lin, B.L.; Gao, X.; Tang, H.; Wu, C.Y. IL-12 promotes HBV-specific central memory CD8+ T cell responses by PBMCs from chronic hepatitis B virus carriers. *Int. Immunopharmacol.* **2007**, *7*, 578–587. [CrossRef]
61. Mollo, S.B.; Ingram, J.T.; Kress, R.L.; Zajac, A.J.; Harrington, L.E. Virus-specific CD4 and CD8 T cell responses in the absence of Th1-associated transcription factors. *J. Leukoc. Biol.* **2014**, *95*, 705–713. [CrossRef] [PubMed]
62. Oja, A.E.; Piet, B.; Helbig, C.; Stark, R.; van der Zwan, D.; Blaauwgeers, H.; Remmerswaal, E.B.M.; Amsen, D.; Jonkers, R.E.; Moerland, P.D.; et al. Trigger-happy resident memory CD4(+) T cells inhabit the human lungs. *Mucosal Immunol.* **2018**, *11*, 654–667. [CrossRef] [PubMed]
63. Raeber, M.E.; Zurbuchen, Y.; Impellizzieri, D.; Boyman, O. The role of cytokines in T-cell memory in health and disease. *Immunol. Rev.* **2018**, *283*, 176–193. [CrossRef] [PubMed]
64. Belz, G.T.; Masson, F. Interleukin-2 tickles T cell memory. *Immunity* **2010**, *32*, 7–9. [CrossRef]
65. Dhume, K.; McKinstry, K.K. Early programming and late-acting checkpoints governing the development of CD4 T-cell memory. *Immunology* **2018**, *155*, 53–62. [CrossRef]
66. Crotty, S.; Johnston, R.J.; Schoenberger, S.P. Effectors and memories: Bcl-6 and Blimp-1 in T and B lymphocyte differentiation. *Nat. Immunol.* **2010**, *11*, 114–120. [CrossRef]
67. McKinstry, K.K.; Alam, F.; Flores-Malavet, V.; Nagy, M.Z.; Sell, S.; Cooper, A.M.; Swain, S.L.; Strutt, T.M. Memory CD4 T cell-derived IL-2 synergizes with viral infection to exacerbate lung inflammation. *Plos Pathog* **2019**, *15*, e1007989. [CrossRef]
68. Prlic, M.; Lefrancois, L.; Jameson, S.C. Multiple choices: Regulation of memory CD8 T cell generation and homeostasis by interleukin (IL)-7 and IL-15. *J. Exp. Med.* **2002**, *195*, F49–F52. [CrossRef]
69. Cui, W.; Kaech, S.M. Generation of effector CD8+ T cells and their conversion to memory T cells. *Immunol. Rev.* **2010**, *236*, 151–166. [CrossRef]
70. Schluns, K.S.; Lefrancois, L. Cytokine control of memory T-cell development and survival. *Nat. Rev. Immunol.* **2003**, *3*, 269–279. [CrossRef]
71. Schenkel, J.M.; Masopust, D. Identification of a resident T-cell memory core transcriptional signature. *Immunol. Cell Biol.* **2014**, *92*, 8–9. [CrossRef] [PubMed]
72. Wang, D.; Yuan, R.; Feng, Y.; El-Asady, R.; Farber, D.L.; Gress, R.E.; Lucas, P.J.; Hadley, G.A. Regulation of CD103 expression by CD8+ T cells responding to renal allografts. *J. Immunol.* **2004**, *172*, 214–221. [CrossRef] [PubMed]
73. Gebhardt, T.; Mackay, L.K. Local immunity by tissue-resident CD8(+) memory T cells. *Front. Immunol.* **2012**, *3*, 340. [CrossRef] [PubMed]
74. Lefrancois, L.; Barrett, T.A.; Havran, W.L.; Puddington, L. Developmental expression of the alpha IEL beta 7 integrin on T cell receptor gamma delta and T cell receptor alpha beta T cells. *Eur. J. Immunol.* **1994**, *24*, 635–640. [CrossRef] [PubMed]

75. Xu, L.; Kitani, A.; Strober, W. Molecular mechanisms regulating TGF-beta-induced Foxp3 expression. *Mucosal Immunol.* **2010**, *3*, 230–238. [CrossRef] [PubMed]
76. Sanchez, A.M.; Zhu, J.; Huang, X.; Yang, Y. The development and function of memory regulatory T cells after acute viral infections. *J. Immunol.* **2012**, *189*, 2805–2814. [CrossRef]
77. Duttagupta, P.A.; Boesteanu, A.C.; Katsikis, P.D. Costimulation signals for memory CD8+ T cells during viral infections. *Crit. Rev. Immunol.* **2009**, *29*, 469–486. [CrossRef]
78. Rahimi, R.A.; Luster, A.D. Chemokines: Critical Regulators of Memory T Cell Development, Maintenance, and Function. *Adv. Immunol.* **2018**, *138*, 71–98. [CrossRef]
79. Negi, S.; Das, D.K.; Pahari, S.; Nadeem, S.; Agrewala, J.N. Potential Role of Gut Microbiota in Induction and Regulation of Innate Immune Memory. *Front. Immunol.* **2019**, *10*, 2441. [CrossRef]
80. Belkaid, Y.; Hand, T.W. Role of the microbiota in immunity and inflammation. *Cell* **2014**, *157*, 121–141. [CrossRef]
81. Thomas, M.L.; Lefrancois, L. Differential expression of the leucocyte-common antigen family. *Immunol. Today* **1988**, *9*, 320–326. [CrossRef]
82. Akbar, A.N.; Salmon, M.; Ivory, K.; Taki, S.; Pilling, D.; Janossy, G. Human CD4+CD45R0+ and CD4+CD45RA+ T cells synergize in response to alloantigens. *Eur. J. Immunol.* **1991**, *21*, 2517–2522. [CrossRef]
83. Lee, W.T.; Vitetta, E.S. Changes in expression of CD45R during the development of Th1 and Th2 cell lines. *Eur. J. Immunol.* **1992**, *22*, 1455–1459. [CrossRef]
84. Merkenschlager, M.; Terry, L.; Edwards, R.; Beverley, P.C. Limiting dilution analysis of proliferative responses in human lymphocyte populations defined by the monoclonal antibody UCHL1: Implications for differential CD45 expression in T cell memory formation. *Eur. J. Immunol.* **1988**, *18*, 1653–1661. [CrossRef]
85. Sanders, M.E.; Makgoba, M.W.; Sharrow, S.O.; Stephany, D.; Springer, T.A.; Young, H.A.; Shaw, S. Human memory T lymphocytes express increased levels of three cell adhesion molecules (LFA-3, CD2, and LFA-1) and three other molecules (UCHL1, CDw29, and Pgp-1) and have enhanced IFN-gamma production. *J. Immunol.* **1988**, *140*, 1401–1407.
86. Akbar, A.N.; Timms, A.; Janossy, G. Cellular events during memory T-cell activation in vitro: The UCHL1 (180,000 MW) determinant is newly synthesized after mitosis. *Immunology* **1989**, *66*, 213–218.
87. Sallusto, F.; Geginat, J.; Lanzavecchia, A. Central memory and effector memory T cell subsets: Function, generation, and maintenance. *Annu Rev Immunol* **2004**, *22*, 745–763. [CrossRef]
88. Bradley, L.M.; Atkins, G.G.; Swain, S.L. Long-term CD4+ memory T cells from the spleen lack MEL-14, the lymph node homing receptor. *J. Immunol.* **1992**, *148*, 324–331.
89. Bradley, L.M.; Duncan, D.D.; Yoshimoto, K.; Swain, S.L. Memory effectors: a potent, IL-4-secreting helper T cell population that develops in vivo after restimulation with antigen. *J. Immunol.* **1993**, *150*, 3119–3130.
90. Budd, R.C.; Cerottini, J.C.; Horvath, C.; Bron, C.; Pedrazzini, T.; Howe, R.C.; MacDonald, H.R. Distinction of virgin and memory T lymphocytes. Stable acquisition of the Pgp-1 glycoprotein concomitant with antigenic stimulation. *J. Immunol.* **1987**, *138*, 3120–3129.
91. Springer, T.A.; Dustin, M.L.; Kishimoto, T.K.; Marlin, S.D. The lymphocyte function-associated LFA-1, CD2, and LFA-3 molecules: Cell adhesion receptors of the immune system. *Annu Rev. Immunol.* **1987**, *5*, 223–252. [CrossRef]
92. Birkeland, M.L.; Johnson, P.; Trowbridge, I.S.; Pure, E. Changes in CD45 isoform expression accompany antigen-induced murine T-cell activation. *Proc. Natl. Acad. Sci. USA* **1989**, *86*, 6734–6738. [CrossRef]
93. Reinhardt, R.L.; Khoruts, A.; Merica, R.; Zell, T.; Jenkins, M.K. Visualizing the generation of memory CD4 T cells in the whole body. *Nature* **2001**, *410*, 101–105. [CrossRef]
94. Ahmadzadeh, M.; Hussain, S.F.; Farber, D.L. Effector CD4 T cells are biochemically distinct from the memory subset: Evidence for long-term persistence of effectors in vivo. *J. Immunol.* **1999**, *163*, 3053–3063.
95. Unsoeld, H.; Pircher, H. Complex memory T-cell phenotypes revealed by coexpression of CD62L and CCR7. *J. Virol.* **2005**, *79*, 4510–4513. [CrossRef]
96. Tough, D.F.; Sprent, J. Turnover of naive- and memory-phenotype T cells. *J. Exp. Med.* **1994**, *179*, 1127–1135. [CrossRef]
97. Yu, X.Z.; Anasetti, C. Memory stem cells sustain disease. *Nat. Med.* **2005**, *11*, 1282–1283. [CrossRef]
98. Rogers, P.R.; Pilapil, S.; Hayakawa, K.; Romain, P.L.; Parker, D.C. CD45 alternative exon expression in murine and human CD4+ T cell subsets. *J. Immunol.* **1992**, *148*, 4054–4065.

99. Klonowski, K.D.; Williams, K.J.; Marzo, A.L.; Blair, D.A.; Lingenheld, E.G.; Lefrancois, L. Dynamics of blood-borne CD8 memory T cell migration in vivo. *Immunity* **2004**, *20*, 551–562. [CrossRef]
100. Jiang, X.; Clark, R.A.; Liu, L.; Wagers, A.J.; Fuhlbrigge, R.C.; Kupper, T.S. Skin infection generates non-migratory memory CD8+ T(RM) cells providing global skin immunity. *Nature* **2012**, *483*, 227–231. [CrossRef]
101. Sheridan, B.S.; Lefrancois, L. Regional and mucosal memory T cells. *Nat. Immunol.* **2011**, *12*, 485–491. [CrossRef] [PubMed]
102. Casey, K.A.; Fraser, K.A.; Schenkel, J.M.; Moran, A.; Abt, M.C.; Beura, L.K.; Lucas, P.J.; Artis, D.; Wherry, E.J.; Hogquist, K.; et al. Antigen-independent differentiation and maintenance of effector-like resident memory T cells in tissues. *J. Immunol.* **2012**, *188*, 4866–4875. [CrossRef] [PubMed]
103. Kohlmeier, J.E.; Cookenham, T.; Roberts, A.D.; Miller, S.C.; Woodland, D.L. Type I interferons regulate cytolytic activity of memory CD8(+) T cells in the lung airways during respiratory virus challenge. *Immunity* **2010**, *33*, 96–105. [CrossRef]
104. Skon, C.N.; Lee, J.Y.; Anderson, K.G.; Masopust, D.; Hogquist, K.A.; Jameson, S.C. Transcriptional downregulation of S1pr1 is required for the establishment of resident memory CD8+ T cells. *Nat. Immunol.* **2013**, *14*, 1285–1293. [CrossRef]
105. Clark, R.A.; Chong, B.F.; Mirchandani, N.; Yamanaka, K.; Murphy, G.F.; Dowgiert, R.K.; Kupper, T.S. a novel method for the isolation of skin resident T cells from normal and diseased human skin. *J. Invest. Derm.* **2006**, *126*, 1059–1070. [CrossRef]
106. Kumar, B.V.; Ma, W.; Miron, M.; Granot, T.; Guyer, R.S.; Carpenter, D.J.; Senda, T.; Sun, X.; Ho, S.H.; Lerner, H.; et al. Human Tissue-Resident Memory T Cells Are Defined by Core Transcriptional and Functional Signatures in Lymphoid and Mucosal Sites. *Cell Rep.* **2017**, *20*, 2921–2934. [CrossRef]
107. Carbone, F.R.; Gebhardt, T. Should I stay or should I go-Reconciling clashing perspectives on CD4(+) tissue-resident memory T cells. *Sci. Immunol.* **2019**, *4*. [CrossRef]
108. Klicznik, M.M.; Morawski, P.A.; Hollbacher, B.; Varkhande, S.R.; Motley, S.J.; Kuri-Cervantes, L.; Goodwin, E.; Rosenblum, M.D.; Long, S.A.; Brachtl, G.; et al. Human CD4(+)CD103(+) cutaneous resident memory T cells are found in the circulation of healthy individuals. *Sci. Immunol.* **2019**, *4*. [CrossRef]
109. Geginat, J.; Sallusto, F.; Lanzavecchia, A. Cytokine-driven proliferation and differentiation of human naive, central memory and effector memory CD4+ T cells. *Pathol. Biol. (Paris)* **2003**, *51*, 64–66. [CrossRef]
110. Park, C.O.; Kupper, T.S. The emerging role of resident memory T cells in protective immunity and inflammatory disease. *Nat. Med.* **2015**, *21*, 688–697. [CrossRef]
111. Swain, S.L.; McKinstry, K.K.; Strutt, T.M. Expanding roles for CD4(+) T cells in immunity to viruses. *Nat. Rev. Immunol.* **2012**, *12*, 136–148. [CrossRef]
112. von Andrian, U.H.; Mackay, C.R. T-cell function and migration. Two sides of the same coin. *N. Engl. J. Med.* **2000**, *343*, 1020–1034. [CrossRef]
113. Kaech, S.M.; Wherry, E.J.; Ahmed, R. Effector and memory T-cell differentiation: Implications for vaccine development. *Nat. Rev. Immunol.* **2002**, *2*, 251–262. [CrossRef]
114. Hengel, R.L.; Thaker, V.; Pavlick, M.V.; Metcalf, J.A.; Dennis, G., Jr.; Yang, J.; Lempicki, R.A.; Sereti, I.; Lane, H.C. Cutting edge: L-selectin (CD62L) expression distinguishes small resting memory CD4+ T cells that preferentially respond to recall antigen. *J. Immunol.* **2003**, *170*, 28–32. [CrossRef]
115. Wang, A.; Chandran, S.; Shah, S.A.; Chiu, Y.; Paria, B.C.; Aghamolla, T.; Alvarez-Downing, M.M.; Lee, C.C.; Singh, S.; Li, T.; et al. The stoichiometric production of IL-2 and IFN-gamma mRNA defines memory T cells that can self-renew after adoptive transfer in humans. *Sci. Transl. Med.* **2012**. [CrossRef]
116. MacLeod, M.K.; David, A.; McKee, A.S.; Crawford, F.; Kappler, J.W.; Marrack, P. Memory CD4 T cells that express CXCR5 provide accelerated help to B cells. *J. Immunol.* **2011**, *186*, 2889–2896. [CrossRef]
117. Fazilleau, N.; Eisenbraun, M.D.; Malherbe, L.; Ebright, J.N.; Pogue-Caley, R.R.; McHeyzer-Williams, L.J.; McHeyzer-Williams, M.G. Lymphoid reservoirs of antigen-specific memory T helper cells. *Nat. Immunol.* **2007**, *8*, 753–761. [CrossRef]
118. Hale, J.S.; Ahmed, R. Memory T follicular helper CD4 T cells. *Front. Immunol.* **2015**, *6*, 16. [CrossRef]
119. Jain, A.; Song, R.; Wakeland, E.K.; Pasare, C. T cell-intrinsic IL-1R signaling licenses effector cytokine production by memory CD4 T cells. *Nat. Commun.* **2018**, *9*, 3185. [CrossRef]
120. Shin, H.; Iwasaki, A. a vaccine strategy that protects against genital herpes by establishing local memory T cells. *Nature* **2012**, *491*, 463–467. [CrossRef]

121. Teijaro, J.R.; Turner, D.; Pham, Q.; Wherry, E.J.; Lefrancois, L.; Farber, D.L. Cutting edge: Tissue-retentive lung memory CD4 T cells mediate optimal protection to respiratory virus infection. *J. Immunol.* **2011**, *187*, 5510–5514. [CrossRef]
122. Sakai, S.; Kauffman, K.D.; Schenkel, J.M.; McBerry, C.C.; Mayer-Barber, K.D.; Masopust, D.; Barber, D.L. Cutting edge: Control of Mycobacterium tuberculosis infection by a subset of lung parenchyma-homing CD4 T cells. *J. Immunol.* **2014**, *192*, 2965–2969. [CrossRef]
123. Watanabe, R.; Gehad, A.; Yang, C.; Scott, L.L.; Teague, J.E.; Schlapbach, C.; Elco, C.P.; Huang, V.; Matos, T.R.; Kupper, T.S.; et al. Human skin is protected by four functionally and phenotypically discrete populations of resident and recirculating memory T cells. *Sci. Transl. Med.* **2015**, *7*, 279ra239. [CrossRef]
124. Stary, G.; Olive, A.; Radovic-Moreno, A.F.; Gondek, D.; Alvarez, D.; Basto, P.A.; Perro, M.; Vrbanac, V.D.; Tager, A.M.; Shi, J.; et al. VACCINES. a mucosal vaccine against Chlamydia trachomatis generates two waves of protective memory T cells. *Science* **2015**, *348*, aaa8205. [CrossRef]
125. Oja, A.E.; Piet, B.; van der Zwan, D.; Blaauwgeers, H.; Mensink, M.; de Kivit, S.; Borst, J.; Nolte, M.A.; van Lier, R.A.W.; Stark, R.; et al. Functional Heterogeneity of CD4(+) Tumor-Infiltrating Lymphocytes With a Resident Memory Phenotype in NSCLC. *Front. Immunol.* **2018**, *9*, 2654. [CrossRef]
126. Glennie, N.D.; Yeramilli, V.A.; Beiting, D.P.; Volk, S.W.; Weaver, C.T.; Scott, P. Skin-resident memory CD4+ T cells enhance protection against Leishmania major infection. *J. Exp. Med.* **2015**, *212*, 1405–1414. [CrossRef]
127. Chapman, T.J.; Topham, D.J. Identification of a unique population of tissue-memory CD4+ T cells in the airways after influenza infection that is dependent on the integrin VLA-1. *J. Immunol.* **2010**, *184*, 3841–3849. [CrossRef]
128. Nogueira, C.V.; Zhang, X.; Giovannone, N.; Sennott, E.L.; Starnbach, M.N. Protective immunity against Chlamydia trachomatis can engage both CD4+ and CD8+ T cells and bridge the respiratory and genital mucosae. *J. Immunol.* **2015**, *194*, 2319–2329. [CrossRef]
129. Liu, W.; Zeng, Z.; Luo, S.; Hu, C.; Xu, N.; Huang, A.; Zheng, L.; Sundberg, E.J.; Xi, T.; Xing, Y. Gastric Subserous Vaccination with Helicobacter pylori Vaccine: An Attempt to Establish Tissue-Resident CD4+ Memory T Cells and Induce Prolonged Protection. *Front. Immunol.* **2019**, *10*, 1115. [CrossRef]
130. Wilk, M.M.; Mills, K.H.G. CD4 TRM Cells Following Infection and Immunization: Implications for More Effective Vaccine Design. *Front. Immunol.* **2018**, *9*, 1860. [CrossRef]
131. Cantero-Perez, J.; Grau-Exposito, J.; Serra-Peinado, C.; Rosero, D.A.; Luque-Ballesteros, L.; Astorga-Gamaza, A.; Castellvi, J.; Sanhueza, T.; Tapia, G.; Lloveras, B.; et al. Resident memory T cells are a cellular reservoir for HIV in the cervical mucosa. *Nat. Commun.* **2019**, *10*, 4739. [CrossRef]
132. Nakanishi, Y.; Lu, B.; Gerard, C.; Iwasaki, A. CD8(+) T lymphocyte mobilization to virus-infected tissue requires CD4(+) T-cell help. *Nature* **2009**, *462*, 510–513. [CrossRef]
133. McKinstry, K.K.; Dutton, R.W.; Swain, S.L.; Strutt, T.M. Memory CD4 T cell-mediated immunity against influenza a virus: More than a little helpful. *Arch. Immunol. Exp. (Warsz)* **2013**, *61*, 341–353. [CrossRef]
134. Shin, H.; Iwasaki, A. Tissue-resident memory T cells. *Immunol. Rev.* **2013**, *255*, 165–181. [CrossRef]
135. Sun, J.C.; Williams, M.A.; Bevan, M.J. CD4+ T cells are required for the maintenance, not programming, of memory CD8+ T cells after acute infection. *Nat. Immunol.* **2004**, *5*, 927–933. [CrossRef]
136. Azadniv, M.; Bowers, W.J.; Topham, D.J.; Crispe, I.N. CD4+ T cell effects on CD8+ T cell location defined using bioluminescence. *PLoS ONE* **2011**, *6*, e16222. [CrossRef]
137. Laidlaw, B.J.; Zhang, N.; Marshall, H.D.; Staron, M.M.; Guan, T.; Hu, Y.; Cauley, L.S.; Craft, J.; Kaech, S.M. CD4+ T cell help guides formation of CD103+ lung-resident memory CD8+ T cells during influenza viral infection. *Immunity* **2014**, *41*, 633–645. [CrossRef]
138. Beura, L.K.; Fares-Frederickson, N.J.; Steinert, E.M.; Scott, M.C.; Thompson, E.A.; Fraser, K.A.; Schenkel, J.M.; Vezys, V.; Masopust, D. CD4(+) resident memory T cells dominate immunosurveillance and orchestrate local recall responses. *J. Exp. Med.* **2019**, *216*, 1214–1229. [CrossRef]
139. Collins, N.; Jiang, X.; Zaid, A.; Macleod, B.L.; Li, J.; Park, C.O.; Haque, A.; Bedoui, S.; Heath, W.R.; Mueller, S.N.; et al. Skin CD4(+) memory T cells exhibit combined cluster-mediated retention and equilibration with the circulation. *Nat. Commun.* **2016**, *7*, 11514. [CrossRef]
140. Turner, D.L.; Farber, D.L. Mucosal resident memory CD4 T cells in protection and immunopathology. *Front. Immunol.* **2014**, *5*, 331. [CrossRef]
141. Clark, R.A. Resident memory T cells in human health and disease. *Sci. Transl. Med.* **2015**, *7*, 269rv261. [CrossRef]

142. Turner, D.L.; Goldklang, M.; Cvetkovski, F.; Paik, D.; Trischler, J.; Barahona, J.; Cao, M.; Dave, R.; Tanna, N.; D'Armiento, J.M.; et al. Biased Generation and In Situ Activation of Lung Tissue-Resident Memory CD4 T Cells in the Pathogenesis of Allergic Asthma. *J. Immunol.* **2018**, *200*, 1561–1569. [CrossRef]
143. Seneschal, J.; Clark, R.A.; Gehad, A.; Baecher-Allan, C.M.; Kupper, T.S. Human epidermal Langerhans cells maintain immune homeostasis in skin by activating skin resident regulatory T cells. *Immunity* **2012**, *36*, 873–884. [CrossRef]
144. Scharschmidt, T.C.; Vasquez, K.S.; Truong, H.A.; Gearty, S.V.; Pauli, M.L.; Nosbaum, A.; Gratz, I.K.; Otto, M.; Moon, J.J.; Liese, J.; et al. a Wave of Regulatory T Cells into Neonatal Skin Mediates Tolerance to Commensal Microbes. *Immunity* **2015**, *43*, 1011–1021. [CrossRef]
145. Burns, J.; Bartholomew, B.; Lobo, S. Isolation of myelin basic protein-specific T cells predominantly from the memory T-cell compartment in multiple sclerosis. *Ann. Neurol.* **1999**, *45*, 33–39. [CrossRef]
146. Allegretta, M.; Nicklas, J.A.; Sriram, S.; Albertini, R.J. T cells responsive to myelin basic protein in patients with multiple sclerosis. *Science* **1990**, *247*, 718–721. [CrossRef]
147. Nielsen, B.R.; Ratzer, R.; Bornsen, L.; von Essen, M.R.; Christensen, J.R.; Sellebjerg, F. Characterization of naive, memory and effector T cells in progressive multiple sclerosis. *J. Neuroimmunol.* **2017**, *310*, 17–25. [CrossRef]
148. Elyaman, W.; Kivisakk, P.; Reddy, J.; Chitnis, T.; Raddassi, K.; Imitola, J.; Bradshaw, E.; Kuchroo, V.K.; Yagita, H.; Sayegh, M.H.; et al. Distinct functions of autoreactive memory and effector CD4+ T cells in experimental autoimmune encephalomyelitis. *Am. J. Pathol.* **2008**, *173*, 411–422. [CrossRef]
149. McKeever, U.; Mordes, J.P.; Greiner, D.L.; Appel, M.C.; Rozing, J.; Handler, E.S.; Rossini, A.A. Adoptive transfer of autoimmune diabetes and thyroiditis to athymic rats. *Proc. Natl. Acad. Sci. USA* **1990**, *87*, 7618–7622. [CrossRef]
150. Mochizuki, M.; Kuwabara, T.; McAllister, C.; Nussenblatt, R.B.; Gery, I. Adoptive transfer of experimental autoimmune uveoretinitis in rats. Immunopathogenic mechanisms and histologic features. *Investig. Ophthalmol. Vis. Sci.* **1985**, *26*, 1–9.
151. Kawakami, N.; Odoardi, F.; Ziemssen, T.; Bradl, M.; Ritter, T.; Neuhaus, O.; Lassmann, H.; Wekerle, H.; Flugel, A. Autoimmune CD4+ T cell memory: Lifelong persistence of encephalitogenic T cell clones in healthy immune repertoires. *J. Immunol.* **2005**, *175*, 69–81. [CrossRef]
152. Swain, S.; Clise-Dwyer, K.; Haynes, L. Homeostasis and the age-associated defect of CD4 T cells. *Semin. Immunol.* **2005**, *17*, 370–377. [CrossRef]
153. Moro-Garcia, M.A.; Alonso-Arias, R.; Lopez-Larrea, C. When Aging Reaches CD4+ T-Cells: Phenotypic and Functional Changes. *Front. Immunol.* **2013**, *4*, 107. [CrossRef]
154. Haynes, L.; Swain, S.L. Why aging T cells fail: Implications for vaccination. *Immunity* **2006**, *24*, 663–666. [CrossRef]
155. Haynes, L.; Eaton, S.M.; Burns, E.M.; Randall, T.D.; Swain, S.L. CD4 T cell memory derived from young naive cells functions well into old age, but memory generated from aged naive cells functions poorly. *Proc. Natl. Acad. Sci. USA* **2003**, *100*, 15053–15058. [CrossRef]
156. Prelog, M. Aging of the immune system: a risk factor for autoimmunity? *Autoimmun. Rev.* **2006**, *5*, 136–139. [CrossRef]
157. Goronzy, J.J.; Weyand, C.M. Immune aging and autoimmunity. *Cell Mol. Life Sci.* **2012**, *69*, 1615–1623. [CrossRef]
158. Bolton, C.; Smith, P.A. The influence and impact of ageing and immunosenescence (ISC) on adaptive immunity during multiple sclerosis (MS) and the animal counterpart experimental autoimmune encephalomyelitis (EAE). *Ageing Res. Rev.* **2018**, *41*, 64–81. [CrossRef]
159. Fulop, T.; Dupuis, G.; Witkowski, J.M.; Larbi, A. The Role of Immunosenescence in the Development of Age-Related Diseases. *Rev. Invest. Clin.* **2016**, *68*, 84–91.
160. Ray, D.; Yung, R. Immune senescence, epigenetics and autoimmunity. *Clin Immunol* **2018**, *196*, 59–63. [CrossRef]
161. Grebenciucova, E.; Berger, J.R. Immunosenescence: The Role of Aging in the Predisposition to Neuro-Infectious Complications Arising from the Treatment of Multiple Sclerosis. *Curr. Neurol. Neurosci. Rep.* **2017**, *17*, 61. [CrossRef]
162. Mills, E.A.; Mao-Draayer, Y. Aging and lymphocyte changes by immunomodulatory therapies impact PML risk in multiple sclerosis patients. *Mult. Scler.* **2018**, *24*, 1014–1022. [CrossRef]

163. Foley, J.; Christensen, A.; Hoyt, T.; Foley, A.; Metzger, R. *Is Aging and Immunosenescence a Risk Factor for Dimethyl Fumarate Induced PML? (P2. 088)*; AAN Enterprises: Karnataka, India, 2016.
164. Elyahu, Y.; Hekselman, I.; Eizenberg-Magar, I.; Berner, O.; Strominger, I.; Schiller, M.; Mittal, K.; Nemirovsky, A.; Eremenko, E.; Vital, A.; et al. Aging promotes reorganization of the CD4 T cell landscape toward extreme regulatory and effector phenotypes. *Sci. Adv.* **2019**, *5*, eaaw8330. [CrossRef]
165. von Delwig, A.; Locke, J.; Robinson, J.H.; Ng, W.F. Response of Th17 cells to a citrullinated arthritogenic aggrecan peptide in patients with rheumatoid arthritis. *Arthritis Rheum* **2010**, *62*, 143–149. [CrossRef]
166. Law, S.C.; Street, S.; Yu, C.H.; Capini, C.; Ramnoruth, S.; Nel, H.J.; van Gorp, E.; Hyde, C.; Lau, K.; Pahau, H.; et al. T-cell autoreactivity to citrullinated autoantigenic peptides in rheumatoid arthritis patients carrying HLA-DRB1 shared epitope alleles. *Arthritis Res. Ther.* **2012**, *14*, R118. [CrossRef]
167. Feitsma, A.L.; van der Voort, E.I.; Franken, K.L.; el Bannoudi, H.; Elferink, B.G.; Drijfhout, J.W.; Huizinga, T.W.; de Vries, R.R.; Toes, R.E.; Ioan-Facsinay, A. Identification of citrullinated vimentin peptides as T cell epitopes in HLA-DR4-positive patients with rheumatoid arthritis. *Arthritis Rheumatol.* **2010**, *62*, 117–125. [CrossRef]
168. Cianciotti, B.C.; Ruggiero, E.; Campochiaro, C.; Oliveira, G.; Magnani, Z.I.; Baldini, M.; Doglio, M.; Tassara, M.; Manfredi, A.A.; Baldissera, E.; et al. CD4(+) memory stem T cells recognizing citrullinated epitopes are expanded in patients with Rheumatoid Arthritis and sensitive to TNF-alpha blockade. *Arthritis Rheumatol.* **2019**. [CrossRef]
169. Snir, O.; Backlund, J.; Bostrom, J.; Andersson, I.; Kihlberg, J.; Buckner, J.H.; Klareskog, L.; Holmdahl, R.; Malmstrom, V. Multifunctional T cell reactivity with native and glycosylated type II collagen in rheumatoid arthritis. *Arthritis Rheumatol.* **2012**, *64*, 2482–2488. [CrossRef]
170. Paroni, M.; Maltese, V.; De Simone, M.; Ranzani, V.; Larghi, P.; Fenoglio, C.; Pietroboni, A.M.; De Riz, M.A.; Crosti, M.C.; Maglie, S.; et al. Recognition of viral and self-antigens by TH1 and TH1/TH17 central memory cells in patients with multiple sclerosis reveals distinct roles in immune surveillance and relapses. *J. Allergy Clin. Immunol.* **2017**, *140*, 797–808. [CrossRef]
171. Venken, K.; Hellings, N.; Hensen, K.; Rummens, J.L.; Stinissen, P. Memory CD4+CD127high T cells from patients with multiple sclerosis produce IL-17 in response to myelin antigens. *J. Neuroimmunol.* **2010**, *226*, 185–191. [CrossRef]
172. Yang, J.; Sundrud, M.S.; Skepner, J.; Yamagata, T. Targeting Th17 cells in autoimmune diseases. *Trends Pharm. Sci* **2014**, *35*, 493–500. [CrossRef]
173. McGeachy, M.J.; Cua, D.J.; Gaffen, S.L. The IL-17 Family of Cytokines in Health and Disease. *Immunity* **2019**, *50*, 892–906. [CrossRef]
174. Kryczek, I.; Zhao, E.; Liu, Y.; Wang, Y.; Vatan, L.; Szeliga, W.; Moyer, J.; Klimczak, A.; Lange, A.; Zou, W. Human TH17 cells are long-lived effector memory cells. *Sci. Transl. Med.* **2011**, *3*, 104ra100. [CrossRef]
175. McGeachy, M.J.; Chen, Y.; Tato, C.M.; Laurence, A.; Joyce-Shaikh, B.; Blumenschein, W.M.; McClanahan, T.K.; O'Shea, J.J.; Cua, D.J. The interleukin 23 receptor is essential for the terminal differentiation of interleukin 17-producing effector T helper cells in vivo. *Nat. Immunol.* **2009**, *10*, 314–324. [CrossRef]
176. Kebir, H.; Ifergan, I.; Alvarez, J.I.; Bernard, M.; Poirier, J.; Arbour, N.; Duquette, P.; Prat, A. Preferential recruitment of interferon-gamma-expressing TH17 cells in multiple sclerosis. *Ann. Neurol.* **2009**, *66*, 390–402. [CrossRef]
177. Haines, C.J.; Chen, Y.; Blumenschein, W.M.; Jain, R.; Chang, C.; Joyce-Shaikh, B.; Porth, K.; Boniface, K.; Mattson, J.; Basham, B.; et al. Autoimmune memory T helper 17 cell function and expansion are dependent on interleukin-23. *Cell Rep* **2013**, *3*, 1378–1388. [CrossRef]
178. Kleinschek, M.A.; Boniface, K.; Sadekova, S.; Grein, J.; Murphy, E.E.; Turner, S.P.; Raskin, L.; Desai, B.; Faubion, W.A.; de Waal Malefyt, R.; et al. Circulating and gut-resident human Th17 cells express CD161 and promote intestinal inflammation. *J. Exp. Med.* **2009**, *206*, 525–534. [CrossRef]
179. Yen, D.; Cheung, J.; Scheerens, H.; Poulet, F.; McClanahan, T.; McKenzie, B.; Kleinschek, M.A.; Owyang, A.; Mattson, J.; Blumenschein, W.; et al. IL-23 is essential for T cell-mediated colitis and promotes inflammation via IL-17 and IL-6. *J Clin Invest* **2006**, *116*, 1310–1316. [CrossRef]
180. Wei, L.; Laurence, A.; Elias, K.M.; O'Shea, J.J. IL-21 is produced by Th17 cells and drives IL-17 production in a STAT3-dependent manner. *J. Biol. Chem.* **2007**, *282*, 34605–34610. [CrossRef]
181. Tian, Y.; Zajac, A.J. IL-21 and T Cell Differentiation: Consider the Context. *Trends Immunol.* **2016**, *37*, 557–568. [CrossRef]

182. Kastirr, I.; Maglie, S.; Paroni, M.; Alfen, J.S.; Nizzoli, G.; Sugliano, E.; Crosti, M.C.; Moro, M.; Steckel, B.; Steinfelder, S.; et al. IL-21 is a central memory T cell-associated cytokine that inhibits the generation of pathogenic Th1/17 effector cells. *J. Immunol.* **2014**, *193*, 3322–3331. [CrossRef]
183. Fritsch, R.D.; Shen, X.; Illei, G.G.; Yarboro, C.H.; Prussin, C.; Hathcock, K.S.; Hodes, R.J.; Lipsky, P.E. Abnormal differentiation of memory T cells in systemic lupus erythematosus. *Arthritis Rheumatol.* **2006**, *54*, 2184–2197. [CrossRef]
184. Mao-Draayer, Y.; Sarazin, J.; Fox, D.; Schiopu, E. The sphingosine-1-phosphate receptor: a novel therapeutic target for multiple sclerosis and other autoimmune diseases. *Clin. Immunol.* **2017**, *175*, 10–15. [CrossRef]
185. Matloubian, M.; Lo, C.G.; Cinamon, G.; Lesneski, M.J.; Xu, Y.; Brinkmann, V.; Allende, M.L.; Proia, R.L.; Cyster, J.G. Lymphocyte egress from thymus and peripheral lymphoid organs is dependent on S1P receptor 1. *Nature* **2004**, *427*, 355–360. [CrossRef]
186. Aoki, M.; Aoki, H.; Ramanathan, R.; Hait, N.C.; Takabe, K. Sphingosine-1-Phosphate Signaling in Immune Cells and Inflammation: Roles and Therapeutic Potential. *Mediat. Inflamm.* **2016**, *2016*, 8606878. [CrossRef]
187. Masopust, D.; Schenkel, J.M. The integration of T cell migration, differentiation and function. *Nat. Rev. Immunol.* **2013**, *13*, 309–320. [CrossRef]
188. Pham, T.H.; Okada, T.; Matloubian, M.; Lo, C.G.; Cyster, J.G. S1P1 receptor signaling overrides retention mediated by G alpha i-coupled receptors to promote T cell egress. *Immunity* **2008**, *28*, 122–133. [CrossRef]
189. Pinschewer, D.D.; Brinkmann, V.; Merkler, D. Impact of sphingosine 1-phosphate modulation on immune outcomes. *Neurology* **2011**, *76*, S15–s19. [CrossRef]
190. Mehling, M.; Brinkmann, V.; Antel, J.; Bar-Or, A.; Goebels, N.; Vedrine, C.; Kristofic, C.; Kuhle, J.; Lindberg, R.L.; Kappos, L. FTY720 therapy exerts differential effects on T cell subsets in multiple sclerosis. *Neurology* **2008**, *71*, 1261–1267. [CrossRef]
191. Mehling, M.; Lindberg, R.; Raulf, F.; Kuhle, J.; Hess, C.; Kappos, L.; Brinkmann, V. Th17 central memory T cells are reduced by FTY720 in patients with multiple sclerosis. *Neurology* **2010**, *75*, 403–410. [CrossRef]
192. Fujii, C.; Kondo, T.; Ochi, H.; Okada, Y.; Hashi, Y.; Adachi, T.; Shin-Ya, M.; Matsumoto, S.; Takahashi, R.; Nakagawa, M.; et al. Altered T cell phenotypes associated with clinical relapse of multiple sclerosis patients receiving fingolimod therapy. *Sci. Rep.* **2016**, *6*, 35314. [CrossRef]
193. Herich, S.; Schneider-Hohendorf, T.; Rohlmann, A.; Khaleghi Ghadiri, M.; Schulte-Mecklenbeck, A.; Zondler, L.; Janoschka, C.; Ostkamp, P.; Richter, J.; Breuer, J.; et al. Human CCR5high effector memory cells perform CNS parenchymal immune surveillance via GZMK-mediated transendothelial diapedesis. *Brain* **2019**, *142*, 3411–3427. [CrossRef]
194. Surh, C.D.; Sprent, J. Homeostasis of naive and memory T cells. *Immunity* **2008**, *29*, 848–862. [CrossRef]
195. Surh, C.D.; Boyman, O.; Purton, J.F.; Sprent, J. Homeostasis of memory T cells. *Immunol. Rev.* **2006**, *211*, 154–163. [CrossRef]
196. Ribas, A.; Shin, D.S.; Zaretsky, J.; Frederiksen, J.; Cornish, A.; Avramis, E.; Seja, E.; Kivork, C.; Siebert, J.; Kaplan-Lefko, P.; et al. PD-1 Blockade Expands Intratumoral Memory T Cells. *Cancer Immunol. Res.* **2016**, *4*, 194–203. [CrossRef]
197. De Sousa Linhares, A.; Leitner, J.; Grabmeier-Pfistershammer, K.; Steinberger, P. Not All Immune Checkpoints Are Created Equal. *Front Immunol.* **2018**, *9*, 1909. [CrossRef]
198. Workman, C.J.; Cauley, L.S.; Kim, I.J.; Blackman, M.A.; Woodland, D.L.; Vignali, D.A. Lymphocyte activation gene-3 (CD223) regulates the size of the expanding T cell population following antigen activation in vivo. *J. Immunol.* **2004**, *172*, 5450–5455. [CrossRef]
199. Avery, L.; Filderman, J.; Szymczak-Workman, A.L.; Kane, L.P. Tim-3 co-stimulation promotes short-lived effector T cells, restricts memory precursors, and is dispensable for T cell exhaustion. *Proc. Natl. Acad. Sci. USA* **2018**, *115*, 2455–2460. [CrossRef]
200. Metz, D.P.; Farber, D.L.; Taylor, T.; Bottomly, K. Differential role of CTLA-4 in regulation of resting memory versus naive CD4 T cell activation. *J. Immunol.* **1998**, *161*, 5855–5861.
201. Wherry, E.J.; Kurachi, M. Molecular and cellular insights into T cell exhaustion. *Nat. Rev. Immunol.* **2015**, *15*, 486–499. [CrossRef]
202. Kim, T.S.; Hufford, M.M.; Sun, J.; Fu, Y.X.; Braciale, T.J. Antigen persistence and the control of local T cell memory by migrant respiratory dendritic cells after acute virus infection. *J. Exp. Med.* **2010**, *207*, 1161–1172. [CrossRef]

203. Vignali, D.A.; Collison, L.W.; Workman, C.J. How regulatory T cells work. *Nat. Rev. Immunol.* **2008**, *8*, 523–532. [CrossRef]
204. Matarese, G.; De Rosa, V.; La Cava, A. Regulatory CD4 T cells: Sensing the environment. *Trends Immunol.* **2008**, *29*, 12–17. [CrossRef]
205. Fehervari, Z.; Sakaguchi, S. CD4+ Tregs and immune control. *J Clin Invest* **2004**, *114*, 1209–1217. [CrossRef]
206. Luckheeram, R.V.; Zhou, R.; Verma, A.D.; Xia, B. CD4(+)T cells: Differentiation and functions. *Clin. Dev. Immunol.* **2012**, *2012*, 925135. [CrossRef]
207. Li, J.; Huston, G.; Swain, S.L. IL-7 promotes the transition of CD4 effectors to persistent memory cells. *J. Exp. Med.* **2003**, *198*, 1807–1815. [CrossRef]
208. Chetoui, N.; Boisvert, M.; Gendron, S.; Aoudjit, F. Interleukin-7 promotes the survival of human CD4+ effector/memory T cells by up-regulating Bcl-2 proteins and activating the JAK/STAT signalling pathway. *Immunology* **2010**, *130*, 418–426. [CrossRef]
209. Bradley, L.M.; Haynes, L.; Swain, S.L. IL-7: Maintaining T-cell memory and achieving homeostasis. *Trends Immunol.* **2005**, *26*, 172–176. [CrossRef]
210. Purton, J.F.; Tan, J.T.; Rubinstein, M.P.; Kim, D.M.; Sprent, J.; Surh, C.D. Antiviral CD4+ memory T cells are IL-15 dependent. *J. Exp. Med.* **2007**, *204*, 951–961. [CrossRef]
211. Picker, L.J.; Reed-Inderbitzin, E.F.; Hagen, S.I.; Edgar, J.B.; Hansen, S.G.; Legasse, A.; Planer, S.; Piatak, M., Jr.; Lifson, J.D.; Maino, V.C.; et al. IL-15 induces CD4 effector memory T cell production and tissue emigration in nonhuman primates. *J. Clin. Invest.* **2006**, *116*, 1514–1524. [CrossRef]
212. Smolders, J.; Heutinck, K.M.; Fransen, N.L.; Remmerswaal, E.B.M.; Hombrink, P.; Ten Berge, I.J.M.; van Lier, R.A.W.; Huitinga, I.; Hamann, J. Tissue-resident memory T cells populate the human brain. *Nat. Commun.* **2018**, *9*, 4593. [CrossRef]
213. Machado-Santos, J.; Saji, E.; Troscher, A.R.; Paunovic, M.; Liblau, R.; Gabriely, G.; Bien, C.G.; Bauer, J.; Lassmann, H. The compartmentalized inflammatory response in the multiple sclerosis brain is composed of tissue-resident CD8+ T lymphocytes and B cells. *Brain* **2018**, *141*, 2066–2082. [CrossRef]
214. Prasad, S.; Hu, S.; Sheng, W.S.; Chauhan, P.; Lokensgard, J.R. Recall Responses from Brain-Resident Memory CD8(+) T Cells (bTRM) Induce Reactive Gliosis. *iScience* **2019**, *20*, 512–526. [CrossRef]
215. Scholler, A.S.; Fonnes, M.; Nazerai, L.; Christensen, J.P.; Thomsen, A.R. Local Antigen Encounter Is Essential for Establishing Persistent CD8(+) T-Cell Memory in the CNS. *Front. Immunol.* **2019**, *10*, 351. [CrossRef]
216. Sabatino, J.J., Jr.; Wilson, M.R.; Calabresi, P.A.; Hauser, S.L.; Schneck, J.P.; Zamvil, S.S. Anti-CD20 therapy depletes activated myelin-specific CD8(+) T cells in multiple sclerosis. *Proc. Natl. Acad. Sci. USA* **2019**. [CrossRef]

© 2020 by the authors. Licensee MDPI, Basel, Switzerland. This article is an open access article distributed under the terms and conditions of the Creative Commons Attribution (CC BY) license (http://creativecommons.org/licenses/by/4.0/).

Review

Kynurenines in the Pathogenesis of Multiple Sclerosis: Therapeutic Perspectives

Tamás Biernacki [1], Dániel Sandi [1], Krisztina Bencsik [1] and László Vécsei [1,2,3,*]

[1] Department of Neurology, Faculty of General Medicine, Albert Szent-Györgyi Clinical Centre, University of Szeged, H-6725 Szeged, Hungary; biernacki.tamas@med.u-szeged.hu (T.B.); sandi.daniel@med.u-szeged.hu (D.S.); krisztina.bencsik@invitel.hu (K.B.)
[2] MTA—SZTE Neuroscience Research Group, H-6725 Szeged, Hungary
[3] Interdisciplinary Excellence Center, University of Szeged, H-6720 Szeged, Hungary
* Correspondence: vecsei.laszlo@med.u-szeged.hu; Tel.: +36-62-545-356; Fax: +36-62-545-597

Received: 1 June 2020; Accepted: 23 June 2020; Published: 26 June 2020

Abstract: Over the past years, an increasing amount of evidence has emerged in support of the kynurenine pathway's (KP) pivotal role in the pathogenesis of several neurodegenerative, psychiatric, vascular and autoimmune diseases. Different neuroactive metabolites of the KP are known to exert opposite effects on neurons, some being neuroprotective (e.g., picolinic acid, kynurenic acid, and the cofactor nicotinamide adenine dinucleotide), while others are toxic to neurons (e.g., 3-hydroxykynurenine, quinolinic acid). Not only the alterations in the levels of the metabolites but also disturbances in their ratio (quinolinic acid/kynurenic acid) have been reported in several diseases. In addition to the metabolites, the enzymes participating in the KP have been unearthed to be involved in modulation of the immune system, the energetic upkeep of neurons and have been shown to influence redox processes and inflammatory cascades, revealing a sophisticated, intertwined system. This review considers various methods through which enzymes and metabolites of the kynurenine pathway influence the immune system, the roles they play in the pathogenesis of neuroinflammatory diseases based on current evidence with a focus on their involvement in multiple sclerosis, as well as therapeutic approaches.

Keywords: kynurenine pathway; kynurenic acid; oxidative stress; quinolinic acid; N-acetylserotonin; IDO; NAD$^+$, multiple sclerosis; laquinimod

1. Introduction

Even though kynurenic acid was discovered roughly 170 years ago, it was not until the 1970s and 1980s that the kynurenine pathway (KP) sparked substantial interest among neuroscientists. This was due to the discovery that the two major products of the pathway, kynurenic acid (KYNA) and quinolinic acid (QUIN), possess significant, yet opposing effects on various neuronal cells and physiological processes [1]. The KP was found to be responsible for the overwhelming majority (>90%) of peripheric tryptophan (TRP) metabolism [2]. In the early days of kynurenine research, the belief held for a long time that the main purpose of the KP is solely the production of nicotinamide adenine dinucleotide (NAD$^+$), a coenzyme already known to be a pivotal molecule in a vast amount of vital biochemical processes including, but not limited to being a key component in several redox reactions and being vital to mitochondrial function [3]. In the past decades, however, significant attention has been directed to the enzymes and metabolites of the KP, after the discovery that an alteration can be found not only in the metabolite levels but in the activity of the enzymes producing them as well in numerous disorders. They have been implicated to play a role in neurodegenerative, psychiatric and developmental diseases, infections, tumors, autism, vascular diseases, allergies, transplant rejections, cancer immunity, immune privilege disorders and also in various autoimmune and neuroinflammatory conditions [4–10].

Kynurenic acid was the first member of the "kynurenine family" derived from the essential amino acid TRP. KYNA has been intensively studied in the past decades, has turned out to be a potent antagonist of excitatory ionotropic glutamate receptor on both the N-methyl-D-aspartate (NMDA) and glycineB site. On the other hand, the other main derivate of the pathway, quinolinic acid, is a selective agonist of the aforementioned receptor. The activation of the NMDA receptors (NMDAR) results in a cationic influx (Na^+, Ca^{2+}, K^+) to the cell; the subsequent increase in intracellular Ca^{2+} level activates several downstream signaling pathways and secondary messenger molecules, which ultimately lead to various synaptic alterations. The superfluous activation of the NMDARs, however, causes an excessive inflow of Ca^{2+} ions, eliciting neurotoxicity and cellular damage, which can even induce neuronal cell death. Neuronal damage due to the excitotoxicity caused by excessive NMDAR activation has been linked to a number of neurodegenerative disorders including Huntington's, Alzheimer's disease, amyotrophic lateral sclerosis (ALS), and multiple sclerosis (MS) [11]. Additional to the confirmed excitotoxicity mediated by the overactivation of the NMDARs, an increased level of QUIN was reported in the pathogenesis of these diseases as well [4–8,11]. In addition to finding an elevation in QUIN levels, a decreased amount of KYNA was observed in some of these diseases. This raised the question, that not simply the increasement of QUIN is essential to the pathogenesis, but a more complex dysregulation is underlying, causing a shift of the KYNA to QUIN ratio. Nowadays, through the advancement of genetic and molecular diagnostics, we gain an increasing amount of insight into the pathomechanism of the diseases burdening humanity; the KP seems to be a key player in many of them. In this review, following a concise introduction about the KP and its two most well-defined neuroactive metabolites, we aim to bring together recent evidence of their diverse effect on immunoregulatory mechanisms and their involvement in MS with a focus on a potential future therapeutic approach.

2. The Production and Metabolism of Kynurenines

Tryptophan is not only one of the most scarcely found amino acid in mammalian organisms (comprising roughly only 1–1.5% of the total protein amino acid content), but is also an essential amino acid for humans and the precursor amine for the synthesis of essential proteins, nicotinic acid, NAD^+, the neurotransmitter serotonin, N-acetylserotonin (NAS), and melatonin [12]. The metabolism of TRP can occur through two major pathways: the methoxyndole pathway (also known as the serotonin pathway, which accounts for ~5% of the metabolism) and the KP. The KP represents the primary route of metabolism for both the periphery and the central nervous system (CNS), accounting for approximately 95% of the TRP metabolism [13]. The metabolism of TRP is conducted by a chain of enzymes mostly found in glial cells after TRP enters the CNS and passes the blood-brain barrier (BBB) via the non-specific and competitive L-type amino acid transporter [14]. Roughly 10% of the total TRP circulating in the blood is bound to albumin; the rest of the unbound TRP can be transported across the BBB by the aforementioned transporter [3]. After entering the CNS, TRP can enter either the kynurenine pathway or the methoxyndole pathway.

2.1. Kynurenine Pathway

All of the intermediary metabolites of the KP are called kynurenines. The first and rate-limiting step of the KP is the conversion of TRP into N-formyl-l-kynurenine by two enzymes. The first enzyme, indoleamine 2,3-dioxygenase (IDO-1) is most prominently expressed and has the highest activity in dendritic cells [15,16]. It has also been shown to be overexpressed in certain neoplastic tissues [17–19] and underexpressed/defective in cases of autoimmune and neuroinflammatory diseases [20–24]. The isoenzyme IDO-2 is mainly expressed in the liver. The second enzyme responsible for the conversion of TRP, tryptophan 2,3-dioxygenase (TDO) is also abundantly expressed in the liver, is responsible for the regulation of the systemic levels of TRP [25]. TDO is also expressed, in lower levels of the CNS: in neurons, endothelial cells, and astrocytes, and similarly to IDO-1, it has been recently found in tumor cells [6,26]. IDO-1 can be potently induced by IFN-γ, TGF-β, Toll-like receptor ligands, and polyamines [27–29]. L-kynurenine is formed afterward from N-formyl-l-kynurenine by a formidase.

The activity of the two rate-limiting enzymes under physiological circumstances are normally quite low in the CNS [30]. Approximately 60% of metabolism along the KP in the brain is mediated by the uptake of the main metabolite, kynurenine from the blood by glial cells through the neutral amino acid carrier [31–33]. The pivotal product, l-kynurenine takes a central position in the pathway, as it can diverge into three very distinct directions from this molecule. L-kynurenine can be further processed by the enzymes kynureninase, kynurenine aminotransferases (KAT; thus far, four isoenzymes have been identified, of which KAT1 and -2 play a capital role in humans [34]), and kynure-nine 3-monooxygenase (KMO). The activity of these enzymes results in the formation of 3-hydroxy-l-kynurenine, KYNA, and anthranilic acid (ANA), respectively. The next molecule, 3-hydroxyanthranilic acid can be either formed from ANA by nonspecific hydroxylases or from 3-hydroxy-l-kynurenine by the action of the kynureninase enzyme. 3-hydroxy-l-kynurenine can be further converted into xanthurenic acid by the KAT enzymes. 3-hydroxyanthranilic acid is further metabolized by 3-hydroxyanthranilate oxidase (3-HAO) into 2-amino-3-carboxymuconate-semialdehyde, an unstable compound. It can be degraded to 2-aminomuconic acid or enzymatically converted into the neuroprotective picolinic acid by the 2-amino-3-carboxymuconate-semialdehyde decarboxylase, or non-enzymatically transformed into QUIN (predominantly in dendritic cells infiltrating the CNS and in microglia), a neurotoxic precursor of NAD^+ and $NADP^+$ (Figure 1).

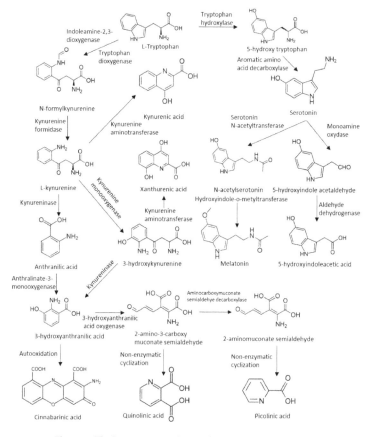

Figure 1. The kynurenine pathway of tryptophan metabolism.

2.2. Methoxyndole Pathway

Another direction of TRP metabolism is through the methoxyndole/serotonin pathway. First, TRP is converted to 5-hydroxy tryptophan by the tryptophan-5-hydroxylase, which afterwards is transformed into 5-hydroxytryptamine (also known as serotonin) by the aromatic L-amino acid decarboxylase enzyme. Second, serotonin is acetylated by an alkylamine N-acetyltransferase forming N-acetylserotonin, which in the end is methylated by an acetylserotonin O-methyltransferase to form melatonin (Figure 1).

3. Neuroactive Metabolites of Tryptophan

3.1. Kynurenic Acid

As mentioned before, KYNA is the end product of one of the three separate branches of the KP. The formation of KYNA is not only segregated from the other pathways (mainly the QUIN branch) enzymatically but physically as well [8,35,36]. In contrast to other products of the KP, KYNA—based on in vitro studies—is mainly considered to be formed in astrocytes, while the other metabolites are produced in microglial cells [37–43]. Additionally, opposed to several other KP metabolites—because of its polar nature and the lack of transporter mechanisms—KYNA produced in the periphery cannot cross the BBB. In order to exert its function in the CNS it has to be de novo synthesized there, for most of which (~75%) the KAT2 enzyme is responsible [44]. The reported concentrations of KYNA in the brain are 0.2–1.5 µM [45–47] (Table 1).

Table 1. Major binding sites and actions of kynurenic acid.

Receptor	Ligand	Action	IC/EC50	Effect
GPR35	cGMP, LPA, T3, rT3, DHICA	Agonist	1–10 µM [48–52]	hyperpolarisation, adenylate cyclase inhibition
AHR	Xenobiotic chemicals	Agonist	10-100 µM [53–55]	migration, proliferation, immunmodulation
NMDAR	Glycine, D-serine (glycine-2 co agonist NR1 site)	Antagonist	~8–10 µM [56–60]	excitation, plasticity, neurodegeneration, depolarization, Ca^{2+} influx
NMDAR	Glutamate, NMDA (glutamate/NMDA NR2 site)	Antagonist	~200–500 µM [57,58,60–65]	excitation, neurodegeneration, depolarization, Ca^{2+} influx
AMPA/Kainate	Glutamate	Antagonist	~250 µM [57,58,64–66]	excitation, depolarization
Free radicals	n/a	n/a	>200 µM [67,68]	hydroxyl, superoxide radical complexation

Abbreviations: GPR35, G protein-coupled receptor 35; AHR, aryl hydrocarbon receptor; NMDAR, N-methyl-D-aspartic acid receptor; AMPA, α-amino-3-hydroxy-5-methyl-4-isoxazolepropionic acid; cGMP, cyclic guanosine monophosphate, LPA, lysophosphatidic acid, T3, triiodothyronine, rT3, reverse triiodothyronine, DHICA, 5,6-dihydroxyindole-2-carboxylic acid. IC/EC50, half maximal inhibitory concentration and half maximal effective concentration respectively.

KYNA possesses neuroprotective properties, as it is an endogenous, broad-spectrum competitive antagonist of all three ionotropic glutamate receptors. It has a stronger affinity for NMDARs and weaker antagonistic effect on kainate and AMPA receptors. KYNA exhibits a particularly high affinity for the strychnine-insensitive glycineB binding NR1 site of the NMDA receptor, as it is able to bind to it and block its activity already in the low micromolar concentrations (IC50 ~ 7.9–15 µM) [69–71]. KYNA can bind to the NMDA recognition site of the receptor as well, albeit much higher concentrations are needed for inhibition (IC_{50} ~ 7.9–15 µM for the NR1 subunit vs. IC_{50} ~ 200–500 µM for the NR2 glutamate-binding site subunit of NMDAR) [70,72]. In low concentrations, the effect of KYNA on AMPA receptor-mediated responses is paradoxically facilitatory (nanomolar to the micromolar range); neuroinhibition is only achieved at high concentrations (micromolar to the millimolar range) of KYNA [73,74]. The inhibition of the NMDA receptors by KYNA has been shown to be able to defend the neurons against the toxic effects of excessive Ca^{2+} influx—which is one of the most investigated and proven processes in neurodegeneration—via the inhibition of extrasynaptic NMDA receptors during excitotoxic insults.

The peripheral blockade of the KMO enzyme has been shown to increase l-kynurenine levels in the blood, which can readily cross the BBB, and is subsequently transformed into KYNA in the CNS. In the CNS, parallel to the rise of KYNA, a drop was observed in the extracellular glutamate level along with a reduced level of synaptic loss, decreased amount of anxiety-related behavior as well as microglial activation in animal models of Alzheimer's disease and Huntington's disease, respectively [75].

In a pivotal study in 2001 it was suggested, that KYNA can noncompetitively bind to the α7 homomeric nicotinic receptors as well [76]. This theory, however, is under heavy dispute nowadays, as to date there is more evidence in support of the fact that KYNA does not influence nicotinic Ach receptors (for a very detailed review on the topic see [73,77,78]). The effects previously attributed to the blockade of the α7 homomeric nicotinic receptors might in fact be due to KYNA's well established and proven actions on other receptors, which are concomitantly expressed on neurons producing α7 homomeric nicotinic receptors.

KYNA is also an endogenous, but—compared to other endogenous ligands—weak activator of the G-protein coupled receptor GPR35 [48,79–81]. GPR35 was discovered at the end of the last century, designated an orphan receptor for almost two decades, thought to be expressed predominantly in the gut and immune cells. Recent studies have shown, however, that GPR35 is in fact the receptor for the mucosal chemokine CXCL17. It is expressed in mucosal tissue, monocytoid cell lines, CD14$^+$ monocytes, T cells, neutrophils, dendritic cells, and in lower levels in B cells, eosinophils, basophils, and on iNKT cells [48,82–84]. In the CNS, GPR35 has been found on astrocytes, and has also been linked to nociception and neurotransmission in the CA1 region of the hippocampus [85,86]. The extent to which the agonistic activity of KYNA is relevant physiologically or under the several pathological conditions that GPR35 has been linked to, unfortunately remains unclear thus far [80].

Furthermore, KYNA acts as a potent endogenous agonist at the aryl hydrocarbon receptor (AHR), a transcription factor in the helix-loop-helix (bHLH) Per/ARNT/Sim family, which is expressed in tumor cells, as well as various cells of the immune system [54,87]. In recent years, it has been shown that not only KYNA, but also other tryptophan metabolites (KYN, NAS), are able to activate AHR responses resulting in different downstream effects [55,88]. The binding of kynurenines to AHR results in the internalization of the receptor into the nucleus where it binds to target genes and activates their transcription [89]. Previously AHR was considered a xenobiotic-sensing receptor that is involved in the metabolism of exogenous toxins. More recent data suggests, however, that it has a function in immune-regulation, carcinogenesis, and tumor growth (through the promotion of the generation of immunosuppressive T cells that support cancer development). It has also been shown to play a role in modulating the synthesis of inflammatory mediators [54,90,91]. The exact role of the AHR in the pathways that it is involved in, however, is yet to be clarified in detail.

Additional to its actions at various receptors, an increasing amount of data shows that KYNA also exhibits potent antioxidant traits in the CNS at physiological conditions [67]. In vivo and in vitro data evidence suggests that KYNA is not only able to halt lipid peroxidation, but can act as a scavenger for hydroxyl and superoxide anions as well as other free radicals, further expanding its diverse neuroprotective properties [92,93].

3.2. N-Acetylserotonin

An intermediate product of the MP of the TRP metabolism, N-acetylserotonin (NAS) has been shown to possess potent antioxidant, antiischemic and antidepressant properties in mice, and was also able to mitigate the neuroinflammation in mice with experimental autoimmune encephalitis (EAE), the most ubiquitously used model of MS [94–96]. As expected, high levels of NAS were isolated from the CNS of mice (from the cervical lymph nodes), especially in the recovery phase of EAE. In fact, NAS has been shown to easily cross the BBB, where it elicited virtually no toxicity. Even though there is an increasing amount of evidence for the presence of the enzyme-producing NAS in organs with function related to the immune system (spleen, bone marrow, and thymus), the exact source of NAS on the periphery remains yet to be pinpointed [97,98].

Recent evidence also suggests that IDO-1 and AHR are functionally intertwined in the modulation of immune responses through a positive feedback loop involving the kynurenine pathway [99]. L-kynurenine is a major endogenous activating ligand of the AHR. The activation of AHR by KYN increases both the expression and the activity of IDO1 in conventional (CD11c+) dendritic cells (where IDO-1 has the highest expression and catalytic activity). This in turn results in the upregulation of TRP breakdown and increased amounts of KYN, which further promotes AHR activity, thus creating a positive feedback loop [15,16,55,88].

A recent study explored the immunomodulatory effect of NAS in EAE. NAS was able to ameliorate the inflammation in EAE in mice with functioning IDO-1 and AHR genes. This effect was lost, however in IDO-1$^{-/-}$ and/or AHR$^{-/-}$ knockout mice [97]. NAS conferred an immunosuppressive effect on dendritic cells via positive allosteric modulation of the IDO1 enzyme. After binding to an allosteric site on the AHR increased IDO-1's catalytic activity, but no changes in gene or protein expression levels were seen. This resulted in increased AHR activation and immunosuppression via the aforementioned feedback loop [97,100].

Not so long ago, decreased IDO-1 expression and KYN levels (but not TDO expression) were found in the peripheral blood monocytes (PBMC) of relapsing-remitting MS (RRMS) patients compared to healthy controls [23]. When stimulated with IFN-γ, a significant increase in kynurenine production and IDO-1 activity was seen in PBMCs from healthy subjects but not RRMS patients. However, the deficient IDO-1 activity seen in MS patients' PBMCs could be increased to levels comparable to those of healthy controls after incubation with NAS.

Up to the present day, the main therapeutic approach in the treatment of MS always has been relatively unselective immunosuppression with drugs that impair the adaptive immune response and the activation and proliferation of T and/or B lymphocytes. The deficiency of IDO-1 enzyme has been reported in a number of autoimmune disorders. Emerging evidence suggests that IDO-1 is a novel type of immune checkpoint molecule—a family of molecular regulators that are pivotal parts of the immune system to obtain self-tolerance and to prevent the development of autoimmune conditions—with diverse effects on both the effector and regulatory arms of the human immune response [101–103]. The discovery of an endogenous, IDO-1-selective positive allosteric modulator, that has the potency to restore the activity of the deficient enzyme to the physiological level may pave the way for the development of novel drugs for MS with a completely different mechanism of action than before.

3.3. Quinolinic Acid

The increased level of quinolinic acid (QUIN) in the CNS has been suggested to be a crucial element in the pathogenesis of several neurological diseases (including, but not limited to, Huntington's disease, Alzheimer's disease, schizophrenia, autism, depression, and epilepsy) [104]. The neurotoxic properties of QUIN are manifold, well-described, and have been investigated in depth in the past decades. QUIN's neuroexcitatory properties can be attributed to several mechanisms, but are mainly due to an extremely specific, though weak (ED50: >100 µM) competitive agonism of the NMDA receptor containing the subunits NR2A, NR2B, and NR2C [5,105].

The NMDA receptors are unequivalently sensitive to QUIN based on their subunit composition, thus QUIN exerts different levels of toxicity in different sites of the brain (because of the different subunit composition of the receptor in different localisations), based on the dominant receptor type in a given anatomical region. It is more than 10-fold more preferential for the NR2B subunit of the NMDA receptors, which are mainly found in the forebrain (abundant in the neocortex and the hippocampus) than it is for the NR2C hindbrain-specific subunit (located mostly in the hindbrain, particularly in the cerebellum and the spinal cord) [106–111]. The undisputable, obligatory role of NMDA receptor activation in QUIN mediated neurotoxicity is supported by the findings that all of its toxic effects studied to date can readily be prevented by NMDA receptor antagonists [1,112]. QUIN can also cause excitotoxicity via the stimulation of synaptosomal glutamate release, inhibition of glutamate reuptake by astrocytes, and the enhancement of reactive oxygen species formation. Reactive oxygen species formation by QUIN requires the presence of Fe^{2+} and the subsequent autooxidation of Fe^{2+}-QUIN complexes, which process can be negated by iron chelation [113]. QUIN also enhances lipid peroxidation for which the presence of Fe^{2+} ions is also obligatory. On the other hand, the presence of FeCl2 deteriorates QUIN's ability to activate the NMDA receptors [114–116]. QUIN is also able to cause damage by the depletion of various endogenous antioxidants and phosphorylate certain proteins implicated in various neurodegenerative disorders [5,117,118].

Furthermore, vast amounts of evidence suggest that QUIN does not simply possess neurotoxic effects, but is tightly involved with the immune system. Certain pro-inflammatory cytokines (TNFα, interleukin-1β) promote the production and the toxicity of QUIN, respectively [119,120]. Certain anti-inflammatory cytokines (such as IL-4), however, can diminish the production of QUIN by inhibiting the IDO and TDO enzymes, while the blockade of certain receptors (adenosine A2A for example) can protect neurons from the toxic effects of QUIN [120,121].

4. The Kynurenine System and Immunoregulation

Various metabolites of the KP have been shown to have profound effects on the functionality of the immune system. Kynurenines have endogenous immunosuppressive attributes through several complex pathways. They modulate the proliferation and function of several immune cells, while, they in turn, are modulated by the cytokines produced by them. In the next couple of paragraphs, we try to give insight into this circle of effects.

4.1. Effects on the Immune Cells

It seems that the kynurenines play a central role in T cell mediated immune responses. TRP metabolites, l-kynurenine in particular, can block antigen-specific T cell proliferation and can even induce apoptosis in these cells [84]. Kynurenines mainly induce negative feed-back loops and cell death in the T-helper-1 (Th1) cell population, while they promote upregulation in the Th2 population. This leads to a relative shift towards Th2 in the Th1–Th2 ratio; thus, kynurenines promote anti-inflammatory responses [84,122].

Their direct effects on the T-helper cells are, however, not the only way through which kynurenines can modulate T cell mediated immune responses. It seems that IDO-1 expression in dendritic cells (DC) induces the generation of a specific subset of T-regulatory (Treg) cells, the FoxP3$^+$ cells [123,124]. The main function of these Treg cells is seemingly to inhibit both the Th1 and Th2 cells and "guide" the immune system back to balance [84].

Kynurenines do not exert their immunomodulatory effects solely through the T cells, but also affect other immune cells. Natural killer (NK) cells are essential effector cells of the innate immune system. Kynurenines have been proven to suppress both the proliferation and the function of these cells. Through the generation of reactive oxygen species kynurenines halt their proliferation, and via the suppression of specific triggering receptors responsible for inducting NK cell-mediated killing function, they impair cell function [125,126]. These effects can lead to serious malfunction of these cells. Some data also suggest that the inhibition of IDO-1 in polymorphonuclear cells (PMN) impair their antifungal capabilities [127].

4.2. The Effect of Cytokines on Kynurenines

Cytokines, modulatory molecules produced (mostly) by immune cells can be divided into pro-inflammatory (e.g., TNF-α, IFN-γ, IL-1, and IL-6) and anti-inflammatory (e.g., IL-4, IL-10) groups. As the names suggest, their effect on the immune response is in direct contrast to each other. The delicate balance of the amount and function of these molecules is essential to the normal functioning of the immune system. It seems that the expression of pro-inflammatory cytokines stimulate the activity of the IDO-1 enzyme; therefore, they increase the rate of kynurenine production [84].

IFN-γ is the most important cytokine in the induction of the IDO-1 enzyme through transcriptomic regulation of the IDO gene [128]. IL-1 and TNF-α act in synergy with interferons in upregulating IDO activity [129]. Thus, the appearance of these cytokines lead indirectly to the increased production of kynurenines. TGF-β is another very potent IDO-1 inductor mainly in Treg and DC cells [130]. IL-23 levels also correlated with kynurenine production in Huntington's disease [131]. On the other hand, IL-4 and IL-13 are strong inhibitors of IFN-γ induced IDO-1 mRNA expression and kynurenine production [132].

4.3. Gut-Microbiome and Kynurenines

In recent years, a plethora of new information has surfaced about the gut microbiome after the discovery that it not only plays a major role in maintaining a healthy state for the host, but its perturbation has been linked to a vast amount of neurodegenerative and immunological diseases [133–139].

Not only are the thousands of bacteria colonizing the gastrointestinal tract in constant flux in response to the changes in the nutrition and diet and drug consumption of the host, but the metabolism of these bacteria and the cytokines and molecules produced by them are in a perpetual shift as well [137,140,141]. As mentioned above, TRP is an essential amino acid for humans, but many bacteria found in the gut can synthesize it. Additionally, bacteria are capable of producing neuroactive kynurenines—which can penetrate the BBB—by various other mechanisms [142–146]. QUIN can be de novo synthesized by gut bacteria from iminoaspartate or metabolized from anthranilic acid, but can also be produced non-enzymatically [147–149].

Numerous studies investigated the change in the composition of MS patients' gut bacteria. Changes in several species were detected; some were depleted, while other species were enriched compared to healthy individuals. The exact role, however, the gut bacteria (which have been associated with the development of MS) play in the regulation of the formation of neuroactive kynurenines or how they shift the balance between the neuroprotective KYNA and neurotoxic QUIN, remains unclear thus far [150–152]. Furthermore, how exactly the microbially regulated TRP metabolism and KP metabolites synthesized in the gut influence the function and dysfunction in the CNS remains elusive to date. How and where the pathway can be targeted is also of great interest and may hold significant therapeutic potential; however, this is yet to be discovered [153,154].

4.4. Disturbances of the Kynurenine System in Neuro-Immunological Conditions

IDO-1 overexpression and tryptophan metabolite disturbances have been found in a number of inflammatory conditions that can affect the CNS. Disturbances in the KP was proven in several acute exogenic inflammatory processes of bacterial or viral origin, as well as in autoimmune conditions such as systemic lupus erythematosus (SLE) and Sjögren's syndrome.

A recent study found major induction of the KP in bacterial and viral CNS infections that correlates strongly with blood-cerebrospinal fluid-barrier dysfunction and the standard measures of neuroinflammation in the cerebrospinal fluid (CSF) [155]. The most prominent finding was in viral meningitis, with the kynurenine/TRP ratio being the most reliable marker [155].

SLE is a systemic autoimmune condition that can affect several organ systems. Classical signs can be seen on the skin (e.g., "butterfly rash"), some forms of the disease are limited to the skin alone. More often than not, however, the disease affects other organ systems (most frequently the kidneys and the respiratory tract), and in a high number of cases, the nervous system as well. Some studies showed, that up to 75% of patients with SLE show some degree of involvement of the CNS with diverse symptoms both in appearance and severity [156]. Kynurenines have been studied particularly in this type of the disease, labeled as neuropsychiatric SLE. Higher levels of QUIN were measured in the plasma and the CSF of SLE patients compared to healthy controls along with lower levels of TRP [157].

Additionally, significant correlation was shown between the kynurenine/TRP ratio and TNF-α levels, demonstrating a connection between the pro-inflammatory pathway and kynurenines [157].

Sjögren's syndrome is classically defined as an autoimmune exocrinopathy: the main targets of the immune response are the salivary and lacrimal glands [158]. However, the involvement of the nervous system (either the periphery or the central) can be present in up to 60–80% of patients, ranging from peripheral neuropathy, to MS-like ("MS-mimic") conditions or neuromyelitis optica spectrum disorder (NMOSD) [159,160]. Disturbance of the kynurenines—not surprisingly—have been found in these patients as well. Analysis of the peripheral blood with flow cytometry of Sjögren's syndrome patients detected higher expression of IDO-1 in the patients' dendritic cells as compared to the dendritic cells of healthy controls [161]. Additionally, overexpression of IDO was demonstrated in both T cells and antigen-presenting cells (APC) of Sjögren's syndrome patients, compared to healthy controls [161,162].

4.5. The Role of the Kynurenine System in Multiple Sclerosis

Multiple sclerosis (MS) is an autoimmune, demyelinating, and neurodegenerative disease of the CNS. As such, kynurenines have been investigated in the pathomechanism of the disease in both pre-clinical and clinical assessments and their important role has been suggested by several studies.

4.5.1. Kynurenines and Animal Models of MS

In spite of vigorous and large-scale research conducted during the past decades into the pathogenesis, causes setting up and initiating MS, the exact trigger(s) remain elusive. Taking into account the complexity of the disease, the fact that the currently available and efficient treatments for MS have different targets (some have an effect on T cells, while some target B cell lines), the notion was raised that potentially different pathomechanisms may be underlying the disease recognized as MS. More recently, it was suggested that in fact not a single, but several intertwining pathological processes may be responsible for the demyelination of neurons and formation of lesions. This results in subsequent tissue injury, the cessation of saltatory nerve conduction, and increased axonal vulnerability followed by an eventual loss of neuronal function and successive cell death.

One hypothesis considers autoreactive T cells and macrophages that infiltrate the CNS and attack oligodendrocytes, which myelinate axons, to be the initiators of acute MS lesions [163]. This theory is supported by the oldest and most frequently used model system for studying MS: the experimental autoimmune encephalomyelitis (EAE) model, which is histologically similar to human MS. In its classic form, EAE mimics the chronic form of MS. In this version, EAE is a mainly cell-mediated condition, in which (most commonly) C57BL/6 mice are immunized agains the myelin sheath with peptides of myelin oligodendrocyte glycoprotein and/or lipopolysaccharides along with an adjuvant (most frequently Complete Freund's Adjuvant, pertussis toxin, or both). The immunization causes autoreactive Th-1 and Th-17 cells [164] to cross the blood-brain barrier which, after a roughly 2-week long incubation period results in inflammation and demyelination, followed by oligodendrocyte and neuronal death that is similar to that seen in MS. The same histopathological results can seen after the adoptive transfer of autoreactive T cells into the CNS [165]. Another frequently used strain of mice (SJL/J) develops a relapsing-remitting form of the disease—the most commonly seen version of MS in humans—after immunization with either intact or fragmented myelin basic protein or proteolipid protein with adjuvants. In this strain the first signs of the disease begin to show 7 to 14 days post-induction. After the initial inflammatory attack subsides, the animals go into remission, after which a remissive phase very similarly to human RRMS subsequent episodes of inflammation, demyelination, and axonal loss occurs [166–168].

A different hypothesis, on the other hand, suggests that the first step in the development of new lesions is in fact oligodendrocyte apoptosis and microglial activation in the presence of only a few or no lymphocytes of phagocytes. This culprit event is in turn followed by primary demyelination and a successive, secondary auto-immune inflammatory process, which leads to subsequent lesion expansion and further oligodendrocyte loss [169]. These histological findings are not present in EAE models, which indeed raises the possibility that multiple processes are responsible for new lesion formation in MS. A most recent study suggests that a specific subset of autoreactive T cells producing IFN-γ and, to a lesser extent, IL-17 targeting the β-synuclein protein (which is abundantly expressed in grey matter, where high density of neuronal processes and synapses are present) may at least in part be responsible for the initiation of grey matter lesion formation, expansion, and subsequent gliosis. They were found to be mainly increased in patients with chronic-progressive MS, whereas myelin-reactive T cells (thought to be mainly responsible for white matter lesion formation) were predominant in patients with relapsing-remitting MS. The recent recognition of this cell type argues for the fact that one of the long speculated alternative pathological mechanisms (such as hypoxia or currently undefined soluble toxic factors) that spark the degenerative grey matter process in MS may be, in fact a previously undiscovered subpopulation of autoreactive T cells [170].

In addition to the aforementioned mechanisms, mitochondrial damage with consequent energy failure was found to be a key component in several aspects that constitute to MS pathogenesis and lesion formation [171–175]. The presence of mitochondrial dysfunction has been linked to oligodendrocyte apoptosis and subsequent demyelination [176], the halting of the differentiation of oligodendrocyte progenitor cells [177], the loss of small-diameter axons and lesion progression [178,179], and astrocytic dysfunction [180].

Several disturbances in the kynurenine system have been observed in the EAE model. Additionally IDO-1 and various KP metabolites have been shown to ameliorate autoimmunity and to promote immune tolerance.

This immunomodulating effect of the KP may be in part responsible for the periodic remissions seen in MS and EAE. Increased kynurenine/tryptophan ratios and microglial/macrophage-derived IDO-1 activity have been shown to be present in the brain and spinal cord of mice at the onset of EAE symptoms; meanwhile, a simultaneous fall in IFN-γ levels was observed [181,182]. This suggests the suppression of IFN-γ producing Th1 cells by increased IDO activity. Encephalitogenic Th1 cells secreting IFN-γ have been shown to induce local IDO expression, which in turn suppresses Th1 and Th17 cells and promotes the expansion of immunoregulatory T cell lines (Th2, Treg cells), terminating its own production, therefore, creating a negative feedback loop. This regulatory mechanism may be at least partly responsible for the cyclicality of relapses and remissions seen in EAE. Furthermore, the decreased activity of IDO-1 in EAE, either by genetic deficiency or pharmacological blockage (by the administration of 1-methyl-tryptophan) has been shown to lead to increased Th1 and Th17 responses with decreased Treg responses resulting in faster disease development, a more severe clinical course, and higher amount of spinal cord inflammation, which argues for this hypothesis [181–183].

The upregulation of IDO-1, successive tryptophane starvation, and the accumulation of its downstream tryptophan metabolite, 3-HAA, on the other hand, had ameliorated the inflammation. The increased amounts of 3-HAA have been shown to directly suppress the activity of Th1 and Th17 cells, and also to reduce their activity indirectly by increasing the amount of TGF-ß secreted by DCs, which subsequently resulted in the increased generation of Treg cells from naive $CD4^+$ cells [181,183]. The impaired immunosuppressive activity of Treg cells on Th1 and Th17 cells and the disruption of regulation between B and T cells, dendritic cells, and natural killer cells have been demonstrated to play a role in the breakdown of self-tolerance and the pathogenesis of MS [184]. Similar results were seen with cinnabarinic acid, a less well-studied endogenous kynurenine metabolite. Systemic administration of cinnabarinic acid was capable of enhancing Treg response at the expense of Th17, proved to be highly protective against EAE. Exogenous cinnabarinic acid was found to enhance endogenous cinnabarinic acid formation in lymphocytes, suggesting the occurrence of a positive feedback loop sustaining immune tolerance [185]. In line with these findings, the protective role of IDO-1 activation in EAE has been demonstrated in another study. Estrogen administration resulted in induced IDO-1 expression in dendritic cells, which in turn has led to concomitant T cell apoptosis and the attenuation of EAE symptoms. This mechanism has been proposed to explain estrogen-mediated EAE suppression and to at least in part underlie the decreased rate of relapses seen in MS patients during pregnancy [186].

Other downstream metabolites of the KP have also been found to have a attenuating effect on EAE symptoms. Both endogenous (3-HAA, 3-KYNA) and orally active synthetic metabolites (N-[3,4-dimethoxycinnamoyl] anthranilic acid) have been shown to suppress the expansion of myelin-specific Th1 and Th17 T cells, inhibit the production of Th1 cytokines and elevate Treg response and ameliorate symptoms in EAE mice demonstrating a pivotal role of downstream kynurenine metabolites in IDO-1 mediated EAE suppression [183,187].

In summary, various metabolites of the KP may suppress auto-immunity and ameliorate symptoms in EAE not only through local tryptophan depletion, but also through influencing T cell differentiation (promoting the expansion of regulatory T cells, while limiting the expansion of autoreactive T cells) mediated through cytokines derived from both T cells and dendritic cells.

In addition to the upregulation of IDO-1, the significantly increased activity of KMO was also observed in the spinal cord of rats with EAE. As a consequence of this increase in KMO activity, neurotoxic levels of QUIN and 3-HK were measured. The addition of a selective KMO inhibitor (Ro 61-8048) resulted in a robust reduction of QUIN and 3-HK levels with a concomitant rise in KYNA concentration. Interestingly, however, this change of balance between neurotoxic and neuroprotective kynurenines did not influence the outcome and severity of EAE. This points against previous findings and suggests that neurotoxicity mediated by QUIN and neuroprotection conveyed by KYNA do not have a vital role in the outcome of EAE [181,188].

Another pathological aspect in EAE was the translocation of the KMO enzyme. In healthy controls, KMO immunoreactivity has been reported to be present in the cytoplasm of both neurons and astroglial cells (most likely inside the mitochondria of these cells). In the case of rats with EAE, however, a very intense KMO immunoreactivity was seen in subependymal, subpial, and perivascular locations in cells that expressed both the inducible nitric oxide synthase enzyme and class II major histocompatibility complex, suggesting these cells to be macrophages. These findings support the notion that the cells of the immune system are responsible for the inflammation are also the source of neurotoxic kynurenines in the CNS of rats with EAE [181,188]. Under pathological conditions (such as in MS and EAE), the breakdown of the blood-brain barrier allows for uncontrolled leukocyte infiltration. Thus, a substantial part of the elevated amount of QUIN measured in the CNS may actually be derived from macrophages originating from the periphery. The release of cytotoxic, pro-inflammatory cytokines by macrophages and microglia concomitant to increased QUIN production by macrophages leads to an amplifying feedback mechanism that further stimulates QUIN synthesis and likely contributes to MS lesion pathology and expansion.

In addition to its elevated levels, the oligodendrocyte apoptosis-inducing properties of QUIN were also observed in the spinal cord of rats with EAE [189,190]. Chronic, low dose exposure to QUIN has been shown, however, to be toxic and cause the destruction of not only oligodendrocytes, but astroglial and neuronal cells as well [190–192]. In addition to the previously mentioned ways, QUIN seems to play a role in neurodegeneration through changing the structure of several important proteins [193], which reduces the cells' ability to neutralize reactive oxygen and nitrogen species and free radicals [194,195], affects the glutathione redox potential [196], depletes superoxide dismutase activity [197], enhances lipid peroxidation [198,199] and disrupts mitochondrial function [200,201].

It is of utmost importance to mention a notable pitfall of EAE. As mentioned previously, in the induction of the disease, the animals are not only immunized against various elements constituting the myelin sheath, but mycobacterium tuberculosis is also a component of the Freund's adjuvant used in the process of auto-immunization in some models. It was demonstrated that the increased levels of QUIN and 3-hydroxy-kynureninase levels observed in the spinal cord of rats with EAE [189] were the consequence of higher IDO-1 and KMO expression levels and activity [188]. The increased activity of IDO-1 and KMO is, however, likely the result of an immune reaction to the presence of bacterial antigens but not myelin proteins. This allegation is supported by the fact that in one study, solely mycobacterium tuberculosis, but not MBP, was found to be a potent activator of IDO-1 [202]. Furthermore, increased IDO-1 activity and significantly lower tryptophan concentrations were found in patients with pulmonary tuberculosis compared to healthy controls [203]. Increased IDO-1 activity was also associated with poor outcomes in patients with bacteremia and cancers [204]. This further underlines the KP's role in xenobiotic sensing immunological processes.

4.5.2. Kynurenine Metabolite Changes in MS

Not only pre-clinical data are available on the kynurenines' role in MS, but a number of studies supplied data from humans as well. In addition to the precise pathomechanism of MS being unknown yet, the exact role the KP plays in it is unclear as well. Therefore, it is not surprising that conflicting data are available regarding the various changes observed in the KP and its metabolites in MS. There is an agreement, however, that a significant dysregulation of the KP is present in MS. Some evidence suggests, that the induction of the kynurenine pathway is mediated by pro-inflammatory cytokine-cascades, as described above [193,205,206]. Many of the previously described proinflammatory factors and cytokines modulating the KP are known to be altered in MS as well. Therefore, it is rational to assume that changes in the KP will be present in patients with MS who show disease activity, e.g., at times of acute lesions formation/expansion, when increased inflammation in the CNS is present. It is also logical to presume that during chronic stages of the disease when little or no CNS inflammation is present only minor or no changes are expected to be seen in the KP. Activation of the kynurenine pathway results in two very distinct and opposite events. Short-term benfits of the KP's activation arise in the form of decreased T cell proliferation (via the previously discussed pathways and feedback loops), leading to immunosuppression, while chronic activation of the KP enzymes induces the production of neurotoxic metabolites and plays a role in preventing the innate repair mechanism of remyelination [207].

The first report exploring the connection between MS and the KP is from 1979, the study has found decreased levels of tryptophan in the serum and CSF of MS patients compared to controls [208]. Later studies in the 1990s, however, have reported controversial results about CSF and serum tryptophan levels in MS [209,210]. Additionall, a negative correlation was found between CSF levels of tryptophan and neopterin during acute relapse, possibly representing IDO-1 activation in CNS-infiltrating macrophages [209]. Following these studies, another group failed, however, to detect a significant baseline difference in the plasma L-kynurenine/tryptophan ratio between relapsing-remitting MS and samples from healthy controls. Interestingly, however, an increased L-kynurenine/tryptophan ratio was detected after treatment with INF-β, implicating IDO-1 activation as a potential mode of action of INF-β products, which were widely used at the time as the first-line treatment of MS (for more details, see the chapter on treatment effects) [211]. These results fall in line with the findings of Rothammer et al., who detected a global decrease of circulating AHR agonists in relapsing-remitting MS patients compared to healthy controls. They have also reported increased global AHR activity during relapse and diminished AHR activity (reflecting decreased AHR agonist levels) in remission in the serum of MS patients implicating the role of the endogenous AHR agonist L-kynurenine. Moreover, AHR ligand levels in patients with mild clinical impairment despite a longstanding disease were unaltered as compared to healthy controls [212].

Several studies since then have succeeded in confirming numerous alterations in the KP in different MS subtypes at various points of the disease, which are in support of the aforementioned ideas. KYNA levels were found to be decreased during the remissive phase and elevated during acute clinical exacerbation in the CSF of RRMS patients compared to healthy controls [7,213,214]. Elevated levels of QUIN were associated with oligodendrocyte, astrocyte, and neuronal loss, while decreased amounts neuroprotective metabolites such as kynurenic acid and picolinic acid were also reported in MS patients [8]. Another study showed that pathological amounts of QUIN might be responsible for the abnormal tau-phosphorylation seen in the progressive phase of the disease [215]. Lower levels of both KAT1 and KAT2 enzymes were also found in MS patients' brain tissue by histopathological processing [216].

A recent study investigated the KP metabolomic profile of MS patients and the balance between its two metabolites, KYNA and QUIN. The ratio of these two metabolites is pivotal, as it determines the overall excitotoxic activity the KP has at the NMDA receptor. Based on the metabolomic analysis and profiling of the KP from the serum of MS, Lim et al. have built a predictive model for the disease subtypes using six predictors. The model evaluated the levels of KYNA, QUIN, tryptophan, picolinic acid, fibroblast growth-factor, and TNF-α (in order of relevance) to predict the disease course with up to 85–91% sensitivity [217]. This points toward the notion that the metabolic profiling of the KP may become a potentially useful biomarker in the future, mainly in separating the different clinical courses of the disease early after disease onset. In the same study KYNA levels were reported to be the highest in the relapsing-remitting group compared to both healthy controls and patients with a progressive disease type. The lowest levels of KYNA were measured in primary/secondary progressive MS patients [217].

Another study investigated the potential difference in IDO-1 expression in and activity in peripheral blood mononuclear cells between healthy controls and RRMS patients in remission and in the acute phase, and whether a change in IDO-1 activity/expression was indicative of an onset of a relapse. IDO-1 expression and activity remained unchanged between healthy controls and patients in the acute phase and between healthy controls and stable RRMS patients. The activity of IDO-1 was shown to be independent of the onset of a relapse. Increased IDO-1 expression and decreased levels of IFN-γ were seen, however, in MS patients with a relapse before corticosteroid treatment compared to patients in remission. Glucocorticoid-induced disease remission resulted in a significant reduction of IDO-1 and IFN-γ gene expression, IDO-1 catalytic activity. Serum neopterin (a protein biomarker for inflammation released by macrophages upon IFN-γ stimulation) concentrations followed the same trend as IDO-1 expression and activity. The pitfalls of this study were the relatively low amount of subjects (15 healthy controls, 21 patients in the acute phase, and 15 in remission), and the fact that different patients were used in the remissive and active groups. The same patient was not examined in both the acute and in remissive phase; therefore, intraindividual changes in IDO-1 activity and its role as a potential relapse indicator could not have been established [218].

A recent study by Rajda et al. investigated the connection and association between biomarkers of inflammation (neopterin), neurodegeneration [neurofilament light chain (NFLc)], tryptophan, and kynurenine metabolites measured at diagnosis in the CSF of MS patients (32 RRMS and five CIS patients) and healthy controls ($n = 22$). Compared to controls, all of the measured markers were elevated in the CSF of MS patients, except for KYNA, which showed no change in MS patients compared to controls. Additionally, a strong positive correlation was found between NFLc normalized for age, neopterin, and QUIN [213,214].

Similarly to the study reported by Rajda et al. [207], Aeinehband et al. [219] failed to show a difference in the levels of KP metabolites (tryptophan, kynurenine, KYNA, and QUIN) in the CSF of MS patients and control subjects with non-inflammatory or inflammatory neurological diseases, when the MS patients were pooled. The study included 71 MS patients, 20 non-inflammatory neurological disease control patients, and 13 control patients with inflammatory neurological disease. After the MS patients were stratified according to their disease subtypes and different phases of the disease significant differences in KP metabolites could be demonstrated. Increased QUIN concentrations and quinolinic acid/kynurenine ratios were seen in RRMS patients during the relapsing phase, whereas patients with secondary progressive MS had lower tryptophan and KYNA levels. Patients with a primary progressive disease similarly to control patients with an inflammatory neurological disease showed increased levels of all evaluated tryptophan metabolites. This further strengthens the findings of Lim et al., in that clinical course and disease activity are reflected by changes in KP metabolites. These findings also raise the possibility that someday clinical course and disease severity may be predicted by profiling the KP's metabolites [219].

4.5.3. The Kynurenine System and Depression in MS

Psychiatric disorders, especially depression is, however, one of the most frequent comorbidity with MS, its prevalence can be as high as ~30.5% [220] even among MS patients. Several studies have linked various disturbances in the kynurenine system and depression. Development of depression is associated with decreased serotonin, melatonin, and N-acetyl-serotonin levels, which in turn have been linked to increased immuno-inflammatory pathway activity in patients with depression [221]. Not so long ago, changes in serotonin transporter levels in MS patients were confirmed. This suggests that alterations in serotonin availability and subsequent disturbances in N-acetyl-serotonin and melatonin production are likely to occur in MS patients in a similar manner to that seen in patients with depression. It was hypothesized by some authors, based on the disease modulating and remyelinating effect of melatonin, that depression may not be in fact a frequently occurring comorbidity, but rather a symptom of MS, a part of the disease itself [222,223]. The levels of several inflammatory cytokines (such as IL-1 β, IL-6, IL-18, TNF-α, and IFN-γ) have been shown to be altered in MS. Many of the same cytokines (especially IFN-γ) also increase IDO-1 activity and expression, and therefore, can cause subsequent depletion of serotonin, N-acetyl-serotonin, and melatonin [224]. Kynurenine is able to cross the blood-brain barrier and can increase tryptophan catabolite levels in the CNS, which is another proposed mechanism via the KP and its metabolites can contribute to the depression, somatization, and fatigue seen in MS [225]. A recent study investigated the correlation between alteration in KP metabolite levels in the CSF of MS patients with short disease duration and an active disease and the presence of neuropsychiatric symptoms. Depressed MS patients were demonstrated to have higher KYNA/tryptophan and kynurenine/tryptophan ratios, which was mainly due to low tryptophan levels.

4.5.4. Treatment Effect on the Kynurenine System

Even though a plethora of disease-modifying treatments are available for the treatment of MS nowadays, the most experience and data exists for one oldest approved drugs, interferon-β-1b (IFN-β). Beta interferon is a standard first line treatment, two shorter (BENEFIT and BEYOND), and one very long term phase IV study (LTF) has provided safety and efficacy data on the effects of IFN-β-1b treatment of MS patients. Amirkhani et al. have found increased plasma and serum concentrations of L-kynurenine and increased kynurenine/tryptophan ratios in patients after treatment with IFN-β compared to baseline [211,226]. In contrast, lower levels of kynurenine and N'-formylkynurenine were detected in the serum of patients receiving IFN-β treatment compared to controls in another study [227]. Depression was one of the most common adverse events reported in patients treated with IFN-β [228–230]. Treatment with interferons in other diseases has been known to increase the risk of depression. One of the most potent inducers of IDO-1 (in addition to the previously mentioned INF-γ) is IFN-α, which is used in the treatment of chronic hepatitis C. Several earlier studies reported a decrease in blood tryptophan concentrations and a concomitant increase of KYN which has been linked to IFN-α induced depression, suggesting the role of IDO-1 in the process [231–234]. This raised the question of whether IFN-β treatment was at least in part responsible for the increased prevalence of depression in MS patients. In vitro studies conducted with human monocyte-derived macrophages support this observation. IFN-β has been shown to be able to induce QUIN production and enhance IDO expression in macrophages in therapeutic doses used in MS [235]. However, it remains unclear whether the changes seen in KP metabolite levels mediated by IFN-β are indeed causatively involved in the development of depressive symptoms in interferon treated MS patients. It is also undetermined whether IFN-β-mediated IDO-1 induction is the reason of the low efficacy of IFN-β treatment in improving MS symptoms [8]. The basis for this hypothesis comes from cell culture experiments conducted inhuman macrophages where IFN-β-treatment resulted in increased levels of QUIN production [235], combined with the fact that QUIN is a weak, but well-established NMDAR agonist [8]. Thus far, however there is no direct evidence demonstrating that the use of IFN-β in therapeutic concentrations result in increased CNS QUIN levels in high enough concentrations to be the sole cause of depressive symptoms in MS patients.

In summary, the results of these studies suggest that the KP, most notably the activity of IDO-1, might be downregulated or unaltered in stable MS, probably contributing to disease pathogenesis, whereas its upregulation can be seen during acute inflammatory relapses, most probably reflecting an endogenous counter-regulatory reaction, which responds to anti-inflammatory therapy. A rise in downstream kynurenine metabolism additional to IDO-1 activity can be seen during an acute inflammatory exacerbation in MS. Furthermore the imbalance between neurotoxic and neuroprotective metabolites of the kynurenine pathway favoring the neurotoxic ones might contribute to neurodegeneration in progressive MS subtypes in part via NMDA receptor-mediated excitotoxicity.

4.5.5. Kynurenic Pathway and Redox Disturbances in Neuroinflammation and Multiple Sclerosis

The de novo, eight-step synthesis of NAD^+ from TRP takes place in the liver, neuronal, and immune cells, and starts with the conversion of quinolinic to nicotinic acid mononucleotide (NaMN) by QUIN phosphoribosyltransferase in the presence of Mg^{2+} The subsequent step in the metabolism takes place in the nucleus, mitochondria, and the Golgi apparatus by the nicotinamide mononucleotide adenyl transferase enzymes (NMNAT1, 2, and 3). NaMN is converted into desamido-NAD^+ with the consumption of ATP. The last step in the cascade is the amidation of desamido-NAD^+ in the presence of glutamine. NAD^+ can be synthesised by at least three additional, TRP-independent routes called salvage pathways. First, it can be produced from nicotinic acid, which is converted to NAD^+ via the three-step Preiss-Handler pathway, taking place in the liver, kidney, intestine, and the heart. A second option is by the nicotinamide salvage pathway (called the two-step Nampt pathway, taking place in adipose tissue, liver, kidneys, and immune cells) in which nicotinamide phosphoribosyltransferase (NMAPT, the rate-limiting, glycosyltransferase enzyme of the pathway) converts nicotine amide and 5-phosphoribosyl-1-pyrophosphate to nicotinamide mononucleotide (NMN) and pyrophosphate, respectively. NMN is afterwards transformed to NAD^+ by the action of NMNAT1, -2, and -3 in the presence of ATP. The third way is the phosphorylation of nicotinamide riboside to NMN by nicotinamide riboside kinases (placed in cardiac and skeletal muscle, neuronal tissue) [236–238]. In physiological conditions from all the possible mechanisms to create NAD^+, the de novo synthesis from TRP seems to be the main source [239]. NAD^+ is not solely a pivotal cofactor in several biochemical pathways but acts as an electron transporter. It also serves as a substrate for the DNA damage sensor and putative nuclear repair enzyme, poly(ADP-ribose) polymerase (PARP). PARP is a nucleotide polymerase abundant in the nucleus, which (in concert with DNA dependent protein kinases) is responsible for upkeeping the integrity of the DNA double-strand. Excessive oxidative stress can damage the DNA causing double-strand breaks, which is the activating signal for the PARP enzyme. In neurons affected by glutamatergic excitotoxicity an increase in intracellular oxidative stress and PARP activity was observed [240]. The activation of PARP results in the poly-ADP-ribosylation of the enzyme itself and other molecules participating in the attempted repair of the damaged DNA segments. To fuel this machinery, NAD^+ and $NADP^+$ are used up, resulting in the depletion of the NAD^+ and $NADP^+$ stores of the cell and the release of nicotinamide as a by-product. The fall in intracellular NAD^+ levels following PARP activation has been observed in many cell lines in the CNS in numerous neurodegenerative disorders, neuroinflammation, aging, and infections, and following the exposure to various excitotoxins and free radicals [241–243]. Excessive activation of PARP has been shown to result in cell lysis and eventual death via the depletion of not only the intracellular NAD^+ but the ATP storage as well, causing energy restriction in the cell and a fall in neurotransmitter levels in the brain. In an attempt to recover the used up NAD^+ from nicotinamide, the aforementioned salvage pathways are activated. To do so, ATP is required, indirectly for the reaction catalyzed by NAMPT and directly for the NMNAT enzyme. The upregulation of the salvage pathways consumes the remaining ATP of the damaged cells. When excessive DNA damage occurs, this becomes a vicious circle, which consumes the NAD^+ and energy reserves of the cell to the brink of energetic collapse. In these situations, the accumulation of poly(ADP)ribose by PARP can induce apoptosis to ensure the timely death of the cell with severely enough damaged DNA, thus reducing the chance for tumor formation [244]. This mechanism is supported by recent data, where it was shown that the blockade of PARP in resulted in the preservation of both NAD$^+$ and ATP levels in cells exposed to oxidant injury, cell lysis was also prevented. DNA damage, however—as expected—was not mitigated, highlighting the crucial role that PARP may play in the pathogenesis of neurodegenerative diseases in which elevated levels of free radicals, excitotoxicity, and pro-oxidants have been confirmed [240].

As mentioned before, to date, only microglia, dendritic cells, astroglia, and macrophages have been shown to express 3-HAO in the CNS, which produces QUIN [245]. IFN-γ has been shown to be the primary activating factor of dendritic and microglial cells, as well as macrophages in both the periphery and in the CNS. Activation by IFN-γ readily modulates these cell line's metabolism and increases their antimicrobial activity through the upregulation of several pathways. It activates the secretion of complement pathway components, induces the production of reactive oxygen species, and upregulates nitric oxide synthase activity. Additionally, it enhances the expression of MHC antigens and upregulates the secretion of several cytokines (including, but not limited to IL-1β, IL-6, TNF-α, platelet-activating factor, macrophage chemotactic protein, etc.), all of which play key roles in the induction of neuroinflammation [239,246]. The rate-limiting enzymes—IDO1,2—of the kynurenine pathway are also potently induced by IFN-γ, leading to the elevated production of neuroactive kynurenines in the CNS by the cells expressing the enzymes. This results in an increased level of all the end products of the KP as well as picolinic acid. Following IDO activation by IFN-γ induction, in addition to other various effects, increased intracellular NAD$^+$ concentrations were measured in astrogliomas, followed by increased TRP catabolism [6,246,247]. How exactly the production of pro-inflammatory and oxidative metabolites vs. anti-inflammatory members of the KP are orchestrated to date is controversial. 3-hydroxyanthranilic acid—a potent antioxidant KP metabolite—inhibits the nitric oxide synthase 2 enzyme in macrophages and can also suppress the inducible nitric oxide synthase, readily inhibiting the NF-κB activation even at sub-millimolar concentrations [248,249]. The increased production of 3-hydroxyanthranilic acid by activated mononuclear phagocytes present in inflamed neuronal tissue may, therefore, mitigate the damage caused by increased oxidative stress and may explain the observed increase in TRP catabolism in these tissues. The increased production of kynurenine and QUIN under these conditions, however, remains controversial [250]. Another kynurenine metabolite, cinnabarinic acid —produced in peripheral mononuclear cells and under oxidative stress via non-enzymatic reactions—has been shown to be a pivotal molecule for immune response. The concomitant expression of IDO and enzymes responsible for the production of free radicals has been demonstrated in immune cells to redirect the kynurenine pathway from the production of quinolinic- or picolinic acid towards cinnabarinic acid. Cinnabaranic acid has been proven to modulate the immune response; recently, it was confirmed to be one of the endogenous ligands of the AHR. Via the stimulation of the AHR, it increases the IL-22 production in human T cells pushing the balance between towards regulatory T cells over IL-17 producing T cells [239,251].

Various alterations in the redox pathways, in the KP, levels of free radicals, and NAD$^+$ production and depletion have been observed in vivo (in both animal models of MS and humans). Neurons in EAE have been shown to be extremely vulnerable to the degenerative characteristics of MS under conditions of NAD$^+$ deficiency [252]. An increased net level of NAD$^+$ was observed in the CNS in the EAE model; however, it is suggested that this was due to the increased NAD$^+$ content of the lymphocytes and APCs infiltrating the CNS, meanwhile, neurons were actually in a state of NAD$^+$ deficiency [238].

Cumulating evidence suggests that TRP metabolism and professional APCs expressing IDO-1 are key components in MS. Accordingly, serum TRP concentrations were found to be low and linked to poor prognosis in autoimmune disorders involving IDO-1 activation and Th1 type cellular immune response [253]. Despite being a key player in immune regulation and TRP metabolism, IDO-1 seems to be a double-edged sword. On one hand, it exerts therapeutically beneficial effects of decreasing autoreactive T cell proliferation; however, persistent activation of IDO-1 has been shown to lead to NAD$^+$ depletion in otherwise healthy neighboring collateral tissues [254]. The suppression of autoreactive T cell proliferation by IDO-1 is in part due to its rapid upregulation in professional APCs—such as microglia, macrophages, and dendritic cells—after being exposed to Th1 type cytokines (most prominently IFN-γ and CD40L, but also TNF-α). The thus induced IDO-1 uses up extracellular TRP, which results in the halting of proliferating T cells at the mid-G1 arrest point. The cells cannot progress further from this point in the absence of another T cell signal, even after the restoration of extracellular TRP levels, which eventually causes autoreactive T cells to diminish [255,256]. One of the most potent modulators of the adaptive immune system, dendritic cells—in which cell line IDO-1 is exceptionally highly expressed—have been linked to relapses/chronicity of neuroinflammation, and to the breakdown of tolerance to CNS autoantigens [257,258]. The importance of immunomodulation exerted by IDO-1 is further underlined by the fact that the ameliorating effect of stem cells in EAE pathogenesis was abrogated by the inhibition of IDO-1. In line with this, other studies not involving stem cells confirmed IDO-1 suppression to exacerbate EAE [181,182,259].

Another way through KYN can exert its immunomodulatory effects is by being a precursor for NAD^+ production. It is proposed that kynurenine is taken up by APCs with induced IDO-1 activation and is readily metabolized into NAD^+ to act as a second messenger, which may be more important to kynurenine's immunomodulatory effects, than the depletion of TRP itself [260,261]. This theory is supported by the confirmed ability of plasmacytoid DCs—a subset of professional APCs—to create a profoundly suppressive microenvironment. This is completely dependent on IDO-1 activity via the constant stimulation of $CD4^+$, $CD25^+$, and $Foxp3^+$ Treg cells, which send tolerogenic signals to other T cells [261]. In light of this information, the production of NAD^+ and its activity as a second messenger may be just as important in immunosuppression in APCs, as is the actual local depletion of its precursors.

In a mammalian EAE model, the depletion of NAD^+ has been linked to PARP activation based on immunostaining findings of its metabolite poly(ADP-ribose). It was abundantly found not only in astrocytes surrounding demyelinated EAE plaques but in microglia, oligodendrocytes, and endothelial cells surrounded by microglia and infiltrating peripheral blood cells as well [262].

TRP levels have been shown to be decreased both in the serum and CSF in MS patients (see before) [8,193,209,263]. In RRMS patients, a more than 50% reduction in serum NAD^+ levels, as well as a two-fold increase in NADH levels and three-fold reduction in the $NAD^+/NADH$ ratio was seen compared to controls. Furthermore, among MS patients NAD^+ levels were the highest among RRMS patients, followed by primary progressive MS patients, and were the lowest in secondary progressive MS [264]. Administration of pharmacological doses of nicotinamide and calorie restriction was, however, able to rise the NAD^+ levels and ameliorate EAE pathogenesis in animal models [265–268].

In summary, during neuroinflammation, the NAD^+ level is elevated in APCs, causing a measurable net increase in NAD^+ content in the CNS; however, at the same time, the adjacent tissue is starved from NAD^+, causing the neurons to be particularly vulnerable to various oxidative and neuroexcitatory insults, resulting in neurodegeneration due to NAD^+ deficiency and subsequent bioenergetic collapse and ultimately cell death. The excessive activation of IDO on one hand can deplete the extracellular TRP—a precursor to NAD^+—on the other hand, it can increase intracellular NAD^+ levels in immune cells. IFN can enhance the viability and endurance of astrocytes against oxidative stress via the increasing intracellular NAD^+ levels [247,269–271]. The increment in intracellular NAD^+ mediated by IFN-γ is abrogated by the inhibition of IDO-1. Additionally, stimulation with IFN-γ along with the concomitant inhibition of PARP-1 results in increased NAD^+ production in glia cells and macrophages [247,269–271]. IFN-γ possesses potent antiviral and antimicrobial activities, which were found to completely rely on its ability to induce IDO-1 activity in various immune cells [272–277].

4.5.6. Treating MS in Light of Kynurenines

As described in detail above, kynurenines are important modulators of the immune system, redox reactions, and excitotoxicity. Different members of the pathway exert opposite effects; thus, some induce neurodegeneration, while others promote neuroprotection. It is therefore rational to think that by modulating this pathway, these effects can be harnessed to develop new therapies in a number of diseases, including MS.

In the past decades, several promising pre-clinical data emerged regarding kynurenines and its analogs as therapeutic targets in experimental models of MS, showing potential benefits. Their in-depth, detailed description, however, exceeds the scope of this study [8]. From many potential candidates, we would like to highlight one molecule, laquinimod, which is of particular interest. In addition to showing remarkable structural similarity to KYNA, it was investigated in several human clinical studies for a number of conditions, including MS. In the next paragraphs, we aim to summarize the mechanism of action and effect on the immune system and potential neuroprotective attributes of laquinimod.

Laquinimod

Laquinimod (or N-ethyl-N-phenyl-5-chloro-1,2-dihydro-4-hydroxy-1-methyl-2-oxo-3-quinolinecarboxamide) (Figure 2) was first synthesized as a structural variant of roquinimex, an initially promising, but later retracted drug for MS (due to severe adverse effects) [278,279]. It displays a remarkable structural similarity to kynurenines, with proposed immunomodulatory and neuroprotective attributes, rather than immunosuppressive effects. It is almost completely (98%) bound to proteins in the plasma, with the ability to diffuse freely across the BBB [280,281]. Estimations put its concentration in the CNS around 13% of the blood concentration [282]. It has a half-life of approximately 80 h and it is metabolized by cytochrome P450 (CYP 450) enzyme family [279].

Even though laquinimod's mechanism of action is still not clear, several studies have shown its strong capabilities to guide the immune system toward the anti-inflammatory paths. In vitro studies established that laquinimod reduces the expression of MHC class II genes and alters the expression of genes involved in the activation of T and B cells [283]. Seemingly it also increases the CD86$^+$ CD25$^+$ and IL10$^+$ CD25$^+$ immunoregulatory subpopulations of B cells, in turn reducing the proliferation and the percentage of INFγ^+ T cells [284].

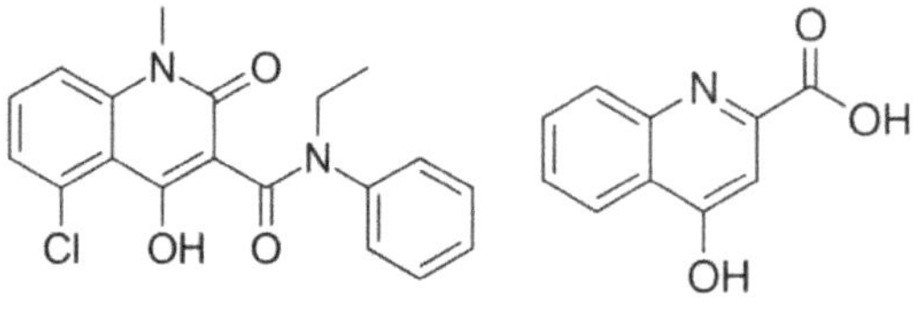

Figure 2. The chemical structure of laquinimod and kynurenic acid.

Animal models demonstrated that laquinimod blocks the release of inflammatory cytokines (TNFα and IL-1) in monocytes and hinders the adhesion and migration of leukocytes by decreasing levels of matrix metalloproteinase 9 (MMP9) and very late antigen-4 (VLA-4) [285–287]. Additionally (by a dose-dependent effect), it inhibits the migration of T cells into the CNS and promotes the production of anti-inflammatory cytokines (TGFβ and IL-4) instead of pro-inflammatory ones (TNFα and IL-12) [281]. It reduces the number of CD4$^+$ dendritic cells and through it, the amount of Th1 and Th17 cells, while increasing the number of Treg cells [288]. Laquinimod was also shown to decrease microglial activation, leading to reduced axonal damage [279]. A recent study has concluded that laquinimod can induce remyelination through the protection of oligodendrocyte progenitors during differentiation [289].

Human data are convincing as well, as decreased secretion of chemokines by mature dendritic cells and reduced number of CD1c$^+$ and plasmacytoid CD303$^+$ DCs in the peripheral blood were found in patients treated with laquinimod [290]. Laquinimod also inhibits T cell secretion of INF-γ, IL-17, granulocyte-macrophage colony-stimulating factor, and TNF-α, while increasing the production of IL-4 by CD4$^+$ T cells [288,290].

However, laquinimod is seemingly not only capable of modulating the immune system, but has neuroprotective properties as well. By increasing the level of brain-derived neurotrophic factor—a protein essential to the development of the CNS—it regulates synaptic plasticity and enhances neuronal and axonal growth [291]. BDNF overexpression was found in the striatum, lateral septal nucleus, nucleus accumbens, and the cortex of EAE mice after laquinimod treatment. Additionally, an increased amount of immunosuppressive Foxp3$^+$ Treg cell was observed, which was associated with reduced astrogliosis, axonal, and myelin damage [292]. In a study evaluating blood samples of 203 MS patients treated with laquinimod a significant increase in serum levels of BDNF was seen in 76% of the patients [293]. Evidence from animal studies suggests that laquinimod may suppress inducible nitric oxide synthase activity; therefore, it can decrease the level of nitric oxide, a potent neurotoxin [287].

Two pivotal, phase III clinical studies were conducted with laquinimod in MS. The ALLEGRO study was placebo-controlled; the BRAVO study had an active comparator (IFN-β i.m.) and placebo arm. In summary: the drug showed a modest effect on the annualized relapse rate (21% in the BRAVO study, 0.30 ± 0.02 vs. 0.39 ± 0.03, $p = 0.002$ against placebo in the ALLEGRO study) [294,295]. A significant reduction in disability progression (40.6%; $p = 0.042$ in the BRAVO, 11.1% vs. 15.7%, $p = 0.01$ against placebo in the ALLEGRO study) was observed [294,295]. Brain atrophy rates were significantly reduced against placebo in the BRAVO study (treatment effect 0.28%, $p < 0.001$) [295]. There was some controversy regarding MRI endpoints: in the ALLEGRO study, laquinimod showed a significant reduction in Gd$^+$ T1-lesion number and the number of new or enlarging T2-lesions ($p < 0.001$), while in BRAVO, it showed no significant benefits [294,295]. The safety profiles were consequently favorable throughout the studies.

Despite this, the Committee for Medicinal Products for Human Use (CHMP) did not approve laquinimod in 2014. The CHMP reasoned that during the animal studies, chronic laquinimod exposure raised the occurrence of malignancies, and even though no malignancies were recorded during human trials, it was deemed too high a risk. Furthermore, it also showed teratogenic tendencies. Combined with the relatively modest effectiveness of the drug, they felt the potential risk of chronic exposure outweighs the benefits of disability reduction; thus, they rejected approval.

In summary, we can conclude that laquinimod has a dual effect by modulating the immune system and—probably even more importantly—shows strong neuroprotective attributes. In spite of these strengths, as described above, it was not approved by the CHMP. However, there are some data that laquinimod may be a beneficial add-on therapy in the future for MS. Additionally, as studies are still ongoing in other diseases—e.g., Huntington's disease—and there are strong pre-clinical results in hemorrhagic stroke, laquinimod may still become an important neuroprotective therapy in the near future [279,296].

5. Conclusions

Alterations can be seen in several cascades and pathways involved in immunomodulation, redox mechanisms, energy management, and the upkeep of genomic integrity of a cell in several neurodegenerative diseases, including MS. This results in diminished resistance and increased vulnerability of neuronal cells to oxidative stress and excitotoxicity. In the past years, intensive research has been invested into the kynurenine system; thus, our understanding of the roles of the metabolites and enzymes participating in the KP has been expanded considerably. Even so, many unsettled issues and controversies remain related to their precise function in immune modulation, neuroprotection, and excitotoxicity. Nonetheless, with recent advances confirming the kynurenine system's pivotal role in the pathogenesis of neurodegenerative diseases, it is highly likely that in the near future additional physiological and pathological roles for neuroactive kynurenines might be unearthed. Additionally, they might prove to be promising targets for drug development based on already known and their yet to be discovered roles. With the rise of targeted drug development and genetic engineering, overcoming past obstacles might indeed become a reality. Targeting the KP in various sites with new, specific agents may be achieved, making the prevention and treatment of several diseases possible by appropriate pharmacological or genetic manipulations.

Author Contributions: T.B. and D.S. wrote the paper. K.B. and L.V. edited and revised the manuscript. All authors read and approved the final manuscript.

Funding: This paper was supported the GINOP 2.3.2-15-2016-00034 and TUDFO/47138-1/2019-ITM programs.

Conflicts of Interest: The authors report no conflict of interest.

Abbreviations

3-HAO	3-hydroxyanthranilate oxidase
3-HK	3-hydroxykynurenine
AHR	aryl hydrocarbon receptor
AMPA	α-amino-3-hydroxy-5-methyl-4-isoxazolepropionic acid
ANA	anthralinic acid
APC	antigen presenting cell
ATP	adenosine triphosphate
BBB	blood-brain barrier
CDxx	cluster of differentiation xx
CHMP	Committee for Medicinal Products for Human Use
CNS	central nervous system
DC	dendritic cell
EAE	experimental autoimmune encephalitis
GRP35	G-protein coupled receptor 35
IDO 1,2	izoenzymes indoleamine 2,3-dioxygenase 1 and 2
IFN-β	interferon beta
IFN-γ	interferon gamma
IL	interleukin
KAT 1, 2	kynurenine aminotransferase 1 and 2
KMO	kynure nine 3-monooxygenase
KP	kynurenine pathway
KYN	kynurenine
KYNA	kynurenic acid
MMP9	matrix metalloproteinase 9 (MMP9)
MP	methoxyndole pathway
MS	multiple sclerosis
NAD$^+$	nicotinamide adenine dinucleotide
NADP$^+$	nicotinamide adenine dinucleotide phosphate
NaMN	nicotinic acid mononucleotide
NAS	N-acetylserotonin
NK	natural killer cell
NMAPT	nicotinamide phosphoribosyltransferase
NMDA	N-methyl-D-aspartate
NMDA	N-methyl-D-aspartate receptor
NMN	nicotinamide mononucleotide
NMNAT	nicotinamide mononucleotide adenyl transferase
PARP	poly(ADP-ribose) polymerase
PBMC	peripherial blood monocytes
PMN	polymorphonuclear cells
PPMS	primary progressive multiple sclerosis
QUIN	quinolinic acid
RRMS	relapsing-remitting multiple sclerosis
SS	Sjögren's syndrome
TDO	tryptophan 2,3-dioxygenase
TGF-β	transforming growth factor beta
TNFα	tumor necrosis factor alpha
Treg	regulatory T cell
TRP	tryptophan
VLA-4	very late antigen-4

References

1. Schwarcz, R.; Bruno, J.P.; Muchowski, P.J.; Wu, H.Q. Kynurenines in the mammalian brain: When physiology meets pathology. *Nat. Revi. Neurosci.* **2012**, *13*, 465–477. [CrossRef] [PubMed]
2. Leklem, J.E. Quantitative aspects of tryptophan metabolism in humans and other species: A review. *Am. J. Clin. Nutr.* **1971**, *24*, 659–672. [CrossRef] [PubMed]
3. Jones, S.P.; Guillemin, G.J.; Brew, B.J. The kynurenine pathway in stem cell biology. *Int. J. Tryptophan Res.* **2013**, *6*, 57–66. [CrossRef]
4. Moroni, F. Tryptophan metabolism and brain function: Focus on kynurenine and other indole metabolites. *Eur. J. Pharmacol.* **1999**, *375*, 87–100. [CrossRef]
5. Guillemin, G.J. Quinolinic acid, the inescapable neurotoxin. *FEBS J.* **2012**, *279*, 1356–1365. [CrossRef]
6. Guillemin, G.J.; Cullen, K.M.; Lim, C.K.; Smythe, G.A.; Garner, B.; Kapoor, V.; Takikawa, O.; Brew, B.J. Characterization of the kynurenine pathway in human neurons. *J. Neurosci.* **2007**, *27*, 12884–12892. [CrossRef]
7. Hartai, Z.; Klivenyi, P.; Janaky, T.; Penke, B.; Dux, L.; Vecsei, L. Kynurenine metabolism in multiple sclerosis. *Acta Neurol. Scand.* **2005**, *112*, 93–96. [CrossRef]
8. Vecsei, L.; Szalardy, L.; Fulop, F.; Toldi, J. Kynurenines in the CNS: Recent advances and new questions. *Nat. Rev. Drug Discov.* **2013**, *12*, 64–82. [CrossRef]
9. Boros, F.A.; Bohar, Z.; Vecsei, L. Genetic alterations affecting the genes encoding the enzymes of the kynurenine pathway and their association with human diseases. *Mutat. Res.* **2018**, *776*, 32–45. [CrossRef] [PubMed]
10. Lovelace, M.D.; Varney, B.; Sundaram, G.; Lennon, M.J.; Lim, C.K.; Jacobs, K.; Guillemin, G.J.; Brew, B.J. Recent evidence for an expanded role of the kynurenine pathway of tryptophan metabolism in neurological diseases. *Neuropharmacology* **2017**, *112*, 373–388. [CrossRef]
11. Braidy, N.; Grant, R.; Adams, S.; Guillemin, G.J. Neuroprotective effects of naturally occurring polyphenols on quinolinic acid-induced excitotoxicity in human neurons. *FEBS J.* **2010**, *277*, 368–382. [CrossRef] [PubMed]
12. Oxenkrug, G.F. Genetic and hormonal regulation of tryptophan kynurenine metabolism: Implications for vascular cognitive impairment, major depressive disorder, and aging. *Ann. N. Y. Acad. Sci.* **2007**, *1122*, 35–49. [CrossRef]
13. Oxenkrug, G.F. Interferon-gamma-inducible kynurenines/pteridines inflammation cascade: Implications for aging and aging-associated psychiatric and medical disorders. *J. Neural Transm.* **2011**, *118*, 75–85. [CrossRef] [PubMed]
14. Pardridge, W.M. Blood-brain barrier carrier-mediated transport and brain metabolism of amino acids. *Neurochem. Res.* **1998**, *23*, 635–644. [CrossRef] [PubMed]
15. Mellor, A.L.; Munn, D.H. IDO expression by dendritic cells: Tolerance and tryptophan catabolism. *Nat. Rev. Immunol.* **2004**, *4*, 762–774. [CrossRef]
16. Grohmann, U.; Fallarino, F.; Puccetti, P. Tolerance, DCs and tryptophan: Much ado about IDO. *Trends Immunol.* **2003**, *24*, 242–248. [CrossRef]
17. Platten, M.; Wick, W.; Van den Eynde, B.J. Tryptophan catabolism in cancer: Beyond IDO and tryptophan depletion. *Cancer Res.* **2012**, *72*, 5435–5440. [CrossRef]
18. Van Baren, N.; Van den Eynde, B.J. Tryptophan-degrading enzymes in tumoral immune resistance. *Front. Immunol.* **2015**, *6*, 34. [CrossRef]
19. Katz, J.B.; Muller, A.J.; Prendergast, G.C. Indoleamine 2,3-dioxygenase in T-cell tolerance and tumoral immune escape. *Immunol. Rev.* **2008**, *222*, 206–221. [CrossRef] [PubMed]
20. Mondanelli, G.; Albini, E.; Pallotta, M.T.; Volpi, C.; Chatenoud, L.; Kuhn, C.; Fallarino, F.; Matino, D.; Belladonna, M.L.; Bianchi, R.; et al. The Proteasome Inhibitor Bortezomib Controls Indoleamine 2,3-Dioxygenase 1 Breakdown and Restores Immune Regulation in Autoimmune Diabetes. *Front. Immunol.* **2017**, *8*, 428. [CrossRef] [PubMed]
21. Llamas-Velasco, M.; Bonay, P.; Jose Concha-Garzon, M.; Corvo-Villen, L.; Vara, A.; Cibrian, D.; Sanguino-Pascual, A.; Sanchez-Madrid, F.; De la Fuente, H.; Dauden, E. Immune cells from patients with psoriasis are defective in inducing indoleamine 2,3-dioxygenase expression in response to inflammatory stimuli. *Br. J. Dermatol.* **2017**, *176*, 695–704. [CrossRef] [PubMed]

22. Cribbs, A.P.; Kennedy, A.; Penn, H.; Read, J.E.; Amjadi, P.; Green, P.; Syed, K.; Manka, S.W.; Brennan, F.M.; Gregory, B.; et al. Treg cell function in rheumatoid arthritis is compromised by ctla-4 promoter methylation resulting in a failure to activate the indoleamine 2,3-dioxygenase pathway. *Arthritis Rheumatol.* **2014**, *66*, 2344–2354. [CrossRef] [PubMed]
23. Negrotto, L.; Correale, J. Amino Acid Catabolism in Multiple Sclerosis Affects Immune Homeostasis. *J. Immunol.* **2017**, *198*, 1900–1909. [CrossRef]
24. Volpi, C.; Mondanelli, G.; Pallotta, M.T.; Vacca, C.; Iacono, A.; Gargaro, M.; Albini, E.; Bianchi, R.; Belladonna, M.L.; Celanire, S.; et al. Allosteric modulation of metabotropic glutamate receptor 4 activates IDO1-dependent, immunoregulatory signaling in dendritic cells. *Neuropharmacology* **2016**, *102*, 59–71. [CrossRef]
25. Miller, C.L.; Llenos, I.C.; Dulay, J.R.; Barillo, M.M.; Yolken, R.H.; Weis, S. Expression of the kynurenine pathway enzyme tryptophan 2,3-dioxygenase is increased in the frontal cortex of individuals with schizophrenia. *Neurobiol. Dis.* **2004**, *15*, 618–629. [CrossRef]
26. Platten, M.; Von Knebel Doeberitz, N.; Oezen, I.; Wick, W.; Ochs, K. Cancer Immunotherapy by Targeting IDO1/TDO and Their Downstream Effectors. *Front. Immunol.* **2014**, *5*, 673. [CrossRef] [PubMed]
27. Grohmann, U.; Bronte, V. Control of immune response by amino acid metabolism. *Immunol. Rev.* **2010**, *236*, 243–264. [CrossRef] [PubMed]
28. Mondanelli, G.; Ugel, S.; Grohmann, U.; Bronte, V. The immune regulation in cancer by the amino acid metabolizing enzymes ARG and IDO. *Curr. Opin. Pharmacol.* **2017**, *35*, 30–39. [CrossRef] [PubMed]
29. Mondanelli, G.; Iacono, A.; Allegrucci, M.; Puccetti, P.; Grohmann, U. Immunoregulatory Interplay between Arginine and Tryptophan Metabolism in Health and Disease. *Front. Immunol.* **2019**, *10*, 1565. [CrossRef] [PubMed]
30. Dang, Y.; Dale, W.E.; Brown, O.R. Comparative effects of oxygen on indoleamine 2,3-dioxygenase and tryptophan 2,3-dioxygenase of the kynurenine pathway. *Free Radic. Biol. Med.* **2000**, *28*, 615–624. [CrossRef]
31. Gal, E.M.; Sherman, A.D. Synthesis and metabolism of L-kynurenine in rat brain. *J. Neurochem.* **1978**, *30*, 607–613. [CrossRef] [PubMed]
32. Speciale, C.; Schwarcz, R. Uptake of kynurenine into rat brain slices. *J. Neurochem.* **1990**, *54*, 156–163. [CrossRef]
33. Fukui, S.; Schwarcz, R.; Rapoport, S.I.; Takada, Y.; Smith, Q.R. Blood-brain barrier transport of kynurenines: Implications for brain synthesis and metabolism. *J. Neurochem.* **1991**, *56*, 2007–2017. [CrossRef] [PubMed]
34. Han, Q.; Li, J.; Li, J. pH dependence, substrate specificity and inhibition of human kynurenine aminotransferase I. *Eur. J. Bochem.* **2004**, *271*, 4804–4814. [CrossRef] [PubMed]
35. Guillemin, G.J.; Smythe, G.; Takikawa, O.; Brew, B.J. Expression of indoleamine 2,3-dioxygenase and production of quinolinic acid by human microglia, astrocytes, and neurons. *Glia* **2005**, *49*, 15–23. [CrossRef] [PubMed]
36. Parrott, J.M.; O'Connor, J.C. Kynurenine 3-Monooxygenase: An Influential Mediator of Neuropathology. *Front. Psychiatry* **2015**, *6*, 116. [CrossRef] [PubMed]
37. Chen, Y.; Brew, B.J.; Guillemin, G.J. Characterization of the kynurenine pathway in NSC-34 cell line: Implications for amyotrophic lateral sclerosis. *J. Neurochem.* **2011**, *118*, 816–825. [CrossRef] [PubMed]
38. Chiarugi, A.; Calvani, M.; Meli, E.; Traggiai, E.; Moroni, F. Synthesis and release of neurotoxic kynurenine metabolites by human monocyte-derived macrophages. *J. Neuroimmunol.* **2001**, *120*, 190–198. [CrossRef]
39. Heyes, M.P.; Chen, C.Y.; Major, E.O.; Saito, K. Different kynurenine pathway enzymes limit quinolinic acid formation by various human cell types. *Biochem. J.* **1997**, *326*, 351–356. [CrossRef]
40. Kiss, C.; Ceresoli-Borroni, G.; Guidetti, P.; Zielke, C.L.; Zielke, H.R.; Schwarcz, R. Kynurenate production by cultured human astrocytes. *J. Neural Transm.* **2003**, *110*, 1–14. [CrossRef]
41. Kocki, T.; Dolinska, M.; Dybel, A.; Urbanska, E.M.; Turski, W.A.; Albrecht, J. Regulation of kynurenic acid synthesis in C6 glioma cells. *J. Neurosci. Res.* **2002**, *68*, 622–626. [CrossRef] [PubMed]
42. Guidetti, P.; Hoffman, G.E.; Melendez-Ferro, M.; Albuquerque, E.X.; Schwarcz, R. Astrocytic localization of kynurenine aminotransferase II in the rat brain visualized by immunocytochemistry. *Glia* **2007**, *55*, 78–92. [CrossRef] [PubMed]
43. Lehrmann, E.; Molinari, A.; Speciale, C.; Schwarcz, R. Immunohistochemical visualization of newly formed quinolinate in the normal and excitotoxically lesioned rat striatum. *Exp. Brain Res.* **2001**, *141*, 389–397. [CrossRef]

44. Zinger, A.; Barcia, C.; Herrero, M.T.; Guillemin, G.J. The involvement of neuroinflammation and kynurenine pathway in Parkinson's disease. *Parkinson's Dis.* **2011**, *2011*, 716859. [CrossRef]
45. Moroni, F.; Russi, P.; Lombardi, G.; Beni, M.; Carla, V. Presence of kynurenic acid in the mammalian brain. *J. Neurochem.* **1988**, *51*, 177–180. [CrossRef] [PubMed]
46. Turski, W.A.; Nakamura, M.; Todd, W.P.; Carpenter, B.K.; Whetsell, W.O., Jr.; Schwarcz, R. Identification and quantification of kynurenic acid in human brain tissue. *Brain Res.* **1988**, *454*, 164–169. [CrossRef]
47. Wu, H.Q.; Guidetti, P.; Goodman, J.H.; Varasi, M.; Ceresoli-Borroni, G.; Speciale, C.; Scharfman, H.E.; Schwarcz, R. Kynurenergic manipulations influence excitatory synaptic function and excitotoxic vulnerability in the rat hippocampus in vivo. *Neuroscience* **2000**, *97*, 243–251. [CrossRef]
48. Wang, J.; Simonavicius, N.; Wu, X.; Swaminath, G.; Reagan, J.; Tian, H.; Ling, L. Kynurenic acid as a ligand for orphan G protein-coupled receptor GPR35. *J. Biol. Chem.* **2006**, *281*, 22021–22028. [CrossRef]
49. Southern, C.; Cook, J.M.; Neetoo-Isseljee, Z.; Taylor, D.L.; Kettleborough, C.A.; Merritt, A.; Bassoni, D.L.; Raab, W.J.; Quinn, E.; Wehrman, T.S.; et al. Screening beta-arrestin recruitment for the identification of natural ligands for orphan G-protein-coupled receptors. *J. Biomol. Screen.* **2013**, *18*, 599–609. [CrossRef]
50. Zhao, P.; Sharir, H.; Kapur, A.; Cowan, A.; Geller, E.B.; Adler, M.W.; Seltzman, H.H.; Reggio, P.H.; Heynen-Genel, S.; Sauer, M.; et al. Targeting of the orphan receptor GPR35 by pamoic acid: A potent activator of extracellular signal-regulated kinase and beta-arrestin2 with antinociceptive activity. *Mol. Pharmacol.* **2010**, *78*, 560–568. [CrossRef]
51. Deng, H.; Hu, H.; Fang, Y. Multiple tyrosine metabolites are GPR35 agonists. *Sci. Rep.* **2012**, *2*, 373. [CrossRef] [PubMed]
52. Guo, J.; Williams, D.J.; Puhl, H.L., 3rd; Ikeda, S.R. Inhibition of N-type calcium channels by activation of GPR35, an orphan receptor, heterologously expressed in rat sympathetic neurons. *J. Pharmacol. Exp. Ther.* **2008**, *324*, 342–351. [CrossRef] [PubMed]
53. Kimura, E.; Kubo, K.I.; Endo, T.; Nakajima, K.; Kakeyama, M.; Tohyama, C. Excessive activation of AhR signaling disrupts neuronal migration in the hippocampal CA1 region in the developing mouse. *J. Toxicol. Sci.* **2017**, *42*, 25–30. [CrossRef] [PubMed]
54. DiNatale, B.C.; Murray, I.A.; Schroeder, J.C.; Flaveny, C.A.; Lahoti, T.S.; Laurenzana, E.M.; Omiecinski, C.J.; Perdew, G.H. Kynurenic acid is a potent endogenous aryl hydrocarbon receptor ligand that synergistically induces interleukin-6 in the presence of inflammatory signaling. *Toxicol. Sci.* **2010**, *115*, 89–97. [CrossRef] [PubMed]
55. Opitz, C.A.; Litzenburger, U.M.; Sahm, F.; Ott, M.; Tritschler, I.; Trump, S.; Schumacher, T.; Jestaedt, L.; Schrenk, D.; Weller, M.; et al. An endogenous tumour-promoting ligand of the human aryl hydrocarbon receptor. *Nature* **2011**, *478*, 197–203. [CrossRef]
56. Henderson, G.; Johnson, J.W.; Ascher, P. Competitive antagonists and partial agonists at the glycine modulatory site of the mouse N-methyl-D-aspartate receptor. *J. Physiol.* **1990**, *430*, 189–212. [CrossRef]
57. Kessler, M.; Baudry, M.; Lynch, G. Quinoxaline derivatives are high-affinity antagonists of the NMDA receptor-associated glycine sites. *Brain Res.* **1989**, *489*, 377–382. [CrossRef]
58. Oliver, M.W.; Kessler, M.; Larson, J.; Schottler, F.; Lynch, G. Glycine site associated with the NMDA receptor modulates long-term potentiation. *Synapse* **1990**, *5*, 265–270. [CrossRef]
59. Kloog, Y.; Lamdani-Itkin, H.; Sokolovsky, M. The glycine site of the N-methyl-D-aspartate receptor channel: Differences between the binding of HA-966 and of 7-chlorokynurenic acid. *J. Neurochem.* **1990**, *54*, 1576–1583. [CrossRef] [PubMed]
60. Mayer, M.L.; Westbrook, G.L.; Vyklicky, L., Jr. Sites of antagonist action on N-methyl-D-aspartic acid receptors studied using fluctuation analysis and a rapid perfusion technique. *J. Neurophysiol.* **1988**, *60*, 645–663. [CrossRef]
61. Olverman, H.J.; Jones, A.W.; Mewett, K.N.; Watkins, J.C. Structure/activity relations of N-methyl-D-aspartate receptor ligands as studied by their inhibition of [3H]D-2-amino-5-phosphonopentanoic acid binding in rat brain membranes. *Neuroscience* **1988**, *26*, 17–31. [CrossRef]
62. Moroni, F.; Pellegrini-Giampietro, D.E.; Alesiani, M.; Cherici, G.; Mori, F.; Galli, A. Glycine and kynurenate modulate the glutamate receptors in the myenteric plexus and in cortical membranes. *Eur. J. Pharmacol.* **1989**, *163*, 123–126. [CrossRef]

63. Robinson, M.B.; Schulte, M.K.; Freund, R.K.; Johnson, R.L.; Koerner, J.F. Structure-function relationships for kynurenic acid analogues at excitatory pathways in the rat hippocampal slice. *Brain Res.* **1985**, *361*, 19–24. [CrossRef]
64. Danysz, W.; Fadda, E.; Wroblewski, J.T.; Costa, E. Different modes of action of 3-amino-1-hydroxy-2-pyrrolidone (HA-966) and 7-chlorokynurenic acid in the modulation of N-methyl-D-aspartate-sensitive glutamate receptors. *Mol. Pharmacol.* **1989**, *36*, 912–916.
65. Danysz, W.; Fadda, E.; Wroblewski, J.T.; Costa, E. Kynurenate and 2-amino-5-phosphonovalerate interact with multiple binding sites of the N-methyl-D-aspartate-sensitive glutamate receptor domain. *Neurosci. Lett.* **1989**, *96*, 340–344. [CrossRef]
66. Fisher, J.L.; Mott, D.D. Distinct functional roles of subunits within the heteromeric kainate receptor. *J. Neurosci.* **2011**, *31*, 17113–17122. [CrossRef]
67. Lugo-Huitron, R.; Blanco-Ayala, T.; Ugalde-Muniz, P.; Carrillo-Mora, P.; Pedraza-Chaverri, J.; Silva-Adaya, D.; Maldonado, P.D.; Torres, I.; Pinzon, E.; Ortiz-Islas, E.; et al. On the antioxidant properties of kynurenic acid: Free radical scavenging activity and inhibition of oxidative stress. *Neurotoxicol. Teratol.* **2011**, *33*, 538–547. [CrossRef]
68. Kubicova, L.; Hadacek, F.; Bachmann, G.; Weckwerth, W.; Chobot, V. Coordination Complex Formation and Redox Properties of Kynurenic and Xanthurenic Acid Can Affect Brain Tissue Homeodynamics. *Antioxidants* **2019**, *8*, 476. [CrossRef]
69. Perkins, M.N.; Stone, T.W. An iontophoretic investigation of the actions of convulsant kynurenines and their interaction with the endogenous excitant quinolinic acid. *Brain Res.* **1982**, *247*, 184–187. [CrossRef]
70. Kessler, M.; Terramani, T.; Lynch, G.; Baudry, M. A glycine site associated with N-methyl-D-aspartic acid receptors: Characterization and identification of a new class of antagonists. *J. Neurochem.* **1989**, *52*, 1319–1328. [CrossRef]
71. Parsons, C.G.; Danysz, W.; Quack, G.; Hartmann, S.; Lorenz, B.; Wollenburg, C.; Baran, L.; Przegalinski, E.; Kostowski, W.; Krzascik, P.; et al. Novel systemically active antagonists of the glycine site of the N-methyl-D-aspartate receptor: Electrophysiological, biochemical and behavioral characterization. *J. Pharmacol. Exp. Ther.* **1997**, *283*, 1264–1275.
72. Szalardy, L.; Zadori, D.; Toldi, J.; Fulop, F.; Klivenyi, P.; Vecsei, L. Manipulating kynurenic acid levels in the brain - on the edge between neuroprotection and cognitive dysfunction. *Curr. Top. Med. Chem.* **2012**, *12*, 1797–1806. [CrossRef] [PubMed]
73. Prescott, C.; Weeks, A.M.; Staley, K.J.; Partin, K.M. Kynurenic acid has a dual action on AMPA receptor responses. *Neurosci. Lett.* **2006**, *402*, 108–112. [CrossRef] [PubMed]
74. Rozsa, E.; Robotka, H.; Vecsei, L.; Toldi, J. The Janus-face kynurenic acid. *J. Neural Transm.* **2008**, *115*, 1087–1091. [CrossRef]
75. Zwilling, D.; Huang, S.Y.; Sathyasaikumar, K.V.; Notarangelo, F.M.; Guidetti, P.; Wu, H.Q.; Lee, J.; Truong, J.; Andrews-Zwilling, Y.; Hsieh, E.W.; et al. Kynurenine 3-monooxygenase inhibition in blood ameliorates neurodegeneration. *Cell* **2011**, *145*, 863–874. [CrossRef]
76. Hilmas, C.; Pereira, E.F.; Alkondon, M.; Rassoulpour, A.; Schwarcz, R.; Albuquerque, E.X. The brain metabolite kynurenic acid inhibits alpha7 nicotinic receptor activity and increases non-alpha7 nicotinic receptor expression: Physiopathological implications. *J. Neurosci.* **2001**, *21*, 7463–7473. [CrossRef]
77. Tanaka, M.; Bohar, Z.; Vecsei, L. Are Kynurenines Accomplices or Principal Villains in Dementia? Maintenance of Kynurenine Metabolism. *Molecules* **2020**, *25*, 564. [CrossRef]
78. Stone, T.W. Kynurenic acid antagonists and kynurenine pathway inhibitors. *Expert Opin. Investig. Drugs* **2001**, *10*, 633–645. [CrossRef]
79. Stone, T.W.; Stoy, N.; Darlington, L.G. An expanding range of targets for kynurenine metabolites of tryptophan. *Trends Pharmacol. Sci.* **2013**, *34*, 136–143. [CrossRef]
80. Mackenzie, A.E.; Lappin, J.E.; Taylor, D.L.; Nicklin, S.A.; Milligan, G. GPR35 as a Novel Therapeutic Target. *Front. Endocrinol.* **2011**, *2*, 68. [CrossRef]
81. Taniguchi, Y.; Tonai-Kachi, H.; Shinjo, K. Zaprinast, a well-known cyclic guanosine monophosphate-specific phosphodiesterase inhibitor, is an agonist for GPR35. *FEBS Lett.* **2006**, *580*, 5003–5008. [CrossRef]
82. Maravillas-Montero, J.L.; Burkhardt, A.M.; Hevezi, P.A.; Carnevale, C.D.; Smit, M.J.; Zlotnik, A. Cutting edge: GPR35/CXCR8 is the receptor of the mucosal chemokine CXCL17. *J. Immunol.* **2015**, *194*, 29–33. [CrossRef]

83. Fallarini, S.; Magliulo, L.; Paoletti, T.; De Lalla, C.; Lombardi, G. Expression of functional GPR35 in human iNKT cells. *Biochem. Biophys. Res. Commun.* **2010**, *398*, 420–425. [CrossRef]
84. Mandi, Y.; Vecsei, L. The kynurenine system and immunoregulation. *J. Neural Transm.* **2012**, *119*, 197–209. [CrossRef]
85. Moroni, F.; Cozzi, A.; Sili, M.; Mannaioni, G. Kynurenic acid: A metabolite with multiple actions and multiple targets in brain and periphery. *J. Neural Transm.* **2012**, *119*, 133–139. [CrossRef]
86. Alkondon, M.; Pereira, E.F.; Todd, S.W.; Randall, W.R.; Lane, M.V.; Albuquerque, E.X. Functional G-protein-coupled receptor 35 is expressed by neurons in the CA1 field of the hippocampus. *Biochem. Pharmacol.* **2015**, *93*, 506–518. [CrossRef]
87. Denison, M.S.; Nagy, S.R. Activation of the aryl hydrocarbon receptor by structurally diverse exogenous and endogenous chemicals. *Ann. Rev. Pharmacol. Toxicol.* **2003**, *43*, 309–334. [CrossRef] [PubMed]
88. Nguyen, N.T.; Kimura, A.; Nakahama, T.; Chinen, I.; Masuda, K.; Nohara, K.; Fujii-Kuriyama, Y.; Kishimoto, T. Aryl hydrocarbon receptor negatively regulates dendritic cell immunogenicity via a kynurenine-dependent mechanism. *Proc. Natl. Acad. Sci. USA* **2010**, *107*, 19961–19966. [CrossRef]
89. Stevens, E.A.; Mezrich, J.D.; Bradfield, C.A. The aryl hydrocarbon receptor: A perspective on potential roles in the immune system. *Immunology* **2009**, *127*, 299–311. [CrossRef]
90. Mezrich, J.D.; Fechner, J.H.; Zhang, X.; Johnson, B.P.; Burlingham, W.J.; Bradfield, C.A. An interaction between kynurenine and the aryl hydrocarbon receptor can generate regulatory T cells. *J. Immunol.* **2010**, *185*, 3190–3198. [CrossRef] [PubMed]
91. Stephens, G.L.; Wang, Q.; Swerdlow, B.; Bhat, G.; Kolbeck, R.; Fung, M. Kynurenine 3-monooxygenase mediates inhibition of Th17 differentiation via catabolism of endogenous aryl hydrocarbon receptor ligands. *Eur. J. Immunol.* **2013**, *43*, 1727–1734. [CrossRef]
92. Perez-Gonzalez, A.; Alvarez-Idaboy, J.R.; Galano, A. Free-radical scavenging by tryptophan and its metabolites through electron transfer based processes. *J. Mol. Model.* **2015**, *21*, 213. [CrossRef] [PubMed]
93. Goda, K.; Hamane, Y.; Kishimoto, R.; Ogishi, Y. Radical scavenging properties of tryptophan metabolites. Estimation of their radical reactivity. *Adv. Exp. Med. Biol.* **1999**, *467*, 397–402. [CrossRef]
94. Zhou, H.; Wang, J.; Jiang, J.; Stavrovskaya, I.G.; Li, M.; Li, W.; Wu, Q.; Zhang, X.; Luo, C.; Zhou, S.; et al. N-acetyl-serotonin offers neuroprotection through inhibiting mitochondrial death pathways and autophagic activation in experimental models of ischemic injury. *J. Neurosci.* **2014**, *34*, 2967–2978. [CrossRef]
95. Jang, S.W.; Liu, X.; Pradoldej, S.; Tosini, G.; Chang, Q.; Iuvone, P.M.; Ye, K. N-acetylserotonin activates TrkB receptor in a circadian rhythm. *Proc. Natl. Acad. Sci. USA* **2010**, *107*, 3876–3881. [CrossRef]
96. Yoo, J.M.; Lee, B.D.; Sok, D.E.; Ma, J.Y.; Kim, M.R. Neuroprotective action of N-acetyl serotonin in oxidative stress-induced apoptosis through the activation of both TrkB/CREB/BDNF pathway and Akt/Nrf2/Antioxidant enzyme in neuronal cells. *Redox Biol.* **2017**, *11*, 592–599. [CrossRef] [PubMed]
97. Mondanelli, G.; Coletti, A.; Greco, F.A.; Pallotta, M.T.; Orabona, C.; Iacono, A.; Belladonna, M.L.; Albini, E.; Panfili, E.; Fallarino, F.; et al. Positive allosteric modulation of indoleamine 2,3-dioxygenase 1 restrains neuroinflammation. *Proc. Natl. Acad. Sci. USA* **2020**, *117*, 3848–3857. [CrossRef]
98. Gomez-Corvera, A.; Cerrillo, I.; Molinero, P.; Naranjo, M.C.; Lardone, P.J.; Sanchez-Hidalgo, M.; Carrascosa-Salmoral, M.P.; Medrano-Campillo, P.; Guerrero, J.M.; Rubio, A. Evidence of immune system melatonin production by two pineal melatonin deficient mice, C57BL/6 and Swiss strains. *J. Pineal Res.* **2009**, *47*, 15–22. [CrossRef]
99. Jaronen, M.; Quintana, F.J. Immunological Relevance of the Coevolution of IDO1 and AHR. *Front. Immunol.* **2014**, *5*, 521. [CrossRef]
100. Lewis-Ballester, A.; Pham, K.N.; Batabyal, D.; Karkashon, S.; Bonanno, J.B.; Poulos, T.L.; Yeh, S.R. Structural insights into substrate and inhibitor binding sites in human indoleamine 2,3-dioxygenase 1. *Nat. Commun.* **2017**, *8*, 1693. [CrossRef] [PubMed]
101. Mondanelli, G.; Iacono, A.; Carvalho, A.; Orabona, C.; Volpi, C.; Pallotta, M.T.; Matino, D.; Esposito, S.; Grohmann, U. Amino acid metabolism as drug target in autoimmune diseases. *Autoimmun. Rev.* **2019**, *18*, 334–348. [CrossRef]
102. Baumeister, S.H.; Freeman, G.J.; Dranoff, G.; Sharpe, A.H. Coinhibitory Pathways in Immunotherapy for Cancer. *Annu. Rev. Immunol.* **2016**, *34*, 539–573. [CrossRef]
103. Orabona, C.; Mondanelli, G.; Puccetti, P.; Grohmann, U. Immune Checkpoint Molecules, Personalized Immunotherapy, and Autoimmune Diabetes. *Trends Mol. Med.* **2018**, *24*, 931–941. [CrossRef] [PubMed]

104. Brundin, L.; Sellgren, C.M.; Lim, C.K.; Grit, J.; Palsson, E.; Landen, M.; Samuelsson, M.; Lundgren, K.; Brundin, P.; Fuchs, D.; et al. An enzyme in the kynurenine pathway that governs vulnerability to suicidal behavior by regulating excitotoxicity and neuroinflammation. *Transl. Psychiatry* **2016**, *6*, e865. [CrossRef] [PubMed]
105. Stone, T.W.; Perkins, M.N. Quinolinic acid: A potent endogenous excitant at amino acid receptors in CNS. *Eur. J. Pharmacol.* **1981**, *72*, 411–412. [CrossRef]
106. Perkins, M.N.; Stone, T.W. Pharmacology and regional variations of quinolinic acid-evoked excitations in the rat central nervous system. *J. Pharmacol. Exp. Ther.* **1983**, *226*, 551–557.
107. Perkins, M.N.; Stone, T.W. Quinolinic acid: Regional variations in neuronal sensitivity. *Brain Res.* **1983**, *259*, 172–176. [CrossRef]
108. De Carvalho, L.P.; Bochet, P.; Rossier, J. The endogenous agonist quinolinic acid and the non endogenous homoquinolinic acid discriminate between NMDAR2 receptor subunits. *Neurochem. Int.* **1996**, *28*, 445–452. [CrossRef]
109. Monaghan, D.T.; Beaton, J.A. Quinolinate differentiates between forebrain and cerebellar NMDA receptors. *Eur. J. Pharmacol.* **1991**, *194*, 123–125. [CrossRef]
110. Scherzer, C.R.; Landwehrmeyer, G.B.; Kerner, J.A.; Standaert, D.G.; Hollingsworth, Z.R.; Daggett, L.P.; Velicelebi, G.; Penney, J.B., Jr.; Young, A.B. Cellular distribution of NMDA glutamate receptor subunit mRNAs in the human cerebellum. *Neurobiol. Dis.* **1997**, *4*, 35–46. [CrossRef]
111. Wang, Y.H.; Bosy, T.Z.; Yasuda, R.P.; Grayson, D.R.; Vicini, S.; Pizzorusso, T.; Wolfe, B.B. Characterization of NMDA receptor subunit-specific antibodies: Distribution of NR2A and NR2B receptor subunits in rat brain and ontogenic profile in the cerebellum. *J. Neurochem.* **1995**, *65*, 176–183. [CrossRef] [PubMed]
112. Santamaria, A.; Rios, C. MK-801, an N-methyl-D-aspartate receptor antagonist, blocks quinolinic acid-induced lipid peroxidation in rat corpus striatum. *Neurosci. Lett.* **1993**, *159*, 51–54. [CrossRef]
113. Platenik, J.; Stopka, P.; Vejrazka, M.; Stipek, S. Quinolinic acid-iron(ii) complexes: Slow autoxidation, but enhanced hydroxyl radical production in the Fenton reaction. *Free Radic. Res.* **2001**, *34*, 445–459. [CrossRef]
114. Rios, C.; Santamaria, A. Quinolinic acid is a potent lipid peroxidant in rat brain homogenates. *Neurochem. Res.* **1991**, *16*, 1139–1143. [CrossRef]
115. Stipek, S.; Stastny, F.; Platenik, J.; Crkovska, J.; Zima, T. The effect of quinolinate on rat brain lipid peroxidation is dependent on iron. *Neurochem. Int.* **1997**, *30*, 233–237. [CrossRef]
116. St'astny, F.; Hinoi, E.; Ogita, K.; Yoneda, Y. Ferrous iron modulates quinolinate-mediated [3H]MK-801 binding to rat brain synaptic membranes in the presence of glycine and spermidine. *Neurosci. Lett.* **1999**, *262*, 105–108. [CrossRef]
117. Pierozan, P.; Zamoner, A.; Soska, A.K.; Silvestrin, R.B.; Loureiro, S.O.; Heimfarth, L.; Mello e Souza, T.; Wajner, M.; Pessoa-Pureur, R. Acute intrastriatal administration of quinolinic acid provokes hyperphosphorylation of cytoskeletal intermediate filament proteins in astrocytes and neurons of rats. *Exp. Neurol.* **2010**, *224*, 188–196. [CrossRef]
118. Rahman, A.; Ting, K.; Cullen, K.M.; Braidy, N.; Brew, B.J.; Guillemin, G.J. The excitotoxin quinolinic acid induces tau phosphorylation in human neurons. *PLoS ONE* **2009**, *4*, e6344. [CrossRef]
119. Guillemin, G.J.; Kerr, S.J.; Smythe, G.A.; Smith, D.G.; Kapoor, V.; Armati, P.J.; Croitoru, J.; Brew, B.J. Kynurenine pathway metabolism in human astrocytes: A paradox for neuronal protection. *J. Neurochem.* **2001**, *78*, 842–853. [CrossRef]
120. Stone, T.W.; Behan, W.M. Interleukin-1beta but not tumor necrosis factor-alpha potentiates neuronal damage by quinolinic acid: Protection by an adenosine A2A receptor antagonist. *J. Neurosci. Res.* **2007**, *85*, 1077–1085. [CrossRef]
121. Musso, T.; Gusella, G.L.; Brooks, A.; Longo, D.L.; Varesio, L. Interleukin-4 inhibits indoleamine 2,3-dioxygenase expression in human monocytes. *Blood* **1994**, *83*, 1408–1411. [CrossRef] [PubMed]
122. Fallarino, F.; Grohmann, U.; Vacca, C.; Bianchi, R.; Orabona, C.; Spreca, A.; Fioretti, M.C.; Puccetti, P. T cell apoptosis by tryptophan catabolism. *Cell Death Differ.* **2002**, *9*, 1069–1077. [CrossRef] [PubMed]
123. Fallarino, F.; Grohmann, U.; You, S.; McGrath, B.C.; Cavener, D.R.; Vacca, C.; Orabona, C.; Bianchi, R.; Belladonna, M.L.; Volpi, C.; et al. The combined effects of tryptophan starvation and tryptophan catabolites down-regulate T cell receptor zeta-chain and induce a regulatory phenotype in naive T cells. *J. Immunol.* **2006**, *176*, 6752–6761. [CrossRef] [PubMed]

124. Hill, M.; Tanguy-Royer, S.; Royer, P.; Chauveau, C.; Asghar, K.; Tesson, L.; Lavainne, F.; Remy, S.; Brion, R.; Hubert, F.X.; et al. IDO expands human CD4+CD25high regulatory T cells by promoting maturation of LPS-treated dendritic cells. *Eur. J. Immunol.* **2007**, *37*, 3054–3062. [CrossRef]
125. Frumento, G.; Rotondo, R.; Tonetti, M.; Damonte, G.; Benatti, U.; Ferrara, G.B. Tryptophan-derived catabolites are responsible for inhibition of T and natural killer cell proliferation induced by indoleamine 2,3-dioxygenase. *J. Exp. Med.* **2002**, *196*, 459–468. [CrossRef]
126. Song, H.; Park, H.; Kim, Y.S.; Kim, K.D.; Lee, H.K.; Cho, D.H.; Yang, J.W.; Hur, D.Y. L-kynurenine-induced apoptosis in human NK cells is mediated by reactive oxygen species. *Int. Immunopharmacol.* **2011**, *11*, 932–938. [CrossRef]
127. Daubener, W.; MacKenzie, C.R. IFN-gamma activated indoleamine 2,3-dioxygenase activity in human cells is an antiparasitic and an antibacterial effector mechanism. *Adv. Exp. Med. Biol.* **1999**, *467*, 517–524. [CrossRef]
128. Robinson, C.M.; Hale, P.T.; Carlin, J.M. The role of IFN-gamma and TNF-alpha-responsive regulatory elements in the synergistic induction of indoleamine dioxygenase. *J. Interferon Cytokine Res.* **2005**, *25*, 20–30. [CrossRef]
129. O'Connor, J.C.; Andre, C.; Wang, Y.; Lawson, M.A.; Szegedi, S.S.; Lestage, J.; Castanon, N.; Kelley, K.W.; Dantzer, R. Interferon-gamma and tumor necrosis factor-alpha mediate the upregulation of indoleamine 2,3-dioxygenase and the induction of depressive-like behavior in mice in response to bacillus Calmette-Guerin. *J. Neurosci.* **2009**, *29*, 4200–4209. [CrossRef]
130. Belladonna, M.L.; Orabona, C.; Grohmann, U.; Puccetti, P. TGF-beta and kynurenines as the key to infectious tolerance. *Trends Mol. Med.* **2009**, *15*, 41–49. [CrossRef]
131. Forrest, C.M.; Mackay, G.M.; Stoy, N.; Spiden, S.L.; Taylor, R.; Stone, T.W.; Darlington, L.G. Blood levels of kynurenines, interleukin-23 and soluble human leucocyte antigen-G at different stages of Huntington's disease. *J. Neurochem.* **2010**, *112*, 112–122. [CrossRef] [PubMed]
132. Chaves, A.C.; Ceravolo, I.P.; Gomes, J.A.; Zani, C.L.; Romanha, A.J.; Gazzinelli, R.T. IL-4 and IL-13 regulate the induction of indoleamine 2,3-dioxygenase activity and the control of Toxoplasma gondii replication in human fibroblasts activated with IFN-gamma. *Eur. J. Immunol.* **2001**, *31*, 333–344. [CrossRef]
133. Gomez, A.; Luckey, D.; Taneja, V. The gut microbiome in autoimmunity: Sex matters. *Clin. Immunol.* **2015**, *159*, 154–162. [CrossRef] [PubMed]
134. Brown, C.T.; Davis-Richardson, A.G.; Giongo, A.; Gano, K.A.; Crabb, D.B.; Mukherjee, N.; Casella, G.; Drew, J.C.; Ilonen, J.; Knip, M.; et al. Gut microbiome metagenomics analysis suggests a functional model for the development of autoimmunity for type 1 diabetes. *PLoS ONE* **2011**, *6*, e25792. [CrossRef]
135. Rosser, E.C.; Mauri, C. A clinical update on the significance of the gut microbiota in systemic autoimmunity. *J. Autoimmun.* **2016**, *74*, 85–93. [CrossRef]
136. Clemente, J.C.; Manasson, J.; Scher, J.U. The role of the gut microbiome in systemic inflammatory disease. *BMJ* **2018**, *360*, j5145. [CrossRef]
137. Ghaisas, S.; Maher, J.; Kanthasamy, A. Gut microbiome in health and disease: Linking the microbiome-gut-brain axis and environmental factors in the pathogenesis of systemic and neurodegenerative diseases. *Pharmacol. Ther.* **2016**, *158*, 52–62. [CrossRef] [PubMed]
138. Westfall, S.; Lomis, N.; Kahouli, I.; Dia, S.Y.; Singh, S.P.; Prakash, S. Microbiome, probiotics and neurodegenerative diseases: Deciphering the gut brain axis. *Cell. Mol. Life Sci.* **2017**, *74*, 3769–3787. [CrossRef] [PubMed]
139. Roy Sarkar, S.; Banerjee, S. Gut microbiota in neurodegenerative disorders. *J. Neuroimmunol.* **2019**, *328*, 98–104. [CrossRef] [PubMed]
140. Bauer, P.V.; Hamr, S.C.; Duca, F.A. Regulation of energy balance by a gut-brain axis and involvement of the gut microbiota. *Cell. Mol. Life Sci.* **2016**, *73*, 737–755. [CrossRef]
141. Ojeda, P.; Bobe, A.; Dolan, K.; Leone, V.; Martinez, K. Nutritional modulation of gut microbiota—The impact on metabolic disease pathophysiology. *J. Nutr. Biochem.* **2016**, *28*, 191–200. [CrossRef] [PubMed]
142. Marinoni, I.; Nonnis, S.; Monteferrante, C.; Heathcote, P.; Hartig, E.; Bottger, L.H.; Trautwein, A.X.; Negri, A.; Albertini, A.M.; Tedeschi, G. Characterization of L-aspartate oxidase and quinolinate synthase from Bacillus subtilis. *FEBS J.* **2008**, *275*, 5090–5107. [CrossRef]
143. Barends, T.R.; Dunn, M.F.; Schlichting, I. Tryptophan synthase, an allosteric molecular factory. *Curr. Opin. Chem. Biol.* **2008**, *12*, 593–600. [CrossRef]

144. Reichmann, D.; Coute, Y.; Ollagnier de Choudens, S. Dual activity of quinolinate synthase: Triose phosphate isomerase and dehydration activities play together to form quinolinate. *Biochemistry* **2015**, *54*, 6443–6446. [CrossRef] [PubMed]
145. Sakuraba, H.; Tsuge, H.; Yoneda, K.; Katunuma, N.; Ohshima, T. Crystal structure of the NAD biosynthetic enzyme quinolinate synthase. *J. Biol. Chem.* **2005**, *280*, 26645–26648. [CrossRef]
146. Saunders, A.H.; Griffiths, A.E.; Lee, K.H.; Cicchillo, R.M.; Tu, L.; Stromberg, J.A.; Krebs, C.; Booker, S.J. Characterization of quinolinate synthases from Escherichia coli, Mycobacterium tuberculosis, and Pyrococcus horikoshii indicates that [4Fe-4S] clusters are common cofactors throughout this class of enzymes. *Biochemistry* **2008**, *47*, 10999–11012. [CrossRef]
147. Garavaglia, S.; Perozzi, S.; Galeazzi, L.; Raffaelli, N.; Rizzi, M. The crystal structure of human alpha-amino-beta-carboxymuconate-epsilon-semialdehyde decarboxylase in complex with 1,3-dihydroxyacetonephosphate suggests a regulatory link between NAD synthesis and glycolysis. *FEBS J.* **2009**, *276*, 6615–6623. [CrossRef] [PubMed]
148. Liu, X.; Dong, Y.; Li, X.; Ren, Y.; Li, Y.; Wang, W.; Wang, L.; Feng, L. Characterization of the anthranilate degradation pathway in Geobacillus thermodenitrificans NG80-2. *Microbiology* **2010**, *156*, 589–595. [CrossRef]
149. Cleaves, H.J.; Miller, S.L. The nicotinamide biosynthetic pathway is a by-product of the RNA world. *J. Mol. Evol.* **2001**, *52*, 73–77. [CrossRef]
150. Shahi, S.K.; Freedman, S.N.; Mangalam, A.K. Gut microbiome in multiple sclerosis: The players involved and the roles they play. *Gut Microbes* **2017**, *8*, 607–615. [CrossRef] [PubMed]
151. Mielcarz, D.W.; Kasper, L.H. The gut microbiome in multiple sclerosis. *Curr. Treat. Options Neurol.* **2015**, *17*, 344. [CrossRef] [PubMed]
152. Jangi, S.; Gandhi, R.; Cox, L.M.; Li, N.; Von Glehn, F.; Yan, R.; Patel, B.; Mazzola, M.A.; Liu, S.; Glanz, B.L.; et al. Alterations of the human gut microbiome in multiple sclerosis. *Nat. Commun.* **2016**, *7*, 12015. [CrossRef] [PubMed]
153. O'Mahony, S.M.; Clarke, G.; Borre, Y.E.; Dinan, T.G.; Cryan, J.F. Serotonin, tryptophan metabolism and the brain-gut-microbiome axis. *Behav. Brain Res.* **2015**, *277*, 32–48. [CrossRef] [PubMed]
154. Kennedy, P.J.; Cryan, J.F.; Dinan, T.G.; Clarke, G. Kynurenine pathway metabolism and the microbiota-gut-brain axis. *Neuropharmacology* **2017**, *112*, 399–412. [CrossRef]
155. Suhs, K.W.; Novoselova, N.; Kuhn, M.; Seegers, L.; Kaever, V.; Muller-Vahl, K.; Trebst, C.; Skripuletz, T.; Stangel, M.; Pessler, F. Kynurenine Is a Cerebrospinal Fluid Biomarker for Bacterial and Viral Central Nervous System Infections. *J. Infect. Dis.* **2019**, *220*, 127–138. [CrossRef]
156. Jeltsch-David, H.; Muller, S. Neuropsychiatric systemic lupus erythematosus: Pathogenesis and biomarkers. *Nat. Rev. Neurol.* **2014**, *10*, 579–596. [CrossRef]
157. Akesson, K.; Pettersson, S.; Stahl, S.; Surowiec, I.; Hedenstrom, M.; Eketjall, S.; Trygg, J.; Jakobsson, P.J.; Gunnarsson, I.; Svenungsson, E.; et al. Kynurenine pathway is altered in patients with SLE and associated with severe fatigue. *Lupus Sci. Med.* **2018**, *5*, e000254. [CrossRef]
158. Vitali, C.; Bombardieri, S.; Jonsson, R.; Moutsopoulos, H.M.; Alexander, E.L.; Carsons, S.E.; Daniels, T.E.; Fox, P.C.; Fox, R.I.; Kassan, S.S.; et al. Classification criteria for Sjogren's syndrome: A revised version of the European criteria proposed by the American-European Consensus Group. *Ann. Rheum. Dis.* **2002**, *61*, 554–558. [CrossRef]
159. Brito-Zeron, P.; Theander, E.; Baldini, C.; Seror, R.; Retamozo, S.; Quartuccio, L.; Bootsma, H.; Bowman, S.J.; Dorner, T.; Gottenberg, J.E.; et al. Early diagnosis of primary Sjogren's syndrome: EULAR-SS task force clinical recommendations. *Exp. Rev. Clin. Immunol.* **2016**, *12*, 137–156. [CrossRef]
160. De Oliveira, F.R.; Fantucci, M.Z.; Adriano, L.; Valim, V.; Cunha, T.M.; Louzada-Junior, P.; Rocha, E.M. Neurological and Inflammatory Manifestations in Sjogren's Syndrome: The Role of the Kynurenine Metabolic Pathway. *Int. J. Mol. Sci.* **2018**, *19*, 3953. [CrossRef]
161. Furuzawa-Carballeda, J.; Hernandez-Molina, G.; Lima, G.; Rivera-Vicencio, Y.; Ferez-Blando, K.; Llorente, L. Peripheral regulatory cells immunophenotyping in primary Sjogren's syndrome: A cross-sectional study. *Arthritis Res. Ther.* **2013**, *15*, R68. [CrossRef] [PubMed]
162. Legany, N.; Berta, L.; Kovacs, L.; Balog, A.; Toldi, G. The role of B7 family costimulatory molecules and indoleamine 2,3-dioxygenase in primary Sjogren's syndrome and systemic sclerosis. *Immunol. Res.* **2017**, *65*, 622–629. [CrossRef] [PubMed]

163. Raine, C.S.; Scheinberg, L.C. On the immunopathology of plaque development and repair in multiple sclerosis. *J. Neuroimmunol.* **1988**, *20*, 189–201. [CrossRef]
164. O'Connor, R.A.; Prendergast, C.T.; Sabatos, C.A.; Lau, C.W.; Leech, M.D.; Wraith, D.C.; Anderton, S.M. Cutting edge: Th1 cells facilitate the entry of Th17 cells to the central nervous system during experimental autoimmune encephalomyelitis. *J. Immunol.* **2008**, *181*, 3750–3754. [CrossRef] [PubMed]
165. Miller, S.D.; Karpus, W.J. Experimental autoimmune encephalomyelitis in the mouse. *Curr. Protoc. Immunol.* **2007**, *15*, 15.1. [CrossRef]
166. Gold, R.; Linington, C.; Lassmann, H. Understanding pathogenesis and therapy of multiple sclerosis via animal models: 70 years of merits and culprits in experimental autoimmune encephalomyelitis research. *Brain* **2006**, *129*, 1953–1971. [CrossRef]
167. Fuvesi, J.; Rajda, C.; Bencsik, K.; Toldi, J.; Vecsei, L. The role of kynurenines in the pathomechanism of amyotrophic lateral sclerosis and multiple sclerosis: Therapeutic implications. *J. Neural Transm.* **2012**, *119*, 225–234. [CrossRef] [PubMed]
168. Lassmann, H.; Ransohoff, R.M. The CD4-Th1 model for multiple sclerosis: A critical [correction of crucial re-appraisal. *Trends Immunol.* **2004**, *25*, 132–137. [CrossRef]
169. Barnett, M.H.; Prineas, J.W. Relapsing and remitting multiple sclerosis: Pathology of the newly forming lesion. *Ann. Neurol.* **2004**, *55*, 458–468. [CrossRef]
170. Lodygin, D.; Hermann, M.; Schweingruber, N.; Flugel-Koch, C.; Watanabe, T.; Schlosser, C.; Merlini, A.; Korner, H.; Chang, H.F.; Fischer, H.J.; et al. beta-Synuclein-reactive T cells induce autoimmune CNS grey matter degeneration. *Nature* **2019**, *566*, 503–508. [CrossRef]
171. Dutta, R.; McDonough, J.; Yin, X.; Peterson, J.; Chang, A.; Torres, T.; Gudz, T.; Macklin, W.B.; Lewis, D.A.; Fox, R.J.; et al. Mitochondrial dysfunction as a cause of axonal degeneration in multiple sclerosis patients. *Ann. Neurol.* **2006**, *59*, 478–489. [CrossRef] [PubMed]
172. Lu, F.; Selak, M.; O'Connor, J.; Croul, S.; Lorenzana, C.; Butunoi, C.; Kalman, B. Oxidative damage to mitochondrial DNA and activity of mitochondrial enzymes in chronic active lesions of multiple sclerosis. *J. Neurol. Sci.* **2000**, *177*, 95–103. [CrossRef]
173. Mahad, D.; Ziabreva, I.; Lassmann, H.; Turnbull, D. Mitochondrial defects in acute multiple sclerosis lesions. *Brain* **2008**, *131*, 1722–1735. [CrossRef]
174. Trapp, B.D.; Stys, P.K. Virtual hypoxia and chronic necrosis of demyelinated axons in multiple sclerosis. *Lancet Neurol.* **2009**, *8*, 280–291. [CrossRef]
175. Witte, M.E.; Bo, L.; Rodenburg, R.J.; Belien, J.A.; Musters, R.; Hazes, T.; Wintjes, L.T.; Smeitink, J.A.; Geurts, J.J.; De Vries, H.E.; et al. Enhanced number and activity of mitochondria in multiple sclerosis lesions. *J. Pathol.* **2009**, *219*, 193–204. [CrossRef]
176. Veto, S.; Acs, P.; Bauer, J.; Lassmann, H.; Berente, Z.; Setalo, G., Jr.; Borgulya, G.; Sumegi, B.; Komoly, S.; Gallyas, F., Jr.; et al. Inhibiting poly(ADP-ribose) polymerase: A potential therapy against oligodendrocyte death. *Brain* **2010**, *133*, 822–834. [CrossRef]
177. Ziabreva, I.; Campbell, G.; Rist, J.; Zambonin, J.; Rorbach, J.; Wydro, M.M.; Lassmann, H.; Franklin, R.J.; Mahad, D. Injury and differentiation following inhibition of mitochondrial respiratory chain complex IV in rat oligodendrocytes. *Glia* **2010**, *58*, 1827–1837. [CrossRef]
178. Mahad, D.; Lassmann, H.; Turnbull, D. Review: Mitochondria and disease progression in multiple sclerosis. *Neuropathol. Appl. Neurobiol.* **2008**, *34*, 577–589. [CrossRef] [PubMed]
179. Mahad, D.J.; Ziabreva, I.; Campbell, G.; Lax, N.; White, K.; Hanson, P.S.; Lassmann, H.; Turnbull, D.M. Mitochondrial changes within axons in multiple sclerosis. *Brain* **2009**, *132*, 1161–1174. [CrossRef]
180. Sharma, R.; Fischer, M.T.; Bauer, J.; Felts, P.A.; Smith, K.J.; Misu, T.; Fujihara, K.; Bradl, M.; Lassmann, H. Inflammation induced by innate immunity in the central nervous system leads to primary astrocyte dysfunction followed by demyelination. *Acta Neuropathol.* **2010**, *120*, 223–236. [CrossRef]
181. Kwidzinski, E.; Bunse, J.; Aktas, O.; Richter, D.; Mutlu, L.; Zipp, F.; Nitsch, R.; Bechmann, I. Indolamine 2,3-dioxygenase is expressed in the CNS and down-regulates autoimmune inflammation. *FASEB J.* **2005**, *19*, 1347–1349. [CrossRef] [PubMed]
182. Sakurai, K.; Zou, J.P.; Tschetter, J.R.; Ward, J.M.; Shearer, G.M. Effect of indoleamine 2,3-dioxygenase on induction of experimental autoimmune encephalomyelitis. *J. Neuroimmunol.* **2002**, *129*, 186–196. [CrossRef]

183. Yan, Y.; Zhang, G.X.; Gran, B.; Fallarino, F.; Yu, S.; Li, H.; Cullimore, M.L.; Rostami, A.; Xu, H. IDO upregulates regulatory T cells via tryptophan catabolite and suppresses encephalitogenic T cell responses in experimental autoimmune encephalomyelitis. *J. Immunol.* **2010**, *185*, 5953–5961. [CrossRef] [PubMed]
184. Gonsette, R.E. Self-tolerance in multiple sclerosis. *Acta Neurol. Belg.* **2012**, *112*, 133–140. [CrossRef]
185. Fazio, F.; Zappulla, C.; Notartomaso, S.; Busceti, C.; Bessede, A.; Scarselli, P.; Vacca, C.; Gargaro, M.; Volpi, C.; Allegrucci, M.; et al. Cinnabarinic acid, an endogenous agonist of type-4 metabotropic glutamate receptor, suppresses experimental autoimmune encephalomyelitis in mice. *Neuropharmacology* **2014**, *81*, 237–243. [CrossRef]
186. Xiao, B.G.; Liu, X.; Link, H. Antigen-specific T cell functions are suppressed over the estrogen-dendritic cell-indoleamine 2,3-dioxygenase axis. *Steroids* **2004**, *69*, 653–659. [CrossRef] [PubMed]
187. Platten, M.; Ho, P.P.; Youssef, S.; Fontoura, P.; Garren, H.; Hur, E.M.; Gupta, R.; Lee, L.Y.; Kidd, B.A.; Robinson, W.H.; et al. Treatment of autoimmune neuroinflammation with a synthetic tryptophan metabolite. *Science* **2005**, *310*, 850–855. [CrossRef] [PubMed]
188. Chiarugi, A.; Cozzi, A.; Ballerini, C.; Massacesi, L.; Moroni, F. Kynurenine 3-mono-oxygenase activity and neurotoxic kynurenine metabolites increase in the spinal cord of rats with experimental allergic encephalomyelitis. *Neuroscience* **2001**, *102*, 687–695. [CrossRef]
189. Flanagan, E.M.; Erickson, J.B.; Viveros, O.H.; Chang, S.Y.; Reinhard, J.F., Jr. Neurotoxin quinolinic acid is selectively elevated in spinal cords of rats with experimental allergic encephalomyelitis. *J. Neurochem.* **1995**, *64*, 1192–1196. [CrossRef] [PubMed]
190. Cammer, W. Oligodendrocyte killing by quinolinic acid in vitro. *Brain Res.* **2001**, *896*, 157–160. [CrossRef]
191. Guillemin, G.J.; Wang, L.; Brew, B.J. Quinolinic acid selectively induces apoptosis of human astrocytes: Potential role in AIDS dementia complex. *J. Neuroinflamm.* **2005**, *2*, 16. [CrossRef] [PubMed]
192. Kerr, S.J.; Armati, P.J.; Guillemin, G.J.; Brew, B.J. Chronic exposure of human neurons to quinolinic acid results in neuronal changes consistent with AIDS dementia complex. *Aids* **1998**, *12*, 355–363. [CrossRef] [PubMed]
193. Rajda, C.; Majlath, Z.; Pukoli, D.; Vecsei, L. Kynurenines and Multiple Sclerosis: The Dialogue between the Immune System and the Central Nervous System. *Int. J. Mol. Sci.* **2015**, *16*, 18270–18282. [CrossRef]
194. Behan, W.M.; McDonald, M.; Darlington, L.G.; Stone, T.W. Oxidative stress as a mechanism for quinolinic acid-induced hippocampal damage: Protection by melatonin and deprenyl. *Br. J. Pharmacol.* **1999**, *128*, 1754–1760. [CrossRef] [PubMed]
195. Santamaria, A.; Jimenez-Capdeville, M.E.; Camacho, A.; Rodriguez-Martinez, E.; Flores, A.; Galvan-Arzate, S. In vivo hydroxyl radical formation after quinolinic acid infusion into rat corpus striatum. *Neuroreport* **2001**, *12*, 2693–2696. [CrossRef] [PubMed]
196. Leipnitz, G.; Schumacher, C.; Scussiato, K.; Dalcin, K.B.; Wannmacher, C.M.; Wyse, A.T.; Dutra-Filho, C.S.; Wajner, M.; Latini, A. Quinolinic acid reduces the antioxidant defenses in cerebral cortex of young rats. *Int. J. Dev. Neurosci.* **2005**, *23*, 695–701. [CrossRef]
197. Rodriguez-Martinez, E.; Camacho, A.; Maldonado, P.D.; Pedraza-Chaverri, J.; Santamaria, D.; Galvan-Arzate, S.; Santamaria, A. Effect of quinolinic acid on endogenous antioxidants in rat corpus striatum. *Brain Res.* **2000**, *858*, 436–439. [CrossRef]
198. Rodriguez, E.; Mendez-Armenta, M.; Villeda-Hernandez, J.; Galvan-Arzate, S.; Barroso-Moguel, R.; Rodriguez, F.; Rios, C.; Santamaria, A. Dapsone prevents morphological lesions and lipid peroxidation induced by quinolinic acid in rat corpus striatum. *Toxicology* **1999**, *139*, 111–118. [CrossRef]
199. Santamaria, A.; Galvan-Arzate, S.; Lisy, V.; Ali, S.F.; Duhart, H.M.; Osorio-Rico, L.; Rios, C.; St'astny, F. Quinolinic acid induces oxidative stress in rat brain synaptosomes. *Neuroreport* **2001**, *12*, 871–874. [CrossRef]
200. Baran, H.; Staniek, K.; Kepplinger, B.; Gille, L.; Stolze, K.; Nohl, H. Kynurenic acid influences the respiratory parameters of rat heart mitochondria. *Pharmacology* **2001**, *62*, 119–123. [CrossRef]
201. Bordelon, Y.M.; Chesselet, M.F.; Nelson, D.; Welsh, F.; Erecinska, M. Energetic dysfunction in quinolinic acid-lesioned rat striatum. *J. Neurochem.* **1997**, *69*, 1629–1639. [CrossRef] [PubMed]
202. Blumenthal, A.; Nagalingam, G.; Huch, J.H.; Walker, L.; Guillemin, G.J.; Smythe, G.A.; Ehrt, S.; Britton, W.J.; Saunders, B.M.M. tuberculosis induces potent activation of IDO-1, but this is not essential for the immunological control of infection. *PLoS ONE* **2012**, *7*, e37314. [CrossRef]

203. Suzuki, Y.; Suda, T.; Asada, K.; Miwa, S.; Suzuki, M.; Fujie, M.; Furuhashi, K.; Nakamura, Y.; Inui, N.; Shirai, T.; et al. Serum indoleamine 2,3-dioxygenase activity predicts prognosis of pulmonary tuberculosis. *Clin. Vaccine Immunol.* **2012**, *19*, 436–442. [CrossRef] [PubMed]
204. Watzlawik, J.O.; Wootla, B.; Rodriguez, M. Tryptophan Catabolites and Their Impact on Multiple Sclerosis Progression. *Curr. Pharma. Des.* **2016**, *22*, 1049–1059. [CrossRef]
205. Pemberton, L.A.; Kerr, S.J.; Smythe, G.; Brew, B.J. Quinolinic acid production by macrophages stimulated with IFN-gamma, TNF-alpha, and IFN-alpha. *J. Interferon Cytokine Res.* **1997**, *17*, 589–595. [CrossRef]
206. Beck, J.; Rondot, P.; Catinot, L.; Falcoff, E.; Kirchner, H.; Wietzerbin, J. Increased production of interferon gamma and tumor necrosis factor precedes clinical manifestation in multiple sclerosis: Do cytokines trigger off exacerbations? *Acta Neurol. Scand.* **1988**, *78*, 318–323. [CrossRef]
207. Rajda, C.; Galla, Z.; Polyak, H.; Maroti, Z.; Babarczy, K.; Pukoli, D.; Vecsei, L. Cerebrospinal Fluid Neurofilament Light Chain Is Associated with Kynurenine Pathway Metabolite Changes in Multiple Sclerosis. *Int. J. Mol. Sci.* **2020**, *21*, 2665. [CrossRef]
208. Monaco, F.; Fumero, S.; Mondino, A.; Mutani, R. Plasma and cerebrospinal fluid tryptophan in multiple sclerosis and degenerative diseases. *J. Neurol. Neurosurg. Psychiatry* **1979**, *42*, 640–641. [CrossRef]
209. Ott, M.; Demisch, L.; Engelhardt, W.; Fischer, P.A. Interleukin-2, soluble interleukin-2-receptor, neopterin, L-tryptophan and beta 2-microglobulin levels in CSF and serum of patients with relapsing-remitting or chronic-progressive multiple sclerosis. *J. Neurol.* **1993**, *241*, 108–114. [CrossRef]
210. Rudzite, V.; Berzinsh, J.; Grivane, I.; Fuchs, D.; Baier-Bitterlich, G.; Wachter, H. Serum tryptophan, kynurenine, and neopterin in patients with Guillain-Barre-syndrome (GBS) and multiple sclerosis (MS). *Adv. Exp. Med. Biol.* **1996**, *398*, 183–187. [CrossRef]
211. Amirkhani, A.; Rajda, C.; Arvidsson, B.; Bencsik, K.; Boda, K.; Seres, E.; Markides, K.E.; Vecsei, L.; Bergquist, J. Interferon-beta affects the tryptophan metabolism in multiple sclerosis patients. *Eur. J. Neurol.* **2005**, *12*, 625–631. [CrossRef]
212. Rothhammer, V.; Borucki, D.M.; Garcia Sanchez, M.I.; Mazzola, M.A.; Hemond, C.C.; Regev, K.; Paul, A.; Kivisakk, P.; Bakshi, R.; Izquierdo, G.; et al. Dynamic regulation of serum aryl hydrocarbon receptor agonists in MS. *Neurol. Neuroimmunol. Neuroinflamm.* **2017**, *4*, e359. [CrossRef] [PubMed]
213. Rejdak, K.; Petzold, A.; Kocki, T.; Kurzepa, J.; Grieb, P.; Turski, W.A.; Stelmasiak, Z. Astrocytic activation in relation to inflammatory markers during clinical exacerbation of relapsing-remitting multiple sclerosis. *J. Neural Transm.* **2007**, *114*, 1011–1015. [CrossRef] [PubMed]
214. Rejdak, K.; Bartosik-Psujek, H.; Dobosz, B.; Kocki, T.; Grieb, P.; Giovannoni, G.; Turski, W.A.; Stelmasiak, Z. Decreased level of kynurenic acid in cerebrospinal fluid of relapsing-onset multiple sclerosis patients. *Neurosci. Lett.* **2002**, *331*, 63–65. [CrossRef]
215. Anderson, J.M.; Hampton, D.W.; Patani, R.; Pryce, G.; Crowther, R.A.; Reynolds, R.; Franklin, R.J.; Giovannoni, G.; Compston, D.A.; Baker, D.; et al. Abnormally phosphorylated tau is associated with neuronal and axonal loss in experimental autoimmune encephalomyelitis and multiple sclerosis. *Brain* **2008**, *131*, 1736–1748. [CrossRef]
216. Lim, C.K.; Brew, B.J.; Sundaram, G.; Guillemin, G.J. Understanding the roles of the kynurenine pathway in multiple sclerosis progression. *Int. J. Tryptophan Res.* **2010**, *3*, 157–167. [CrossRef] [PubMed]
217. Lim, C.K.; Bilgin, A.; Lovejoy, D.B.; Tan, V.; Bustamante, S.; Taylor, B.V.; Bessede, A.; Brew, B.J.; Guillemin, G.J. Kynurenine pathway metabolomics predicts and provides mechanistic insight into multiple sclerosis progression. *Sci. Rep.* **2017**, *7*, 41473. [CrossRef]
218. Mancuso, R.; Hernis, A.; Agostini, S.; Rovaris, M.; Caputo, D.; Fuchs, D.; Clerici, M. Indoleamine 2,3 Dioxygenase (IDO) Expression and Activity in Relapsing-Remitting Multiple Sclerosis. *PLoS ONE* **2015**, *10*, e0130715. [CrossRef]
219. Aeinehband, S.; Brenner, P.; Stahl, S.; Bhat, M.; Fidock, M.D.; Khademi, M.; Olsson, T.; Engberg, G.; Jokinen, J.; Erhardt, S.; et al. Cerebrospinal fluid kynurenines in multiple sclerosis; relation to disease course and neurocognitive symptoms. *Brain Behav. Immun.* **2016**, *51*, 47–55. [CrossRef]
220. Boeschoten, R.E.; Braamse, A.M.J.; Beekman, A.T.F.; Cuijpers, P.; Van Oppen, P.; Dekker, J.; Uitdehaag, B.M.J. Prevalence of depression and anxiety in Multiple Sclerosis: A systematic review and meta-analysis. *J. Neurol. Sci.* **2017**, *372*, 331–341. [CrossRef]
221. Anderson, G.; Maes, M. Oxidative/nitrosative stress and immuno-inflammatory pathways in depression: Treatment implications. *Curr. Pharm. Des.* **2014**, *20*, 3812–3847. [CrossRef]

222. Anderson, G.; Maes, M. TRYCAT pathways link peripheral inflammation, nicotine, somatization and depression in the etiology and course of Parkinson's disease. *CNS Neurol. Disord. Drug Targets* **2014**, *13*, 137–149. [CrossRef] [PubMed]
223. Maes, M.; Kubera, M.; Obuchowiczwa, E.; Goehler, L.; Brzeszcz, J. Depression's multiple comorbidities explained by (neuro)inflammatory and oxidative & nitrosative stress pathways. *Neuro Endocrinol. Lett.* **2011**, *32*, 7–24. [PubMed]
224. Campbell, B.M.; Charych, E.; Lee, A.W.; Moller, T. Kynurenines in CNS disease: Regulation by inflammatory cytokines. *Front. Neurosci.* **2014**, *8*, 12. [CrossRef]
225. Anderson, G.; Maes, M.; Berk, M. Inflammation-related disorders in the tryptophan catabolite pathway in depression and somatization. *Adv. Protein Chem. Struct. Biol.* **2012**, *88*, 27–48. [CrossRef] [PubMed]
226. Durastanti, V.; Lugaresi, A.; Bramanti, P.; Amato, M.; Bellantonio, P.; De Luca, G.; Picconi, O.; Fantozzi, R.; Locatelli, L.; Solda, A.; et al. Neopterin production and tryptophan degradation during 24-months therapy with interferon beta-1a in multiple sclerosis patients. *J. Transl. Med.* **2011**, *9*, 42. [CrossRef]
227. Sadowska-Bartosz, I.; Adamczyk-Sowa, M.; Gajewska, A.; Bartosz, G. Oxidative modification of blood serum proteins in multiple sclerosis after interferon or mitoxantrone treatment. *J. Neuroimmunol.* **2014**, *266*, 67–74. [CrossRef]
228. Kappos, L.; Polman, C.H.; Freedman, M.S.; Edan, G.; Hartung, H.P.; Miller, D.H.; Montalban, X.; Barkhof, F.; Bauer, L.; Jakobs, P.; et al. Treatment with interferon beta-1b delays conversion to clinically definite and McDonald MS in patients with clinically isolated syndromes. *Neurology* **2006**, *67*, 1242–1249. [CrossRef]
229. O'Connor, P.; Filippi, M.; Arnason, B.; Comi, G.; Cook, S.; Goodin, D.; Hartung, H.P.; Jeffery, D.; Kappos, L.; Boateng, F.; et al. 250 microg or 500 microg interferon beta-1b versus 20 mg glatiramer acetate in relapsing-remitting multiple sclerosis: A prospective, randomised, multicentre study. *Lancet Neurol.* **2009**, *8*, 889–897. [CrossRef]
230. Reder, A.T.; Ebers, G.C.; Traboulsee, A.; Li, D.; Langdon, D.; Goodin, D.S.; Bogumil, T.; Beckmann, K.; Konieczny, A. Investigators of the 16-Year Long-Term Follow-Up, S. Cross-sectional study assessing long-term safety of interferon-beta-1b for relapsing-remitting MS. *Neurology* **2010**, *74*, 1877–1885. [CrossRef]
231. Smith, A.K.; Simon, J.S.; Gustafson, E.L.; Noviello, S.; Cubells, J.F.; Epstein, M.P.; Devlin, D.J.; Qiu, P.; Albrecht, J.K.; Brass, C.A.; et al. Association of a polymorphism in the indoleamine- 2,3-dioxygenase gene and interferon-alpha-induced depression in patients with chronic hepatitis C. *Mol. Psychiatry* **2012**, *17*, 781–789. [CrossRef] [PubMed]
232. Capuron, L.; Neurauter, G.; Musselman, D.L.; Lawson, D.H.; Nemeroff, C.B.; Fuchs, D.; Miller, A.H. Interferon-alpha-induced changes in tryptophan metabolism. relationship to depression and paroxetine treatment. *Biol. Psychiatry* **2003**, *54*, 906–914. [CrossRef]
233. Wichers, M.C.; Koek, G.H.; Robaeys, G.; Verkerk, R.; Scharpe, S.; Maes, M. IDO and interferon-alpha-induced depressive symptoms: A shift in hypothesis from tryptophan depletion to neurotoxicity. *Mol. Psychiatry* **2005**, *10*, 538–544. [CrossRef] [PubMed]
234. Galvao-de Almeida, A.; Quarantini, L.C.; Sampaio, A.S.; Lyra, A.C.; Parise, C.L.; Parana, R.; De Oliveira, I.R.; Koenen, K.C.; Miranda-Scippa, A.; Guindalini, C. Lack of association of indoleamine 2,3-dioxygenase polymorphisms with interferon-alpha-related depression in hepatitis C. *Brain Behav. Immun.* **2011**, *25*, 1491–1497. [CrossRef]
235. Guillemin, G.J.; Kerr, S.J.; Pemberton, L.A.; Smith, D.G.; Smythe, G.A.; Armati, P.J.; Brew, B.J. IFN-beta1b induces kynurenine pathway metabolism in human macrophages: Potential implications for multiple sclerosis treatment. *J. Interferon Cytokine Res.* **2001**, *21*, 1097–1101. [CrossRef]
236. Ratajczak, J.; Joffraud, M.; Trammell, S.A.; Ras, R.; Canela, N.; Boutant, M.; Kulkarni, S.S.; Rodrigues, M.; Redpath, P.; Migaud, M.E.; et al. NRK1 controls nicotinamide mononucleotide and nicotinamide riboside metabolism in mammalian cells. *Nat. Commun.* **2016**, *7*, 13103. [CrossRef]
237. Revollo, J.R.; Grimm, A.A.; Imai, S. The regulation of nicotinamide adenine dinucleotide biosynthesis by Nampt/PBEF/visfatin in mammals. *Curr. Opin. Gastroenterol.* **2007**, *23*, 164–170. [CrossRef]
238. Penberthy, W.T.; Tsunoda, I. The importance of NAD in multiple sclerosis. *Curr. Pharm. Des.* **2009**, *15*, 64–99. [CrossRef]
239. Braidy, N.; Grant, R. Kynurenine pathway metabolism and neuroinflammatory disease. *Neural Regen. Res.* **2017**, *12*, 39–42. [CrossRef]

240. Massudi, H.; Grant, R.; Guillemin, G.J.; Braidy, N. NAD$^+$ metabolism and oxidative stress: The golden nucleotide on a crown of thorns. *Redox Rep. Commun. Free Radic. Res.* **2012**, *17*, 28–46. [CrossRef]
241. Abeti, R.; Duchen, M.R. Activation of PARP by oxidative stress induced by beta-amyloid: Implications for Alzheimer's disease. *Neurochem. Res.* **2012**, *37*, 2589–2596. [CrossRef] [PubMed]
242. Braidy, N.; Poljak, A.; Grant, R.; Jayasena, T.; Mansour, H.; Chan-Ling, T.; Guillemin, G.J.; Smythe, G.; Sachdev, P. Mapping NAD(+) metabolism in the brain of ageing Wistar rats: Potential targets for influencing brain senescence. *Biogerontology* **2014**, *15*, 177–198. [CrossRef] [PubMed]
243. Verdin, E. NAD(+) in aging, metabolism, and neurodegeneration. *Science* **2015**, *350*, 1208–1213. [CrossRef]
244. Chiarugi, A. Intrinsic mechanisms of poly(ADP-ribose) neurotoxicity: Three hypotheses. *Neurotoxicology* **2005**, *26*, 847–855. [CrossRef]
245. Braidy, N.; Rossez, H.; Lim, C.K.; Jugder, B.E.; Brew, B.J.; Guillemin, G.J. Characterization of the Kynurenine Pathway in CD8(+) Human Primary Monocyte-Derived Dendritic Cells. *Neurotox. Res.* **2016**, *30*, 620–632. [CrossRef]
246. Schwarcz, R.; Stone, T.W. The kynurenine pathway and the brain: Challenges, controversies and promises. *Neuropharmacology* **2017**, *112*, 237–247. [CrossRef] [PubMed]
247. Grant, R.S.; Passey, R.; Matanovic, G.; Smythe, G.; Kapoor, V. Evidence for increased de novo synthesis of NAD in immune-activated RAW264.7 macrophages: A self-protective mechanism? *Arch. Biochem. Biophys.* **1999**, *372*, 1–7. [CrossRef]
248. Oh, G.S.; Pae, H.O.; Choi, B.M.; Chae, S.C.; Lee, H.S.; Ryu, D.G.; Chung, H.T. 3-Hydroxyanthranilic acid, one of metabolites of tryptophan via indoleamine 2,3-dioxygenase pathway, suppresses inducible nitric oxide synthase expression by enhancing heme oxygenase-1 expression. *Biochem. Biophys. Res. Commun.* **2004**, *320*, 1156–1162. [CrossRef] [PubMed]
249. Sekkai, D.; Guittet, O.; Lemaire, G.; Tenu, J.P.; Lepoivre, M. Inhibition of nitric oxide synthase expression and activity in macrophages by 3-hydroxyanthranilic acid, a tryptophan metabolite. *Arch. Biochem. Biophys.* **1997**, *340*, 117–123. [CrossRef]
250. Krause, D.; Suh, H.S.; Tarassishin, L.; Cui, Q.L.; Durafourt, B.A.; Choi, N.; Bauman, A.; Cosenza-Nashat, M.; Antel, J.P.; Zhao, M.L.; et al. The tryptophan metabolite 3-hydroxyanthranilic acid plays anti-inflammatory and neuroprotective roles during inflammation: Role of hemeoxygenase-1. *Am. J. Pathol.* **2011**, *179*, 1360–1372. [CrossRef]
251. Lowe, M.M.; Mold, J.E.; Kanwar, B.; Huang, Y.; Louie, A.; Pollastri, M.P.; Wang, C.; Patel, G.; Franks, D.G.; Schlezinger, J.; et al. Identification of cinnabarinic acid as a novel endogenous aryl hydrocarbon receptor ligand that drives IL-22 production. *PLoS ONE* **2014**, *9*, e87877. [CrossRef]
252. Suh, S.W.; Hamby, A.M.; Swanson, R.A. Hypoglycemia, brain energetics, and hypoglycemic neuronal death. *Glia* **2007**, *55*, 1280–1286. [CrossRef] [PubMed]
253. Schrocksnadel, K.; Wirleitner, B.; Winkler, C.; Fuchs, D. Monitoring tryptophan metabolism in chronic immune activation. *Clin. Chim. Acta* **2006**, *364*, 82–90. [CrossRef]
254. Penberthy, W.T. Pharmacological targeting of IDO-mediated tolerance for treating autoimmune disease. *Curr. Drug Metabol.* **2007**, *8*, 245–266. [CrossRef] [PubMed]
255. Munn, D.H.; Shafizadeh, E.; Attwood, J.T.; Bondarev, I.; Pashine, A.; Mellor, A.L. Inhibition of T cell proliferation by macrophage tryptophan catabolism. *J. Exp. Med.* **1999**, *189*, 1363–1372. [CrossRef] [PubMed]
256. Munn, D.H.; Sharma, M.D.; Mellor, A.L. Ligation of B7-1/B7-2 by human CD4+ T cells triggers indoleamine 2,3-dioxygenase activity in dendritic cells. *J. Immunol.* **2004**, *172*, 4100–4110. [CrossRef]
257. Iribarren, P.; Cui, Y.H.; Le, Y.; Wang, J.M. The role of dendritic cells in neurodegenerative diseases. *Archiv. Immunol. Ther. Exp.* **2002**, *50*, 187–196.
258. Pashenkov, M.; Teleshova, N.; Link, H. Inflammation in the central nervous system: The role for dendritic cells. *Brain Pathol.* **2003**, *13*, 23–33. [CrossRef]
259. Matysiak, M.; Stasiolek, M.; Orlowski, W.; Jurewicz, A.; Janczar, S.; Raine, C.S.; Selmaj, K. Stem cells ameliorate EAE via an indoleamine 2,3-dioxygenase (IDO) mechanism. *J. Neuroimmunol.* **2008**, *193*, 12–23. [CrossRef]
260. Belladonna, M.L.; Grohmann, U.; Guidetti, P.; Volpi, C.; Bianchi, R.; Fioretti, M.C.; Schwarcz, R.; Fallarino, F.; Puccetti, P. Kynurenine pathway enzymes in dendritic cells initiate tolerogenesis in the absence of functional IDO. *J. Immunol.* **2006**, *177*, 130–137. [CrossRef]

261. Sharma, M.D.; Baban, B.; Chandler, P.; Hou, D.Y.; Singh, N.; Yagita, H.; Azuma, M.; Blazar, B.R.; Mellor, A.L.; Munn, D.H. Plasmacytoid dendritic cells from mouse tumor-draining lymph nodes directly activate mature Tregs via indoleamine 2,3-dioxygenase. *J. Clin. Investig.* **2007**, *117*, 2570–2582. [CrossRef] [PubMed]
262. Kauppinen, T.M.; Suh, S.W.; Genain, C.P.; Swanson, R.A. Poly(ADP-ribose) polymerase-1 activation in a primate model of multiple sclerosis. *J. Neurosci. Res.* **2005**, *81*, 190–198. [CrossRef] [PubMed]
263. Opitz, C.A.; Wick, W.; Steinman, L.; Platten, M. Tryptophan degradation in autoimmune diseases. *Cell. Mol. Life Sci.* **2007**, *64*, 2542–2563. [CrossRef]
264. Braidy, N.; Lim, C.K.; Grant, R.; Brew, B.J.; Guillemin, G.J. Serum nicotinamide adenine dinucleotide levels through disease course in multiple sclerosis. *Brain Res.* **2013**, *1537*, 267–272. [CrossRef]
265. Kaneko, S.; Wang, J.; Kaneko, M.; Yiu, G.; Hurrell, J.M.; Chitnis, T.; Khoury, S.J.; He, Z. Protecting axonal degeneration by increasing nicotinamide adenine dinucleotide levels in experimental autoimmune encephalomyelitis models. *J. Neurosci.* **2006**, *26*, 9794–9804. [CrossRef] [PubMed]
266. Esquifino, A.I.; Cano, P.; Jimenez, V.; Cutrera, R.A.; Cardinali, D.P. Experimental allergic encephalomyelitis in male Lewis rats subjected to calorie restriction. *J. Physiol. Biochem.* **2004**, *60*, 245–252. [CrossRef] [PubMed]
267. Piccio, L.; Stark, J.L.; Cross, A.H. Chronic calorie restriction attenuates experimental autoimmune encephalomyelitis. *J. Leukocyte Biol.* **2008**, *84*, 940–948. [CrossRef]
268. Esquifino, A.I.; Cano, P.; Jimenez-Ortega, V.; Fernandez-Mateos, M.P.; Cardinali, D.P. Immune response after experimental allergic encephalomyelitis in rats subjected to calorie restriction. *J. Neuroinflamm.* **2007**, *4*, 6. [CrossRef]
269. Grant, R.S.; Naif, H.; Espinosa, M.; Kapoor, V. IDO induction in IFN-gamma activated astroglia: A role in improving cell viability during oxidative stress. *Redox Rep. Communic. Free Radic. Res.* **2000**, *5*, 101–104. [CrossRef]
270. Grant, R.; Kapoor, V. Inhibition of indoleamine 2,3-dioxygenase activity in IFN-gamma stimulated astroglioma cells decreases intracellular NAD levels. *Biochem. Pharmacol.* **2003**, *66*, 1033–1036. [CrossRef]
271. Kujundzic, R.N.; Lowenthal, J.W. The role of tryptophan metabolism in iNOS transcription and nitric oxide production by chicken macrophage cells upon treatment with interferon gamma. *Immunol. Lett.* **2008**, *115*, 153–159. [CrossRef]
272. Adams, O.; Besken, K.; Oberdorfer, C.; MacKenzie, C.R.; Russing, D.; Daubener, W. Inhibition of human herpes simplex virus type 2 by interferon gamma and tumor necrosis factor alpha is mediated by indoleamine 2,3-dioxygenase. *Microbes Infect.* **2004**, *6*, 806–812. [CrossRef] [PubMed]
273. Adams, O.; Besken, K.; Oberdorfer, C.; MacKenzie, C.R.; Takikawa, O.; Daubener, W. Role of indoleamine-2,3-dioxygenase in alpha/beta and gamma interferon-mediated antiviral effects against herpes simplex virus infections. *J. Virol.* **2004**, *78*, 2632–2636. [CrossRef] [PubMed]
274. Bodaghi, B.; Goureau, O.; Zipeto, D.; Laurent, L.; Virelizier, J.L.; Michelson, S. Role of IFN-gamma-induced indoleamine 2,3 dioxygenase and inducible nitric oxide synthase in the replication of human cytomegalovirus in retinal pigment epithelial cells. *J. Immunol.* **1999**, *162*, 957–964.
275. Sanni, L.A.; Thomas, S.R.; Tattam, B.N.; Moore, D.E.; Chaudhri, G.; Stocker, R.; Hunt, N.H. Dramatic changes in oxidative tryptophan metabolism along the kynurenine pathway in experimental cerebral and noncerebral malaria. *Am. Pathol.* **1998**, *152*, 611–619.
276. Pfefferkorn, E.R. Interferon gamma blocks the growth of Toxoplasma gondii in human fibroblasts by inducing the host cells to degrade tryptophan. *Proc. Natl. Acad. Sci. USA* **1984**, *81*, 908–912. [CrossRef] [PubMed]
277. Byrne, G.I.; Lehmann, L.K.; Landry, G.J. Induction of tryptophan catabolism is the mechanism for gamma-interferon-mediated inhibition of intracellular Chlamydia psittaci replication in T24 cells. *Infect. Immun.* **1986**, *53*, 347–351. [CrossRef] [PubMed]
278. Jonsson, S.; Andersson, G.; Fex, T.; Fristedt, T.; Hedlund, G.; Jansson, K.; Abramo, L.; Fritzson, I.; Pekarski, O.; Runstrom, A.; et al. Synthesis and biological evaluation of new 1,2-dihydro-4-hydroxy-2-oxo-3-quinolinecarboxamides for treatment of autoimmune disorders: Structure-activity relationship. *J. Med. Chem.* **2004**, *47*, 2075–2088. [CrossRef]
279. Majlath, Z.; Annus, A.; Vecsei, L. Kynurenine System and Multiple Sclerosis, Pathomechanism and Drug Targets with an Emphasis on Laquinimod. *Curr. Drug Targ.* **2018**, *19*, 805–814. [CrossRef]
280. Varrin-Doyer, M.; Zamvil, S.S.; Schulze-Topphoff, U. Laquinimod, an up-and-coming immunomodulatory agent for treatment of multiple sclerosis. *Exp. Neurol.* **2014**, *262*, 66–71. [CrossRef]

281. Bruck, W.; Wegner, C. Insight into the mechanism of laquinimod action. *J. Neurol. Sci.* **2011**, *306*, 173–179. [CrossRef] [PubMed]
282. Ali, R.; Nicholas, R.S.; Muraro, P.A. Drugs in development for relapsing multiple sclerosis. *Drugs* **2013**, *73*, 625–650. [CrossRef]
283. Gurevich, M.; Gritzman, T.; Orbach, R.; Tuller, T.; Feldman, A.; Achiron, A. Laquinimod suppress antigen presentation in relapsing-remitting multiple sclerosis: In-vitro high-throughput gene expression study. *J. Neuroimmunol.* **2010**, *221*, 87–94. [CrossRef] [PubMed]
284. Toubi, E.; Nussbaum, S.; Staun-Ram, E.; Snir, A.; Melamed, D.; Hayardeny, L.; Miller, A. Laquinimod modulates B cells and their regulatory effects on T cells in multiple sclerosis. *J. Neuroimmunol.* **2012**, *251*, 45–54. [CrossRef] [PubMed]
285. Bjork, P.; Bjork, A.; Vogl, T.; Stenstrom, M.; Liberg, D.; Olsson, A.; Roth, J.; Ivars, F.; Leanderson, T. Identification of human S100A9 as a novel target for treatment of autoimmune disease via binding to quinoline-3-carboxamides. *PLoS Biol.* **2009**, *7*, e97. [CrossRef]
286. Mishra, M.K.; Wang, J.; Silva, C.; Mack, M.; Yong, V.W. Kinetics of proinflammatory monocytes in a model of multiple sclerosis and its perturbation by laquinimod. *Am. J. Pathol.* **2012**, *181*, 642–651. [CrossRef]
287. Wegner, C.; Stadelmann, C.; Pfortner, R.; Raymond, E.; Feigelson, S.; Alon, R.; Timan, B.; Hayardeny, L.; Bruck, W. Laquinimod interferes with migratory capacity of T cells and reduces IL-17 levels, inflammatory demyelination and acute axonal damage in mice with experimental autoimmune encephalomyelitis. *J. Neuroimmunol.* **2010**, *227*, 133–143. [CrossRef]
288. Schulze-Topphoff, U.; Shetty, A.; Varrin-Doyer, M.; Molnarfi, N.; Sagan, S.A.; Sobel, R.A.; Nelson, P.A.; Zamvil, S.S. Laquinimod, a quinoline-3-carboxamide, induces type II myeloid cells that modulate central nervous system autoimmunity. *PLoS ONE* **2012**, *7*, e33797. [CrossRef]
289. Nyamoya, S.; Steinle, J.; Chrzanowski, U.; Kaye, J.; Schmitz, C.; Beyer, C.; Kipp, M. Laquinimod Supports Remyelination in Non-Supportive Environments. *Cells* **2019**, *8*, 1363. [CrossRef]
290. Jolivel, V.; Luessi, F.; Masri, J.; Kraus, S.H.; Hubo, M.; Poisa-Beiro, L.; Klebow, S.; Paterka, M.; Yogev, N.; Tumani, H.; et al. Modulation of dendritic cell properties by laquinimod as a mechanism for modulating multiple sclerosis. *Brain* **2013**, *136*, 1048–1066. [CrossRef]
291. Kalb, R. The protean actions of neurotrophins and their receptors on the life and death of neurons. *Trends Neurosci.* **2005**, *28*, 5–11. [CrossRef] [PubMed]
292. Aharoni, R.; Saada, R.; Eilam, R.; Hayardeny, L.; Sela, M.; Arnon, R. Oral treatment with laquinimod augments regulatory T-cells and brain-derived neurotrophic factor expression and reduces injury in the CNS of mice with experimental autoimmune encephalomyelitis. *J. Neuroimmunol.* **2012**, *251*, 14–24. [CrossRef] [PubMed]
293. Thone, J.; Ellrichmann, G.; Seubert, S.; Peruga, I.; Lee, D.H.; Conrad, R.; Hayardeny, L.; Comi, G.; Wiese, S.; Linker, R.A.; et al. Modulation of autoimmune demyelination by laquinimod via induction of brain-derived neurotrophic factor. *Am. J. Pathol.* **2012**, *180*, 267–274. [CrossRef] [PubMed]
294. Comi, G.; Jeffery, D.; Kappos, L.; Montalban, X.; Boyko, A.; Rocca, M.A.; Filippi, M.; Group, A.S. Placebo-controlled trial of oral laquinimod for multiple sclerosis. *N. Engl. J. Med.* **2012**, *366*, 1000–1009. [CrossRef]
295. Vollmer, T.L.; Sorensen, P.S.; Selmaj, K.; Zipp, F.; Havrdova, E.; Cohen, J.A.; Sasson, N.; Gilgun-Sherki, Y.; Arnold, D.L.; Group, B.S. A randomized placebo-controlled phase III trial of oral laquinimod for multiple sclerosis. *J. Neurol.* **2014**, *261*, 773–783. [CrossRef] [PubMed]
296. Matsumoto, K.; Kinoshita, K.; Yoshimizu, A.; Kurauchi, Y.; Hisatsune, A.; Seki, T.; Katsuki, H. Laquinimod and 3,3′-diindolylmethane alleviate neuropathological events and neurological deficits in a mouse model of intracerebral hemorrhage. *J. Neuroimmunol.* **2020**, *342*, 577195. [CrossRef]

© 2020 by the authors. Licensee MDPI, Basel, Switzerland. This article is an open access article distributed under the terms and conditions of the Creative Commons Attribution (CC BY) license (http://creativecommons.org/licenses/by/4.0/).

Review

The Role of Granulocyte-Macrophage Colony-Stimulating Factor in Murine Models of Multiple Sclerosis

Kelly L. Monaghan [1] and Edwin C.K. Wan [1,2,3,*]

1. Department of Microbiology, Immunology, and Cell Biology, West Virginia University, Morgantown, WV 26506, USA; klm0031@mix.wvu.edu
2. Department of Neuroscience, West Virginia University, Morgantown, WV 26506, USA
3. Rockefeller Neuroscience Institute, West Virginia University, Morgantown, WV 26506, USA
* Correspondence: edwin.wan@hsc.wvu.edu; Tel.:+1-304-293-6293

Received: 13 February 2020; Accepted: 3 March 2020; Published: 4 March 2020

Abstract: Multiple sclerosis (MS) is an immune-mediated disease that predominantly impacts the central nervous system (CNS). Animal models have been used to elucidate the underpinnings of MS pathology. One of the most well-studied models of MS is experimental autoimmune encephalomyelitis (EAE). This model was utilized to demonstrate that the cytokine granulocyte-macrophage colony-stimulating factor (GM-CSF) plays a critical and non-redundant role in mediating EAE pathology, making it an ideal therapeutic target. In this review, we will first explore the role that GM-CSF plays in maintaining homeostasis. This is important to consider, because any therapeutics that target GM-CSF could potentially alter these regulatory processes. We will then focus on current findings related to the function of GM-CSF signaling in EAE pathology, including the cell types that produce and respond to GM-CSF and the role of GM-CSF in both acute and chronic EAE. We will then assess the role of GM-CSF in alternative models of MS and comment on how this informs the understanding of GM-CSF signaling in the various aspects of MS immunopathology. Finally, we will examine what is currently known about GM-CSF signaling in MS, and how this has promoted clinical trials that directly target GM-CSF.

Keywords: multiple sclerosis; experimental autoimmune encephalomyelitis; monocytes; granulocyte-macrophage colony-stimulating factor

1. Introduction

Multiple sclerosis (MS) is a chronic immune-mediated disease that impacts approximately 2.3 million people world-wide [1]. MS is characterized by the formation of demyelinating lesions, which are disseminated in both time and space. The location of the lesions correlates with the manifestation of physical disease symptoms [2]. In addition to demyelination, peripheral immune cell infiltration to the CNS is associated with inflammation, tissue damage, and axonal loss [3]. There are three major subtypes of MS: (1) relapsing remitting MS (RRMS), (2) secondary progressive MS (SPMS), and primary progressive MS (PPMS) [4,5]. RRMS is the most common subtype. This disease course is defined by periods of exacerbation followed by periods of clinical recovery, although new lesions can develop in clinically silent areas during periods of remission without the presentation of overt clinical symptoms [5]. A majority of RRMS patients will develop SPMS, which is defined as the progressive worsening of neurological dysfunction, without remission [5]. PPMS is less common and is defined as the accumulation of neurological dysfunction following onset of clinical symptoms with no remission [5]. While some studies have suggested that these three subtypes are one disease with differing clinical manifestations, it is important to distinguish between these subtypes. This is because the current disease-modifying agents that are used to treat MS are efficacious at treating neuroinflammation and abrogating some of the tissue damage and demyelination associated with the active phase of the disease, when patients exhibit overt clinical symptoms [6–8]. However, these same disease-modifying agents are not efficacious at impeding disease progression [6–8]. Consequently, the major focus in the field of MS research is to develop novel therapeutic strategies to dampen neuroinflammation and prevent MS progression.

Animal model systems of MS have provided insight into the immunopathology of MS. Studies in these models have directly and indirectly contributed to the development of disease-modifying agents that are utilized in the clinic [8]. The most widely studied murine model of multiple sclerosis is experimental autoimmune encephalomyelitis (EAE). This animal model closely recapitulates the neuroinflammatory process that is associated with MS [9]. Consequently, this model has been used to identify novel therapeutic targets by ascertaining those mediators that are critical for potentiating neuroinflammation. One such mediator that has gained attention for its role in promoting EAE-associated inflammation is the cytokine granulocyte-macrophage colony-stimulating factor (GM-CSF). This cytokine first drew attention when a clinical report in 1998, which assessed cytokine concentrations in the cerebral spinal fluid of MS patients with active disease, found that the levels of GM-CSF are significantly increased in MS patients compared to healthy controls [10]. Based on this observation, McQualter and colleagues wanted to determine whether GM-CSF played a critical and non-redundant role in promoting EAE pathology. [11]. This study, which will be discussed in detail later in this review, is the first to underscore the critical role of GM-CSF in potentiating EAE pathology. Based on their findings and the aforementioned clinical study, McQualter and colleagues posited that GM-CSF is a putative therapeutic target for MS treatment [11]. Since then, much information has been gleaned about the role of GM-CSF in EAE pathology, including the cells types that produce and respond to this cytokine. It is evident from recent studies that GM-CSF plays a dynamic role in mediating EAE pathology. In this review, we will explore the current findings related to the function of GM-CSF signaling in EAE pathology. We will then assess the role of GM-CSF in alternative models of MS and comment on how this informs the understanding of GM-CSF signaling in the various aspects of MS immunopathology. Finally, we will explore the studies that have directly ascertained the function of GM-CSF in MS, and what implications these findings have for developing novel therapies that target GM-CSF and its downstream mediators.

2. GM-CSF

2.1. Protein Structure, Receptor Structure, and Signaling

GM-CSF is a 114 amino acid polypeptide that is secreted as a monomeric 23kDA glycosylated small glycoprotein protein, though the molecular weight can vary depending on the extent of glycosylation [12]. Human *CSF2* is encoded by 2.5kb mRNA that consists of four exons on the chromosome region 5q31 [12,13]. Murine and human GM-CSF share 70% nucleotide and 56% sequence homolog, suggesting that while cross-reactivity between human and murine GM-CSF does not occur, murine models can be utilized to study the role of GM-CSF in the context of human diseases [12]. The GM-CSF receptor is a heterodimer that consists of an α subunit and a common beta chain (βc) subunit, which is shared with IL-3 and IL-5 [14]. Interestingly, functional mutagenesis studies and crystal structure analysis of the GM-CSF receptor demonstrate that receptor activation is predicated on the assembly of the GM-CSF receptor into a dodecamer or higher order structure [15]. Activation of the GM-CSF receptor requires both the α subunit and βc subunit. The βc subunit is associated with Janus kinase 2 (JAK2); however, the βc subunit keeps its tails far enough apart that transphosphorylation of JAK2 cannot occur [16,17]. When GM-CSF binds to the receptor, the higher order dodecamer complex brings the subunit tails close enough together to mediate the interaction between the JAK2 molecules, resulting in functional dimerization and transphosphorylation [15,17]. The activation of JAK2 results in the activation of the signal transducer and activator of transcription 5 (STAT5). STAT5 can then translocate to the nucleus and regulate the expression of target genes [18]. GM-CSF is known to play an indispensable role of JAK2-STAT5 signaling [19]. GM-CSF can also activate the interferon regulatory factor 4 (IRF4)-CCL17 pathway which is associated with pain [20]. GM-CSF signaling activates IRF4 by enhancing the activity of JMJD3 demethylase [20]. The upregulation of IRF4 results in an increased expression of MHC II by differentiating monocytes and an increase in the production of CCL17 [20]. Additionally, GM-CSF signaling is implicated in the AKT-ERK mediated activation of NF-κB [21]. Given the pleiotropic nature of GM-CSF, it is unsurprising that this cytokine plays a major role in both maintaining homeostasis and promoting inflammation.

2.2. Cellular Source and Function of GM-CSF during Homeostasis

GM-CSF is a pleiotropic cytokine that is known to be a major mediator in inflammation; however, GM-CSF also functions in maintaining homeostasis. In the lungs, GM-CSF is abundantly produced by epithelial cells. Murine studies utilizing GM-CSF-deficient mice ($Csf^{-/-}$) reveal that GM-CSF is required for the development of functional alveolar macrophages through the regulation of the transcription factor PU.1 [22,23]. Given that alveolar macrophages play a major role in facilitating the clearance of surfactant from the alveolar space, GM-CSF-deficient mice develop a condition known as pulmonary alveolar proteinosis (PAP), which is characterized by the accumulation of surfactant in the alveolar space [23,24]. Further investigation posited that GM-CSF signaling directly regulates the differentiation of liver-derived fetal monocytes into immature alveolar macrophages during embryonic development [23]. GM-CSF signaling also promotes the differentiation of immature alveolar macrophages into mature alveolar macrophages, postnatally [23]. Intriguingly, immunocompromised patients that develop cryptococcal meningitis have circulating anti-GM-CSF autoantibodies. These patients exhibit reduced surfactant clearance, and a number of these patients subsequently developed PAP [25].

In addition to promoting the development of alveolar macrophages, GM-CSF also appears to play a minor role in the development of tissue-resident conventional dendritic cells (cDCs). $Csf2^{-/-}$ or $Csfr2^{-/-}$ mice have fewer CD103+ cDCs in the lung, dermis, and intestine [24,26,27]. In other lymphoid tissues, however, tissue-resident cDC development appears to be normal [28]. This is an interesting observation given that, under inflammatory conditions, GM-CSF is a major cytokine that promotes monocyte differentiation into dendritic cells, and a more critical role of this cytokine in cDC development is anticipated [29]. Since GM-CSF and its downstream mediators are potential therapeutic targets, it is necessary to consider the role that GM-CSF plays in the development of both alveolar macrophages and cDCs to prevent undesirable and potentially dangerous off-target effects.

2.3. GM-CSF in Murine Models of Multiple Sclerosis

GM-CSF in Experimental Autoimmune Encephalomyelitis

Experimental autoimmune encephalomyelitis (EAE) is the most well-studied model of multiple sclerosis. This model was established in 1933 by Rivers and colleagues in an attempt to address human encephalitis resulting from rabbit spinal cord contamination in the human rabies vaccine [30]. Since its development, rodent and primate models have utilized some variation of this model to generate acute monophasic, relapsing–remitting, and chronic inflammatory phenotypes [31]. Given that the role of GM-CSF has been elucidated in murine EAE models, we will focus on murine models for the remainder of this review. EAE can be induced through two mechanisms [32]. The first is active EAE induction, whereby myelin or brain tissue peptides such as myelin oligodendrocyte glycoprotein amino acid 35-55 ($MOG_{(35-55)}$), myelin basic protein (MBP), or proteolipid protein (PLP) are emulsified in complete Freund's adjuvant (CFA) and subcutaneously injected into naïve recipient mice [33]. This is followed by two intraperitoneal injections (IP) of pertussis toxin at 2- and 48-h post induction. The pertussis toxin is thought to increase the permeability of the blood–brain barrier, thereby facilitating peripheral immune cell infiltration into the CNS parenchyma [34]. The resulting clinical presentation of active EAE induction is contingent on the strain of mice being utilized. For example, when EAE is induced via active induction with MOG $_{(35-55)}$ in CFA in mice on a C57BL/6J background, the mice develop a monophasic and chronic disease pattern that is characterized by white matter demyelination and peripheral CD4+ T cell and myeloid cell infiltration [35]. The onset of clinical symptoms usually appears between days 9–10, and the symptoms reach peak severity between days 13–15 [35]. Active EAE induction in C57BL/6 mice is a valuable tool for recapitulating the immune cell infiltration and resulting neuroinflammation that mediate MS pathology [31]. In addition, EAE is commonly induced in SJL/J mice using PLP$_{(139-151)}$. Active EAE induction in the SJL/J mice results in a relapsing–remitting disease course which is characterized by peripheral immune cell infiltration, inflammation, and demyelination (relapses), followed by the resolution of inflammation but the progression of white matter damage and axonal damage with no overt clinical symptoms (remission) [31]. This model is a useful tool to study relapsing–remitting MS [32]. The other major mechanism to induce EAE is through the adoptive transfer of pathogenic CD4+ T cells. In this model, antigen-specific CD4+ T cells are transferred to naïve recipient mice to induce EAE. In this model, the priming phase of EAE that occurs in the periphery is bypassed, therefore the in vitro manipulation of CD4+ T cells prior to transfer can allow researchers to study the role of various cytokines during the effector phase of EAE [33]. Neither active nor passive EAE induction completely recapitulates all aspects of MS immunopathology; however, EAE is a useful tool to study various aspects of the immune-mediate response. This is evidenced by the successful development of standard-of-care MS disease-modifying agents utilizing EAE models, including interferon beta, glatiramer acetate, and natalizumab (anti-alpha 4 beta 1 integrin) [36–38]. Though the exact mechanism has not been fully elucidated, interferon beta is thought to act as an immunomodulatory agent that dampens inflammation in the CNS [39]. Additionally, interferon beta is also thought to prevent the migration of proinflammatory immune cells into the CNS [39]. Glatiramer acetate is a synthetic amino acid copolymer that is thought to expand the regulatory T cell population in the periphery, which can migrate into the CNS parenchyma and produce anti-inflammatory mediators that abrogate the activation of immune cells that are reactive against myelin [40]. Natalizumab binds to the α_4 subunit of $\alpha_4\beta_1$ integrin on the surface of lymphocytes, which prevents binding to the vascular cell adhesion molecule 1 (VCAM-1). This prevents T cells from migrating into the CNS parenchyma [41,42]. Consequently, EAE is currently the best model to understand the role of GM-CSF in MS pathogenesis and its therapeutic implications.

The first study to assess the role of GM-CSF in EAE pathology was conducted by McQualter and colleagues in 2001. Their goal was to determine whether GM-CSF played a critical and non-redundant role in promoting EAE pathology, which was based on previous findings suggesting that that concentration of GM-CSF was increased in the cerebral spinal fluid of patients with MS compared to healthy controls [10]. To this end, they generated a GM-CSF-deficient mouse that was backcrossed to an EAE-suspectable NOD/Lt background. EAE was induced through active induction with MOG $_{(35-55)}$ and the clinical presentation in this particular stain of mice was a relapsing–remitting biphasic phenotype. The study found that, although functionally normal in terms of hematopoiesis, these mice are resistant to the EAE, which was demonstrated by the lack of immune cell infiltration into the CNS in addition to the absence of clinical symptoms, suggesting that GM-CSF is important for the development of demyelinating lesions and the migration and/or expansion of immune cells within the CNS [11]. These findings suggest that GM-CSF is a conceivable threptic target for MS. In order to develop these novel therapies, it is necessary to understand the cell types and subsequent signaling pathways that regulate the production and response to GM-CSF. The proposed role of GM-CSF during EAE is detailed in Figure 1.

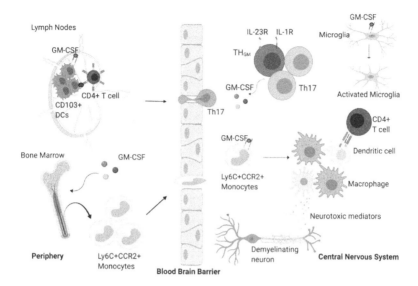

Figure 1. The proposed role of granulocyte-macrophage colony-stimulating factor (GM-CSF) during experimental autoimmune encephalomyelitis (EAE). GM-CSF promotes the accumulation of CD103+ dendritic cells (DCs) in the lymph nodes which can present myelin antigen to CD4+ T cells. These CD4+ T cells can then migrate into the central nervous system (CNS) parenchyma where they begin to produce GM-CSF exclusively, or GM-CSF and IL-17. GM-CSF production by the CD4+ T cells promotes the migration of Ly6C+CCR2+ cells from the bone marrow to the CNS. Once in the CNS, GM-CSF signaling promotes the differentiation of monocytes into a heterogenous population of monocyte-derived macrophages and monocyte derived dendritic cells. Monocyte-derived dendritic cells can interact with and promote the activation of infiltrating CD4+ T cells. In addition, these differentiated cells can secrete mediators that directly promote demyelination, tissue damage and axonal loss. GM-CSF can also promote the activation of CNS-resident microglia. These reactive microglia can potentiate the inflammatory milieu by producing proinflammatory mediators.

2.4. T cells Are the Predominant Source of GM-CSF during EAE

In 2007, a study published by Ponomarev and colleagues identified the cellular source of GM-CSF during EAE as T cells and not CNS-resident microglia or other infiltrated peripheral immune cells [43]. This study suggests that Th1 CD4+ T cells are the major T cell subset that produce GM-CSF. This idea that Th1 cells are the major source of inflammation in EAE is due to the fact that both IL-12 and IL-23 share the p40 subunit, therefore any efficacious strategies that blocked IL-12p40 subsequently block both IL-12 and IL-23 activity [44]. With the discovery of IL-23 and Th17 cells, however, the notion that Th1 cells are the predominant source of inflammation during EAE was quickly challenged [45,46]. One study that challenges this notion demonstrates that while it is true that the passive transfer of IL-12p70- and IL-23-polarized cells can cause EAE, treatment with anti–GM-CSF can ameliorate EAE induced in the mice that receive IL-23 polarized Th17 cells, but not IL-12p70 polarized Th1 cells [47]. This suggests that Th17, but not Th1 cells, are the major source of GM-CSF during EAE. The role of Th17 cells in EAE is further supported by an elegant study that demonstrated that the upregulation of both IL-23 and RORγt license the Th17 cells to produce GM-CSF. IL-12 and IFNγ on the other hand, are negative regulators of GM-CSF production by these cells [48]. Furthermore, GM-CSF secretion from $Ifng^{-/-}Il17a^{-/-}$ mice was sufficient to induce EAE; however, $Csf2^{-/-}$ mice, which lack GM-CSF, do no develop EAE, suggesting that other inflammatory mediators are not sufficient to induce pathology [48]. A study by Mangalam and colleagues found that IFNγ sequesters infiltrating immune cells to the spinal cord during EAE, and it partially suppresses the production of GM-CSF by

Th17 cells, rendering them less pathogenic [49]. Additional studies demonstrate that targeting other Th17-associated cytokines including IL-17F, IL-22, and IL-21 does not confer resistance to EAE [50–52]. These findings indicate that GM-CSF is the major Th17-associated cytokine that licenses the CD4+ T cells to become encephalitogenic. In fact, GM-CSF is now thought to be the only Th17-associated cytokine that had a non-redundant function in promoting EAE pathology [53]. A recent study found that GM-CSF production by Th17 cells is not restricted to upstream regulation by IL-23. This study showed that STAT5 deficiency in CD4+ T cells confers resistance to EAE by impairing the expression of GM-CSF [54]. Further investigation found that IL-7 acts upstream of STAT5. The study posited that the CD4+ T cells regulated by the IL-7-STAT5 axis are a distinct subset of Th cells, which they named Th$_{GM}$ cells. These cells have minimal expression of the master gene regulators of Th1 and Th17 cells, T-bet and RORγt. Microarray analysis revealed that the Th$_{GM}$ cells co-express GM-CSF and IL-3, which is not the case in either Th1 or Th17 cells. Additionally, when the three Th subtypes were adoptively transferred into $Rag2^{-/-}$ mice, the Th$_{GM}$ cells caused more severe EAE compared to EAE induced by the transfer of Th1 or Th17 cells [54]. A recent study further supported the notion that Th$_{GM}$ are a distinct subset of pathogenic Th cells. To this end, they generated a fate-map and reporter of GM-CSF expression mouse stain, whereby they were able to identify a subset of Th cells that required IL-23R and IL-1R signaling but not IL-6R signaling, to promote pathogenesis [55]. Furthermore, when this subset of Th cells was ablated, the inflammatory cascade was perturbed; however, the accumulation of Th1 and Th17 cells were not impacted, further underscoring the notion that these cells are a distinct subset of GM-CSF Th cells [55]. Interestingly, the production of GM-CSF may be dependent on a subset of CCR4-expressing dendritic cells [56]. When CCR4 expression was ablated in this cell subset, these cells showed a significant decrease in the expression of IL-23 [56]. Consequently, these mice were protected against EAE and had less GM-CSF overall in the spinal cords, suggesting that CCR4 expression on DCs maintains the Th17 population, thereby regulating the production of GM-CSF [56]. It is evident from the aforementioned findings that CD4+ T cells are the major cellular source of GM-CSF during EAE. Additional studies are required to confirm that Th$_{GM}$ cells are, in fact, a distinct subset of Th cells during EAE. These cells may serve as novel therapeutic targets. A list of cell types that produce GM-CSF during EAE is summarized in Table 1.

Table 1. Immune cell types that produce or respond to GM-CSF during EAE.

Cell Types that Produce GM-CSF	Cellular and Molecular Signals Involved
Th17 cells	IL-23-mediated expression of RORγt [48,57].
Th$_{GM}$	IL-7-mediated activation of STAT5 [54]; IL-23R and IL-1R signaling [55].
CD8+ T cells	IL-23 induces but IFN-β suppresses GM-CSF production [58,59].
B cells	B-cell receptor, CD40, and IL-4-mediated STAT5/6 activation [60].
Dendritic cells	CCL17/CCL22-mediated expression of GM-CSF via CCR4 [56].
CNS endothelial cells	Monocyte-produced, IL-1β-mediated expression of GM-CSF [61].
Cell Types that Respond to GM-CSF	**Cell Type-Specific Biological Function of GM-CSF during EAE**
Monocytes	Stimulates CNS migration; induces the production of proinflammatory cytokines and neurotoxic mediators; promotes cell differentiation [62–65].
Dendritic cells	Induces the production of IL-23 that promotes EAE [56].
CD103+ dendritic cells	Induces cell accumulation in the skin and peripheral lymph nodes that can then present antigen to pathogenetic CD4+ T cells [66].
Neutrophils	Promotes cell accumulation in the brain that causes atypical EAE [67].
Microglia	Induces activation and promotes onset of EAE [43,68].
Astrocytes	Promotes the upregulation of proinflammatory gene expression [69].

2.5. Many Immune Cells Respond to GM-CSF during EAE

Once CD4+ T cells were found to be the major cellular source of GM-CSF during EAE, there was a need to understand which immune cell types were responding to the high level of GM-CSF that was being produced by these cells. One of the first studies to address this question suggested that myeloid cells are a major component of the inflammatory infiltrate, and these cells must migrate into the CNS prior to EAE relapses [70]. Using GM-CS-deficient mice, King and colleagues demonstrated that GM-CSF promotes $CD11b^{hi}$ $Ly6C^{hi}$ egress from the bone marrow, across the blood–brain barrier, and into the CNS parenchyma, where these cells will upregulate the expression of proinflammatory cytokines [70]. Additional studies demonstrate that GM-CSF deletion results in fewer monocyte-derived cells in the CNS parenchyma following EAE induction, and the overexpression of GM-CSF results in increased monocyte migration, which underscores the role of GM-CSF in mediating monocyte migration from the bone marrow into the CNS parenchyma [11,71,72]. The conditional deletion of the *Csf2r* on various immune cells, including CD103+ conventional dendritic cells, CNS-resident microglia, and neutrophils, does not alter the progression of EAE [64]. However, when *Csf2r* is deleted on CCR2+Ly6C+ monocytes, the mice are resistant to EAE and have a similar phenotype to the complete $Csf2^{-/-}$ mice, suggesting that this particular subset of monocytes responds to GM-CSF and is critical in mediating EAE pathology [64]. It is thought that, in addition to promoting the migration of CCR2+Ly6C+ cells into the CNS, GM-CSF is required to promote the differentiation of these specific infiltrated monocytes into antigen-presenting cells which can subsequently produce proinflammatory cytokines and present antigen to and maintain the pathogenic CD4+ T cell population [62,73]. In fact, Helft and colleagues found that when bone marrow monocytes are treated with GM-CSF in vitro, the resulting population is heterogenous in nature, and is comprised of monocyte-derived dendritic cells and macrophages, supporting the idea that GM-CSF promotes monocyte differentiation [29]. There is also evidence to suggest that, once in the parenchyma, the monocyte-derived cells can produce mediators that directly promote tissue damage, demyelination, and axonal loss [63,65].

In addition to monocytes, there are other cell-types that can respond to GM-CSF during EAE. The accumulation of CD103+ dendric cells in the lymph nodes is dependent on the presence of GM-CSF [66]. The CD103+ dendric cells present myelin antigen to, and subsequently activate, naïve CD4+ T cells, and therefore contribute to the onset of EAE [66]. However, when the *Csf2r* is conditionally deleted in CD103+ dendritic cells, severe EAE could still be observed in these mice, suggesting that *Csf2r* expression on the CD103+ dendric cells is not exclusively required for EAE initiation and/or progression [64]. Neutrophils are an additional myeloid cell type that is known to respond to GM-CSF [74]. Studies using anti-CXCR2, a major chemoattractant for neutrophils, demonstrate that inhibiting the activity of this chemokine confers protection against EAE [75]. Furthermore, GM-CSF is thought to promote the accumulation of neutrophils in the brain of mice with atypical EAE, wherein mice exhibit extensive inflammation in both the brain and spinal cord [67]. Therefore, neutrophils may be an important cell type that respond to GM-CSF and subsequently promote EAE pathology. In addition to myeloid cells, CNS-resident microglia become activated in response to GM-CSF produced by infiltrating CD4+ T cells prior to the onset of clinical symptoms, suggesting that GM-CSF-dependent microglial activation is required for the progression of EAE [43,68]. However, there have been very few studies to assess how GM-CSF promotes the activation of microglia during EAE. This will require further investigation. In addition, astrocytes are known to promote EAE pathology [76]. In a recent study, Wheeler and colleagues utilized transcriptome analyses to characterize astrocyte activation during EAE in response to GM-CSF signaling [69]. They found that GM-CSF stimulation promoted the expression of MAFG and MAT2α which are thought to repress anti-inflammatory transcriptional programs. In addition, GM-CSF stimulation in astrocytes promoted proinflammatory transcriptional programs, suggesting that GM-CSF signaling in astrocytes renders them pathogenic in the context of EAE [69]. It is evident that monocytes are the predominant cell type that respond to GM-CSF during EAE. Consequently, targeting GM-CSF and/or the downstream mediators of GM-CSF signaling in these cells may be a promising therapeutic approach to curtail pathogenic monocyte infiltration and differentiation in the CNS. A list of cell types that respond to GM-CSF during EAE is summarized in Table 1.

2.6. GM-CSF in Other Murine Models of MS

The role of GM-CSF in less commonly used models of MS has not been well elucidated. One model that may depend on GM-CSF signaling is the Theiler's murine encephalomyelitis virus-induced demyelinating disease (TMEV-IDD) [77]. Theiler's murine virus is an enteric commensal in most mouse stains; however, when injected via intracranial injection into susceptible mice, such as SJL/J mice, the result is a chronic and progressive demyelinating disease [78,79]. The chronic phase of TMEV-IDD is characterized by inflammation, demyelination, axonal degeneration, and astrogliosis, making this a suitable model to study MS progression [79]. One study suggested that GM-CSF may play a role in promoting pathology in this model. Bone marrow cells stimulated with GM-CSF were infected with TMEV, and the presence of GM-CSF was found to promote virus replication and the production of proinflammatory cytokines, indicating that GM-CSF is important in inducing TMEV-IDD [77]. The importance of GM-CSF in this model further highlights the important role of GM-CSF in promoting neuroinflammation associated with immune cell infiltration into the CNS parenchyma, although additional in vivo studies need to be performed to further characterize the role of GM-CSF in this model of demyelinating disease. Interestingly, in the Cuprizone model, which is a non-inflammatory model of MS that promotes demyelination by promoting the death of mature myelin-producing oligodendrocytes, there is no literature to support the notion that GM-CSF plays a role in promoting pathology [80,81]. This suggests that GM-CSF does not directly facilitate demyelination, rather it promotes the differentiation and activation of immune cells that can then directly promote demyelination. Consequently, as therapies are being developed, co-treatment with a mediator that prevents demyelination by protecting mature oligodendrocytes should be considered.

2.7. Controversy over GM-CSF in Murine Models of MS

Studies from animal models have convincingly demonstrated that GM-CSF plays a critical role in promoting EAE pathology. However, recent studies have brought the importance of this cytokine in the onset of disease into question. The first study performed by Pierson and Goverman sought to determine the role of GM-CSF in EAE that is induced in C3HeB/FeJ mice, which develop an inflammatory disease in both the brain and spinal cord [67]. This model is unique because the inflammation resulting from EAE induced in mice on a C57BL/6 background has a strong predilection for the spinal cord [31]. GM-CSF-deficient C3HeB/FeJ mice develop EAE because IL-17 is able to compensate for the loss of GM-CSF, and is able to promote neutrophil accumulation, inflammation, and demyelination [67]. Interestingly, this study also determined that the co-expression of IL-17 and GM-CSF is required to promote immune cell migration into the brain, which is normally inhibited by IFNγ [79]. This suggests that, while GM-CSF may be important for promoting EAE pathology, it is not the only cytokine that is essential. Therefore, when therapeutics are being generated, an approach that involves targeting multiple cytokines, including GM-CSF, should be considered. This notion is further supported by a study that posited that GM-CSF is required for the accumulation of pathogenic CD4+ T cells in the lymph nodes, but is not required for the activation of these cells, and is therefore not required for the onset of EAE [71]. To determine whether or not this was the case, active EAE was induced in $Csf2^{-/-}$ mice, and the number of pathogenic CD4+ T cells in the lymph nodes was significantly decreased, suggesting that the accumulation of CD4+ T cells in the lymph nodes is dependent on GM-CSF [71]. However, when these T cells were expanded in vitro under Th17-polarizing conditions, and were adoptively transferred to wild-type recipient mice, these cells were able to induce EAE, suggesting that GM-CSF is not exclusively required for the development of encephalitogenic CD4+ T cells [71]. Additionally, when wild-type pathogenic CD4+ T cells were adoptively transferred to $Csf2^{-/-}$ mice, the onset of clinical EAE symptoms was unaltered; however, perturbing GM-CSF signaling did alter the immune cell profile in the CNS, thereby decreasing disease severity and preventing the progression to chronic disability [71]. This study suggests that GM-CSF is indispensable for promoting EAE onset; however, GM-CSF signaling is compulsory for EAE progression. These findings emphasize the importance of considering the role of additional cytokines that can be therapeutically targeted in conjunction with GM-CSF.

3. GM-CSF in MS

3.1. Immune Cells that Produce and Respond to GM-CSF during MS

Animal models of MS have allowed us to gain significant insight into the role of GM-CSF in promoting this immune-mediated disease. Despite the breakthroughs in our understanding of this cytokines in murine models, the role of GM-CSF in MS is still not completely elucidated. It has been known for some time that the concentration of GM-CSF is significantly increased in the cerebral spinal fluid (CSF) of patients with active MS compared to healthy controls [10]. Similar to murine EAE models, GM-CSF is thought to be produced by CD4+ T cells that contain an MS-associated polymorphism in the IL-2 receptor alpha gene [82]. Similar to EAE, a distinct subset of CCR6-expressing Th cells that exclusively produce GM-CSF have been found in high numbers in the CSF of patients with active disease, suggesting that CD4+ T cells are the major cellular source of GM-CSF during MS [83]. In fact, efficacious treatment with the disease-modifying agent interferon beta significantly decreases the number of GM-CSF-producing CD4+ T cells in the peripheral blood and in the CSF of patients with MS compared to untreated patients [59]. In addition to CD4+ T cells, GM-CSF is produced by a subset of B cells and CD8+ T cells during active MS [59,60]. The recent success of the disease-modifying agent Ocrelizumab, which depletes CD20-expressing B cells, for the treatment of progressive MS, suggests that pathogenic B cells may play an important role in mediating MS pathology [84]. Additional mechanistic studies are needed to address the role of GM-CSF-producing B and T cells in MS. Such mechanistic studies can inform more efficacious therapeutics to prevent MS progression.

Similar to EAE, monocytes appear to be the major cell types that respond to GM-CSF in MS [85]. GM-CSF increases the migration of monocytes across the blood–brain barrier and, once in the parenchyma, promotes the differentiation of monocytes into monocyte-derived antigen-presenting cells [85]. Analyses of postmortem brain tissue obtained from patients with MS demonstrate that these monocyte-derived antigen-presenting cells are the predominant cell type found at the site of active demyelinating lesions [85–88]. Moreover, these same cells have been found to persist at the sites of chronic demyelinating lesions [89]. Since GM-CSF plays a critical role in promoting the migration of these cells across the blood–brain barrier, targeting GM-CSF therapeutically may prevent the migration of pathogenic monocyte-derived cells into the CNS. Additionally, an analysis of active and chronic demyelinating lesions found that the expression of the GM-CSF receptor is highly upregulated on the lesion-associated astrocytes and microglia, suggesting that both of these CNS-resident cells may upregulate proinflammatory genes in response to GM-CSF signaling [90]. Additional studies are required to ascertain the role of GM-CSF signaling in these CNS-resident cells, and how this signaling contributes to MS pathology.

3.2. Clinical Trials Therapeutically Targeting GM-CSF

There are numerous clinical trials that are attempting to target GM-CSF or the GM-CSF receptor for the treatment of autoimmune diseases [91]. One biologic that has been tested in clinical trials as an MS therapy is MOR103 [91]. MOR103 is a humanized monoclonal antibody against GM-CSF [92]. In a 20-week, randomized, double-blind, placebo-controlled phase 1b dose-escalation trial, patients with relapsing–remitting MS or secondary progressive MS with less than 10 gadolinium-enhancing lesions were administered through intravenous infusions of either MOR103 or a placebo control [92]. The primary objective of this study was safety, and although MOR103 demonstrated only modest efficacy, it was well-tolerated in patients with MS, and overall had a favorable safety profile [92]. This study is important, because there are some risks associated with blocking the biological activity of GM-CSF, which are important to keep in mind when developing a therapy against GM-CSF. Targeting GM-CSF and its receptor has been associated with exacerbations in pre-existing intestinal inflammation and the onset of colitis [93,94]. Additionally, as was previously mentioned, GM-CSF signaling is critical for the development of alveolar macrophages [22,23]. The accumulation of autoantibodies against GM-CSF is associated with an increased risk of developing pulmonary alveolar proteinosis, which is characterized by decreases in alveolar macrophages which result in the abnormal accumulation of surfactant in the lungs [95]. Additional clinical trials will need to further evaluate the efficacy of MOR103. Although this is the only clinical trial that has assessed the use of anti-GM-CSF or GM-CSF receptor inhibitors to treat MS, there are numerous clinical trials that are attempting to utilize these biologics to treat Rheumatoid Arthritis [91]. Many of these therapeutics will likely also be tested for efficacy to treat MS in the future.

4. Conclusions

Murine models of MS have allowed us to gain insight into the important role that GM-CSF plays in mediating neuroinflammation. These models have proven to be a useful tool for studying immune cell infiltration and the resulting inflammatory milieu, given that many of the major underpinnings of EAE pathology have been validated in patients with MS. These models have established that CD4+ T cells are the major cellular source of GM-CSF during EAE, and the monocytes are the major cell type that responds to that GM-CSF. These monocytes can then infiltrate into the CNS and promote inflammation, demyelination, and axonal loss. These cells and their downstream mediators are ideal targets for MS therapies. However, there is still much to be learned about the role of GM-CSF in mediating EAE pathology. Additional studies need to access the importance of Th$_{GM}$ cells, the role of GM-CSF signaling in lesion-associated microglia and astrocytes, and the functional importance of GM-CSF-producing B cells in patients with MS. Despite these shortcomings in the literature, GM-CSF appears to be a promising therapeutic target for treating MS progression. Evidence from the literature also strongly alludes to the notion that a combined therapy approach that includes inhibiting the biological function of GM-CSF will likely be the most efficacious approach to treat MS.

Funding: This work was supported by NIH grant P20 GM109098 and the Innovation Award Program from Praespero to Edwin Wan.

Conflicts of Interest: The authors declare no conflict of interest. The funders had no role in the design of the study; in the collection, analyses, or interpretation of data; in the writing of the manuscript, or in the decision to publish the results.

References

1. Wallin, M.T.; Culpepper, W.J.; Campbell, J.D.; Nelson, L.M.; Langer-Gould, A.; Marrie, R.A.; Cutter, G.R.; Kaye, W.E.; Wagner, L.; Tremlett, H.; et al. The prevalence of MS in the United States. *A Popul. Based Estim. Using Health Claims Data* **2019**, *92*, e1029–e1040. [CrossRef] [PubMed]
2. Dendrou, C.A.; Fugger, L.; Friese, M.A. Immunopathology of multiple sclerosis. *Nat. Rev. Immunol.* **2015**, *15*, 545–558. [CrossRef] [PubMed]
3. Popescu, B.F.G.; Pirko, I.; Lucchinetti, C.F. Pathology of multiple sclerosis: Where do we stand? *CONTINUUM Lifelong Learn. Neurol.* **2013**, *19*, 901–921. [CrossRef] [PubMed]
4. Katz Sand, I. Classification, diagnosis, and differential diagnosis of multiple sclerosis. *Curr. Opin. Neurol.* **2015**, *28*, 193–205. [CrossRef]
5. Krieger, S.C.; Cook, K.; De Nino, S.; Fletcher, M. The topographical model of multiple sclerosis: A dynamic visualization of disease course. *Neurol. (R) Neuroimmunol. Neuroinflammation* **2016**, *3*, e279. [CrossRef]
6. Shull, C.; Hoyle, B.; Iannotta, C.; Fletcher, E.; Curan, M.; Cipollone, V. A current understanding of multiple sclerosis. *Jaapa* **2020**. [CrossRef]
7. Klineova, S.; Lublin, F.D. Clinical Course of Multiple Sclerosis. *Cold Spring Harb. Perspect. Med.* **2018**, *8*. [CrossRef]

8. Finkelsztejn, A. Multiple sclerosis: Overview of disease-modifying agents. *Perspect. Med. Chem.* **2014**, *6*, 65–72. [CrossRef]
9. Denic, A.; Johnson, A.J.; Bieber, A.J.; Warrington, A.E.; Rodriguez, M.; Pirko, I. The relevance of animal models in multiple sclerosis research. *Pathophysiology* **2011**, *18*, 21–29. [CrossRef]
10. Carrieri, P.B.; Provitera, V.; De Rosa, T.; Tartaglia, G.; Gorga, F.; Perrella, O. Profile of cerebrospinal fluid and serum cytokines in patients with relapsing-remitting multiple sclerosis: A correlation with clinical activity. *Immunopharmacol. Immunotoxicol.* **1998**, *20*, 373–382. [CrossRef]
11. McQualter, J.L.; Darwiche, R.; Ewing, C.; Onuki, M.; Kay, T.W.; Hamilton, J.A.; Reid, H.H.; Bernard, C.C. Granulocyte macrophage colony-stimulating factor: A new putative therapeutic target in multiple sclerosis. *J. Exp. Med.* **2001**, *194*, 873–882. [CrossRef] [PubMed]
12. Shi, Y.; Liu, C.H.; Roberts, A.I.; Das, J.; Xu, G.; Ren, G.; Zhang, Y.; Zhang, L.; Yuan, Z.R.; Tan, H.S.; et al. Granulocyte-macrophage colony-stimulating factor (GM-CSF) and T-cell responses: What we do and don't know. *Cell Res.* **2006**, *16*, 126–133. [CrossRef] [PubMed]
13. Bowers, S.R.; Mirabella, F.; Calero-Nieto, F.J.; Valeaux, S.; Hadjur, S.; Baxter, E.W.; Merkenschlager, M.; Cockerill, P.N. A conserved insulator that recruits CTCF and cohesin exists between the closely related but divergently regulated interleukin-3 and granulocyte-macrophage colony-stimulating factor genes. *Mol. Cell. Biol.* **2009**, *29*, 1682–1693. [CrossRef] [PubMed]
14. Matsuguchi, T.; Zhao, Y.; Lilly, M.B.; Kraft, A.S. The cytoplasmic domain of granulocyte-macrophage colony-stimulating factor (GM-CSF) receptor alpha subunit is essential for both GM-CSF-mediated growth and differentiation. *J. Biol. Chem.* **1997**, *272*, 17450–17459. [CrossRef]
15. Hansen, G.; Hercus, T.R.; McClure, B.J.; Stomski, F.C.; Dottore, M.; Powell, J.; Ramshaw, H.; Woodcock, J.M.; Xu, Y.; Guthridge, M.; et al. The Structure of the GM-CSF Receptor Complex Reveals a Distinct Mode of Cytokine Receptor Activation. *Cell* **2008**, *134*, 496–507. [CrossRef]
16. Brizzi, M.F.; Zini, M.G.; Aronica, M.G.; Blechman, J.M.; Yarden, Y.; Pegoraro, L. Convergence of signaling by interleukin-3, granulocyte-macrophage colony-stimulating factor, and mast cell growth factor on JAK2 tyrosine kinase. *J. Biol. Chem.* **1994**, *269*, 31680–31684.
17. Hercus, T.R.; Thomas, D.; Guthridge, M.A.; Ekert, P.G.; King-Scott, J.; Parker, M.W.; Lopez, A.F. The granulocyte-macrophage colony-stimulating factor receptor: Linking its structure to cell signaling and its role in disease. *Blood* **2009**, *114*, 1289–1298. [CrossRef]
18. Yeh, J.E.; Toniolo, P.A.; Frank, D.A. JAK2-STAT5 signaling: A novel mechanism of resistance to targeted PI3K/mTOR inhibition. *Jak-Stat.* **2013**, *2*, e24635. [CrossRef]
19. Lehtonen, A.; Matikainen, S.; Miettinen, M.; Julkunen, I. Granulocyte-macrophage colony-stimulating factor (GM-CSF)-induced STAT5 activation and target-gene expression during human monocyte/macrophage differentiation. *J. Leukoc Biol* **2002**, *71*, 511–519.
20. Cook, A.D.; Lee, M.C.; Saleh, R.; Khiew, H.W.; Christensen, A.D.; Achuthan, A.; Fleetwood, A.J.; Lacey, D.C.; Smith, J.E.; Forster, I.; et al. TNF and granulocyte macrophage-colony stimulating factor interdependence mediates inflammation via CCL17. *Jci Insight* **2018**, *3*. [CrossRef]
21. Bozinovski, S.; Jones, J.E.; Vlahos, R.; Hamilton, J.A.; Anderson, G.P. Granulocyte/macrophage-colony-stimulating factor (GM-CSF) regulates lung innate immunity to lipopolysaccharide through Akt/Erk activation of NFkappa B and AP-1 in vivo. *J. Biol. Chem.* **2002**, *277*, 42808–42814. [CrossRef]
22. Shibata, Y.; Berclaz, P.Y.; Chroneos, Z.C.; Yoshida, M.; Whitsett, J.A.; Trapnell, B.C. GM-CSF regulates alveolar macrophage differentiation and innate immunity in the lung through PU.1. *Immunity* **2001**, *15*, 557–567. [CrossRef]
23. Guilliams, M.; De Kleer, I.; Henri, S.; Post, S.; Vanhoutte, L.; De Prijck, S.; Deswarte, K.; Malissen, B.; Hammad, H.; Lambrecht, B.N. Alveolar macrophages develop from fetal monocytes that differentiate into long-lived cells in the first week of life via GM-CSF. *J. Exp. Med.* **2013**, *210*, 1977–1992. [CrossRef] [PubMed]
24. Becher, B.; Tugues, S.; Greter, M. GM-CSF: From Growth Factor to Central Mediator of Tissue Inflammation. *Immunity* **2016**, *45*, 963–973. [CrossRef] [PubMed]
25. Rosen, L.B.; Freeman, A.F.; Yang, L.M.; Jutivorakool, K.; Olivier, K.N.; Angkasekwinai, N.; Suputtamongkol, Y.; Bennett, J.E.; Pyrgos, V.; Williamson, P.R.; et al. Anti-GM-CSF autoantibodies in patients with cryptococcal meningitis. *J. Immunol.* **2013**, *190*, 3959–3966. [CrossRef]

26. Greter, M.; Helft, J.; Chow, A.; Hashimoto, D.; Mortha, A.; Agudo-Cantero, J.; Bogunovic, M.; Gautier, E.L.; Miller, J.; Leboeuf, M.; et al. GM-CSF Controls Nonlymphoid Tissue Dendritic Cell Homeostasis but Is Dispensable for the Differentiation of Inflammatory Dendritic Cells. *Immunity* **2012**, *36*, 1031–1046. [CrossRef]
27. Kingston, D.; Schmid, M.A.; Onai, N.; Obata-Onai, A.; Baumjohann, D.; Manz, M.G. The concerted action of GM-CSF and Flt3-ligand on in vivo dendritic cell homeostasis. *Blood* **2009**, *114*, 835–843. [CrossRef]
28. Vremec, D.; Lieschke, G.J.; Dunn, A.R.; Robb, L.; Metcalf, D.; Shortman, K. The influence of granulocyte/macrophage colony-stimulating factor on dendritic cell levels in mouse lymphoid organs. *Eur. J. Immunol.* **1997**, *27*, 40–44. [CrossRef]
29. Helft, J.; Böttcher, J.; Chakravarty, P.; Zelenay, S.; Huotari, J.; Schraml, B.U.; Goubau, D.; Reis e Sousa, C. GM-CSF Mouse Bone Marrow Cultures Comprise a Heterogeneous Population of CD11c+MHCII+ Macrophages and Dendritic Cells. *Immunity* **2015**, *42*, 1197–1211. [CrossRef]
30. Rivers, T.M.; Sprunt, D.H.; Berry, G.P. OBSERVATIONS ON ATTEMPTS TO PRODUCE ACUTE DISSEMINATED ENCEPHALOMYELITIS IN MONKEYS. *J. Exp. Med.* **1933**, *58*, 39–53. [CrossRef]
31. Constantinescu, C.S.; Farooqi, N.; O'Brien, K.; Gran, B. Experimental autoimmune encephalomyelitis (EAE) as a model for multiple sclerosis (MS). *Br. J. Pharm.* **2011**, *164*, 1079–1106. [CrossRef]
32. McCarthy, D.P.; Richards, M.H.; Miller, S.D. Mouse models of multiple sclerosis: Experimental autoimmune encephalomyelitis and Theiler's virus-induced demyelinating disease. *Methods Mol. Biol* **2012**, *900*, 381–401. [CrossRef] [PubMed]
33. Stromnes, I.M.; Goverman, J.M. Active induction of experimental allergic encephalomyelitis. *Nat. Protoc.* **2006**, *1*, 1810–1819. [CrossRef] [PubMed]
34. Linthicum, D.S.; Munoz, J.J.; Blaskett, A. Acute experimental autoimmune encephalomyelitis in mice. I. Adjuvant action of Bordetella pertussis is due to vasoactive amine sensitization and increased vascular permeability of the central nervous system. *Cell. Immunol.* **1982**, *73*, 299–310. [CrossRef]
35. Sun, D.; Whitaker, J.N.; Huang, Z.; Liu, D.; Coleclough, C.; Wekerle, H.; Raine, C.S. Myelin Antigen-Specific CD8+ T Cells Are Encephalitogenic and Produce Severe Disease in C57BL/6 Mice. *J. Immunol.* **2001**, *166*, 7579–7587. [CrossRef] [PubMed]
36. Abreu, S.L. Suppression of experimental allergic encephalomyelitis by interferon. *Immunol. Commun.* **1982**, *11*, 1–7. [CrossRef]
37. Teitelbaum, D.; Meshorer, A.; Hirshfeld, T.; Arnon, R.; Sela, M. Suppression of experimental allergic encephalomyelitis by a synthetic polypeptide. *Eur. J. Immunol* **1971**, *1*, 242–248. [CrossRef]
38. Yednock, T.A.; Cannon, C.; Fritz, L.C.; Sanchez-Madrid, F.; Steinman, L.; Karin, N. Prevention of experimental autoimmune encephalomyelitis by antibodies against alpha 4 beta 1 integrin. *Nature* **1992**, *356*, 63–66. [CrossRef]
39. Yong, V.W. Differential mechanisms of action of interferon-β and glatiramer acetate in MS. *Neurology* **2002**, *59*, 802–808. [CrossRef]
40. Ziemssen, T.; Schrempf, W. Glatiramer acetate: Mechanisms of action in multiple sclerosis. *Int. Rev. Neurobiol.* **2007**, *79*, 537–570. [CrossRef]
41. Singer, B.A. The role of natalizumab in the treatment of multiple sclerosis: Benefits and risks. *Adv. Neurol Disord.* **2017**, *10*, 327–336. [CrossRef] [PubMed]
42. Rice, G.P.; Hartung, H.P.; Calabresi, P.A. Anti-alpha4 integrin therapy for multiple sclerosis: Mechanisms and rationale. *Neurology* **2005**, *64*, 1336–1342. [CrossRef] [PubMed]
43. Ponomarev, E.D.; Shriver, L.P.; Maresz, K.; Pedras-Vasconcelos, J.; Verthelyi, D.; Dittel, B.N. GM-CSF production by autoreactive T cells is required for the activation of microglial cells and the onset of experimental autoimmune encephalomyelitis. *J. Immunol* **2007**, *178*, 39–48. [CrossRef]
44. Kreymborg, K.; Böhlmann, U.; Becher, B. IL-23: Changing the verdict on IL-12 function in inflammation and autoimmunity. *Expert Opin. Ther. Targets* **2005**, *9*, 1123–1136. [CrossRef] [PubMed]
45. Harrington, L.E.; Hatton, R.D.; Mangan, P.R.; Turner, H.; Murphy, T.L.; Murphy, K.M.; Weaver, C.T. Interleukin 17–producing CD4+ effector T cells develop via a lineage distinct from the T helper type 1 and 2 lineages. *Nat. Immunol.* **2005**, *6*, 1123–1132. [CrossRef]
46. Park, H.; Li, Z.; Yang, X.O.; Chang, S.H.; Nurieva, R.; Wang, Y.-H.; Wang, Y.; Hood, L.; Zhu, Z.; Tian, Q.; et al. A distinct lineage of CD4 T cells regulates tissue inflammation by producing interleukin 17. *Nat. Immunol.* **2005**, *6*, 1133–1141. [CrossRef] [PubMed]

47. Kroenke, M.A.; Carlson, T.J.; Andjelkovic, A.V.; Segal, B.M. IL-12– and IL-23–modulated T cells induce distinct types of EAE based on histology, CNS chemokine profile, and response to cytokine inhibition. *J. Exp. Med.* **2008**, *205*, 1535–1541. [CrossRef]
48. Codarri, L.; Gyulveszi, G.; Tosevski, V.; Hesske, L.; Fontana, A.; Magnenat, L.; Suter, T.; Becher, B. RORgammat drives production of the cytokine GM-CSF in helper T cells, which is essential for the effector phase of autoimmune neuroinflammation. *Nat. Immunol* **2011**, *12*, 560–567. [CrossRef]
49. Mangalam, A.K.; Luo, N.; Luckey, D.; Papke, L.; Hubbard, A.; Wussow, A.; Smart, M.; Giri, S.; Rodriguez, M.; David, C. Absence of IFN-gamma increases brain pathology in experimental autoimmune encephalomyelitis-susceptible DRB1*0301.DQ8 HLA transgenic mice through secretion of proinflammatory cytokine IL-17 and induction of pathogenic monocytes/microglia into the central nervous system. *J. Immunol.* **2014**, *193*, 4859–4870. [CrossRef]
50. Haak, S.; Croxford, A.L.; Kreymborg, K.; Heppner, F.L.; Pouly, S.; Becher, B.; Waisman, A. IL-17A and IL-17F do not contribute vitally to autoimmune neuro-inflammation in mice. *J. Clin. Investig.* **2009**, *119*, 61–69. [CrossRef]
51. Kreymborg, K.; Etzensperger, R.; Dumoutier, L.; Haak, S.; Rebollo, A.; Buch, T.; Heppner, F.L.; Renauld, J.C.; Becher, B. IL-22 is expressed by Th17 cells in an IL-23-dependent fashion, but not required for the development of autoimmune encephalomyelitis. *J. Immunol.* **2007**, *179*, 8098–8104. [CrossRef] [PubMed]
52. Coquet, J.M.; Chakravarti, S.; Smyth, M.J.; Godfrey, D.I. Cutting Edge: IL-21 Is Not Essential for Th17 Differentiation or Experimental Autoimmune Encephalomyelitis. *J. Immunol.* **2008**, *180*, 7097–7101. [CrossRef] [PubMed]
53. Becher, B.; Segal, B.M. T(H)17 cytokines in autoimmune neuro-inflammation. *Curr. Opin. Immunol.* **2011**, *23*, 707–712. [CrossRef] [PubMed]
54. Sheng, W.; Yang, F.; Zhou, Y.; Yang, H.; Low, P.Y.; Kemeny, D.M.; Tan, P.; Moh, A.; Kaplan, M.H.; Zhang, Y.; et al. STAT5 programs a distinct subset of GM-CSF-producing T helper cells that is essential for autoimmune neuroinflammation. *Cell Res.* **2014**, *24*, 1387–1402. [CrossRef] [PubMed]
55. Komuczki, J.; Tuzlak, S.; Friebel, E.; Hartwig, T.; Spath, S.; Rosenstiel, P.; Waisman, A.; Opitz, L.; Oukka, M.; Schreiner, B.; et al. Fate-Mapping of GM-CSF Expression Identifies a Discrete Subset of Inflammation-Driving T Helper Cells Regulated by Cytokines IL-23 and IL-1beta. *Immunity* **2019**, *50*, 1289–1304. [CrossRef]
56. Poppensieker, K.; Otte, D.M.; Schurmann, B.; Limmer, A.; Dresing, P.; Drews, E.; Schumak, B.; Klotz, L.; Raasch, J.; Mildner, A.; et al. CC chemokine receptor 4 is required for experimental autoimmune encephalomyelitis by regulating GM-CSF and IL-23 production in dendritic cells. *Proc. Natl. Acad. Sci. USA* **2012**, *109*, 3897–3902. [CrossRef]
57. El-Behi, M.; Ciric, B.; Dai, H.; Yan, Y.; Cullimore, M.; Safavi, F.; Zhang, G.X.; Dittel, B.N.; Rostami, A. The encephalitogenicity of T(H)17 cells is dependent on IL-1- and IL-23-induced production of the cytokine GM-CSF. *Nat. Immunol.* **2011**, *12*, 568–575. [CrossRef]
58. Ciric, B.; El-behi, M.; Cabrera, R.; Zhang, G.X.; Rostami, A. IL-23 drives pathogenic IL-17-producing CD8+ T cells. *J. Immunol.* **2009**, *182*, 5296–5305. [CrossRef]
59. Rasouli, J.; Ciric, B.; Imitola, J.; Gonnella, P.; Hwang, D.; Mahajan, K.; Mari, E.R.; Safavi, F.; Leist, T.P.; Zhang, G.X.; et al. Expression of GM-CSF in T Cells Is Increased in Multiple Sclerosis and Suppressed by IFN-beta Therapy. *J. Immunol.* **2015**, *194*, 5085–5093. [CrossRef]
60. Li, R.; Rezk, A.; Miyazaki, Y.; Hilgenberg, E.; Touil, H.; Shen, P.; Moore, C.S.; Michel, L.; Althekair, F.; Rajasekharan, S.; et al. Proinflammatory GM-CSF-producing B cells in multiple sclerosis and B cell depletion therapy. *Sci. Transl. Med.* **2015**, *7*, 310ra166. [CrossRef]
61. Pare, A.; Mailhot, B.; Levesque, S.A.; Juzwik, C.; Ignatius Arokia Doss, P.M.; Lecuyer, M.A.; Prat, A.; Rangachari, M.; Fournier, A.; Lacroix, S. IL-1beta enables CNS access to CCR2(hi) monocytes and the generation of pathogenic cells through GM-CSF released by CNS endothelial cells. *Proc. Natl. Acad. Sci. USA* **2018**, *115*, E1194–E1203. [CrossRef] [PubMed]
62. Croxford, A.L.; Spath, S.; Becher, B. GM-CSF in Neuroinflammation: Licensing Myeloid Cells for Tissue Damage. *Trends Immunol.* **2015**, *36*, 651–662. [CrossRef] [PubMed]
63. Segal, B.M. Modulation of the Innate Immune System: A Future Approach to the Treatment of Neurological Disease. *Clin. Immunol.* **2018**, *189*, 1–3. [CrossRef] [PubMed]

64. Croxford, A.L.; Lanzinger, M.; Hartmann, F.J.; Schreiner, B.; Mair, F.; Pelczar, P.; Clausen, B.E.; Jung, S.; Greter, M.; Becher, B. The Cytokine GM-CSF Drives the Inflammatory Signature of CCR2+ Monocytes and Licenses Autoimmunity. *Immunity* **2015**, *43*, 502–514. [CrossRef] [PubMed]
65. Spath, S.; Komuczki, J.; Hermann, M.; Pelczar, P.; Mair, F.; Schreiner, B.; Becher, B. Dysregulation of the Cytokine GM-CSF Induces Spontaneous Phagocyte Invasion and Immunopathology in the Central Nervous System. *Immunity* **2017**, *46*, 245–260. [CrossRef]
66. King, I.L.; Kroenke, M.A.; Segal, B.M. GM-CSF-dependent, CD103+ dermal dendritic cells play a critical role in Th effector cell differentiation after subcutaneous immunization. *J. Exp. Med.* **2010**, *207*, 953–961. [CrossRef]
67. Pierson, E.R.; Goverman, J.M. GM-CSF is not essential for experimental autoimmune encephalomyelitis but promotes brain-targeted disease. *Jci Insight* **2017**, *2*, e92362. [CrossRef]
68. Ponomarev, E.D.; Shriver, L.P.; Maresz, K.; Dittel, B.N. Microglial cell activation and proliferation precedes the onset of CNS autoimmunity. *J. Neurosci Res.* **2005**, *81*, 374–389. [CrossRef]
69. Wheeler, M.A.; Clark, I.C.; Tjon, E.C.; Li, Z.; Zandee, S.E.J.; Couturier, C.P.; Watson, B.R.; Scalisi, G.; Alkwai, S.; Rothhammer, V.; et al. MAFG-driven astrocytes promote CNS inflammation. *Nature* **2020**, *578*, 593–599. [CrossRef]
70. King, I.L.; Dickendesher, T.L.; Segal, B.M. Circulating Ly-6C+ myeloid precursors migrate to the CNS and play a pathogenic role during autoimmune demyelinating disease. *Blood* **2009**, *113*, 3190–3197. [CrossRef]
71. Duncker, P.C.; Stoolman, J.S.; Huber, A.K.; Segal, B.M. GM-CSF Promotes Chronic Disability in Experimental Autoimmune Encephalomyelitis by Altering the Composition of Central Nervous System-Infiltrating Cells, but Is Dispensable for Disease Induction. *J. Immunol.* **2018**, *200*, 966–973. [CrossRef] [PubMed]
72. Grifka-Walk, H.M.; Giles, D.A.; Segal, B.M. IL-12-polarized Th1 cells produce GM-CSF and induce EAE independent of IL-23. *Eur. J. Immunol.* **2015**, *45*, 2780–2786. [CrossRef] [PubMed]
73. Ko, H.-J.; Brady, J.L.; Ryg-Cornejo, V.; Hansen, D.S.; Vremec, D.; Shortman, K.; Zhan, Y.; Lew, A.M. GM-CSF–Responsive Monocyte-Derived Dendritic Cells Are Pivotal in Th17 Pathogenesis. *J. Immunol.* **2014**, *192*, 2202–2209. [CrossRef] [PubMed]
74. Kobayashi, S.D.; Voyich, J.M.; Whitney, A.R.; DeLeo, F.R. Spontaneous neutrophil apoptosis and regulation of cell survival by granulocyte macrophage-colony stimulating factor. *J. Leukoc. Biol.* **2005**, *78*, 1408–1418. [CrossRef] [PubMed]
75. Kroenke, M.A.; Chensue, S.W.; Segal, B.M. EAE mediated by a non-IFN-γ/non-IL-17 pathway. *Eur. J. Immunol.* **2010**, *40*, 2340–2348. [CrossRef] [PubMed]
76. Baecher-Allan, C.; Kaskow, B.J.; Weiner, H.L. Multiple Sclerosis: Mechanisms and Immunotherapy. *Neuron* **2018**, *97*, 742–768. [CrossRef]
77. Schneider, K.M.; Watson, N.B.; Minchenberg, S.B.; Massa, P.T. The influence of macrophage growth factors on Theiler's Murine Encephalomyelitis Virus (TMEV) infection and activation of macrophages. *Cytokine* **2018**, *102*, 83–93. [CrossRef]
78. Clatch, R.J.; Melvold, R.W.; Miller, S.D.; Lipton, H.L. Theiler's murine encephalomyelitis virus (TMEV)-induced demyelinating disease in mice is influenced by the H-2D region: Correlation with TEMV-specific delayed-type hypersensitivity. *J. Immunol.* **1985**, *135*, 1408–1414.
79. Gerhauser, I.; Hansmann, F.; Ciurkiewicz, M.; Löscher, W.; Beineke, A. Facets of Theiler's Murine Encephalomyelitis Virus-Induced Diseases: An Update. *Int. J. Mol. Sci.* **2019**, *20*, 448. [CrossRef]
80. Lassmann, H.; Bradl, M. Multiple sclerosis: Experimental models and reality. *Acta Neuropathol.* **2017**, *133*, 223–244. [CrossRef]
81. Palle, P.; Monaghan, K.L.; Milne, S.M.; Wan, E.C.K. Cytokine Signaling in Multiple Sclerosis and Its Therapeutic Applications. *Med. Sci.* **2017**, *5*, 23. [CrossRef] [PubMed]
82. Hartmann, F.J.; Khademi, M.; Aram, J.; Ammann, S.; Kockum, I.; Constantinescu, C.; Gran, B.; Piehl, F.; Olsson, T.; Codarri, L.; et al. Multiple sclerosis-associated IL2RA polymorphism controls GM-CSF production in human TH cells. *Nat. Commun.* **2014**, *5*, 5056. [CrossRef] [PubMed]
83. Restorick, S.M.; Durant, L.; Kalra, S.; Hassan-Smith, G.; Rathbone, E.; Douglas, M.R.; Curnow, S.J. CCR6(+) Th cells in the cerebrospinal fluid of persons with multiple sclerosis are dominated by pathogenic non-classic Th1 cells and GM-CSF-only-secreting Th cells. *Brain Behav. Immun.* **2017**, *64*, 71–79. [CrossRef] [PubMed]
84. Sellebjerg, F.; Blinkenberg, M.; Sorensen, P.S. Anti-CD20 Monoclonal Antibodies for Relapsing and Progressive Multiple Sclerosis. *Cns Drugs* **2020**, *10*. [CrossRef]

85. Vogel, D.Y.; Kooij, G.; Heijnen, P.D.; Breur, M.; Peferoen, L.A.; van der Valk, P.; de Vries, H.E.; Amor, S.; Dijkstra, C.D. GM-CSF promotes migration of human monocytes across the blood brain barrier. *Eur. J. Immunol.* **2015**, *45*, 1808–1819. [CrossRef]
86. Lucchinetti, C.; Bruck, W.; Parisi, J.; Scheithauer, B.; Rodriguez, M.; Lassmann, H. Heterogeneity of multiple sclerosis lesions: Implications for the pathogenesis of demyelination. *Ann. Neurol.* **2000**, *47*, 707–717. [CrossRef]
87. Vogel, D.Y.; Vereyken, E.J.; Glim, J.E.; Heijnen, P.D.; Moeton, M.; van der Valk, P.; Amor, S.; Teunissen, C.E.; van Horssen, J.; Dijkstra, C.D. Macrophages in inflammatory multiple sclerosis lesions have an intermediate activation status. *J. Neuroinflamm.* **2013**, *10*, 35. [CrossRef]
88. Mishra, M.K.; Yong, V.W. Myeloid cells—targets of medication in multiple sclerosis. *Nat. Rev. Neurol.* **2016**, *12*, 539–551. [CrossRef]
89. Prineas, J.W.; Kwon, E.E.; Cho, E.S.; Sharer, L.R.; Barnett, M.H.; Oleszak, E.L.; Hoffman, B.; Morgan, B.P. Immunopathology of secondary-progressive multiple sclerosis. *Ann. Neurol.* **2001**, *50*, 646–657. [CrossRef]
90. Imitola, J.; Rasouli, J.; Watanabe, F.; Mahajan, K.; Sharan, A.D.; Ciric, B.; Zhang, G.X.; Rostami, A. Elevated expression of granulocyte-macrophage colony-stimulating factor receptor in multiple sclerosis lesions. *J. Neuroimmunol.* **2018**, *317*, 45–54. [CrossRef]
91. Shiomi, A.; Usui, T.; Mimori, T. GM-CSF as a therapeutic target in autoimmune diseases. *Inflamm. Regen.* **2016**, *36*, 8. [CrossRef] [PubMed]
92. Constantinescu, C.S.; Asher, A.; Fryze, W.; Kozubski, W.; Wagner, F.; Aram, J.; Tanasescu, R.; Korolkiewicz, R.P.; Dirnberger-Hertweck, M.; Steidl, S.; et al. Randomized phase 1b trial of MOR103, a human antibody to GM-CSF, in multiple sclerosis. *Neurol Neuroimmunol Neuroinflamm* **2015**, *2*, e117. [CrossRef] [PubMed]
93. Biondo, M.; Nasa, Z.; Marshall, A.; Toh, B.H.; Alderuccio, F. Local transgenic expression of granulocyte macrophage-colony stimulating factor initiates autoimmunity. *J. Immunol.* **2001**, *166*, 2090–2099. [CrossRef] [PubMed]
94. Hirata, Y.; Egea, L.; Dann, S.M.; Eckmann, L.; Kagnoff, M.F. GM-CSF-facilitated dendritic cell recruitment and survival govern the intestinal mucosal response to a mouse enteric bacterial pathogen. *Cell Host Microbe* **2010**, *7*, 151–163. [CrossRef] [PubMed]
95. Carey, B.; Trapnell, B.C. The molecular basis of pulmonary alveolar proteinosis. *Clin. Immunol.* **2010**, *135*, 223–235. [CrossRef] [PubMed]

© 2020 by the authors. Licensee MDPI, Basel, Switzerland. This article is an open access article distributed under the terms and conditions of the Creative Commons Attribution (CC BY) license (http://creativecommons.org/licenses/by/4.0/).

MDPI
St. Alban-Anlage 66
4052 Basel
Switzerland
Tel. +41 61 683 77 34
Fax +41 61 302 89 18
www.mdpi.com

Cells Editorial Office
E-mail: cells@mdpi.com
www.mdpi.com/journal/cells